D1243395

North American
COMBUSTION HANDBOOK

North American
COMBUSTION HANDBOOK

A Practical Basic Reference on the
Art and Science of Industrial Heating
with Gaseous and Liquid Fuels

Volume II: Combustion Equipment, Controls, Systems,
Heat Recovery, Process Control Optimization, Pollution Reduction,
Noise Minimization, Oxygen Enrichment and Oxy-Fuel Firing

Third Edition

NORTH AMERICAN Mfg. Co.
CLEVELAND, OH 44105-5600 USA

NORTH AMERICAN COMBUSTION HANDBOOK
3rd Edition, Volume II

First Edition, 1952, 1957, 1965   55 000 copies
Second Edition, 1978   17 562 copies
Second Edition, second printing, 1982   5 154 copies
Third Edition, Volume I, 1986   25 000 copies
Third Edition, Volume II, 1997   20 000 copies

Library of Congress Cataloging in Publication Data

Reed, Richard J. (Richard James), 1924—
 North American Combustion Handbook.

 Includes bibliographies, appendix, glossary,
and index.
   1. Combustion engineering -- Handbooks, manuals,
 etc. 2. Controls -- Handbooks, manuals, etc.
 I. North American Mfg. Co. II. Title. III. Title:
 Combustion Handbook.

ISBN 0-9601596-3-0 (v.II)

PREFACE TO THE FIRST EDITION

The purpose of this handbook is to provide users of combustion equipment with:

a basic explanation of the theory of combustion,

outlines for combustion, heat transfer, and fluid calculations,

charts and data to simplify and speed these calculations, and

a discussion of combustion equipment, its selection and operation.

The first chapter of this handbook is intentionally very elementary. It is designed specifically for the newcomer to the field of combustion engineering.

Although the stated scope of this handbook is *industrial heating with gaseous and liquid fuels*, the reader will find that the book also contains considerable information applicable to commercial heating and to solid fuels.

We have endeavored to collect enough information under one cover to permit easy computation of typical combustion problems with a choice of several degrees of accuracy. Frequent use has been made of analogies and examples in order to make this handbook as generally useful as possible.

PREFACE TO THE THIRD EDITION, Volume II

North American Mfg. Co., is grateful to the users and builders of industrial process heating systems for their widespread acceptance of earlier editions of this handbook as their "bible." Also appreciated is the fact that many universities, libraries, and training schools around the world have adopted it as a textbook and/or reference.

Requests for more copies of this handbook after the first 77 000 copies encouraged us to enlarge and improve it in this 2-volume edition, which includes many updates, and five new parts (Heat Recovery, Process Control Optimization, Pollution Control, Noise Minimization, and Oxygen Enrichment/Oxy-Fuel Firing). We thank you for your patience and understanding during this major rework.

I and contributors Richard C. Riccardi, Thomas F. Robertson, Robert E. Schreter, and Hisashi Kobayashi thank North American Mfg. Co. for the freedom and assistance granted us in the planning and preparation of this book. The ideas, experience, and skills of many knowledgeable and loyal North American employees are embodied in it.

Richard J. Reed, Technical Information Director
NORTH AMERICAN Mfg. Co.

March, 1997

CONTENTS

Part 6. FUEL-BURNING EQUIPMENT

Part 7. COMBUSTION CONTROL

Part 10. PROCESS CONTROL OPTIMIZATION

Part 11. POLLUTION CONTROL

Part 12. NOISE MINIMIZATION

Part 13. OXYGEN ENRICHMENT AND OXY-FUEL FIRING

This is a handbook of engineering practice and recommendations—all subject to local, state, and federal codes, and insurance requirements....and good common sense.

WARNING: Situations dangerous to personnel and property can develop from incorrect operation of combustion equipment. North American urges compliance with safety standards and insurance underwriters' recommendations, and care in operation.

INTRODUCTION

Fire has been known to man since the dawn of time, but for many ages, it was only a thing of danger -- as forest fires, volcanic eruptions, cooking accidents -- because man could not control it.

Humans' first step in controlling fire was starting it; and the next was stopping it. After people learned these two procedures, they began to make good use of fire for heat and light. But many centuries passed before there was any further progress in the art of burning fuels. A fire was simply a heap of wood, a pile of coal, or a pot of oil. These all burned slowly -- at nature's own pace. If man wanted more heat, he simply had to build a *bigger fire*. It was not until fairly recent times that man learned to build a *faster fire*.

This matter of speed in burning was a new element of control that permitted large amounts of heat output within a small space -- not a simple trick. It required a knowledge of the burning process and thoughtful design of burning equipment. Although burning, or combustion, is really a chemical process, practical control of it is mainly a matter of fluid dynamics and heat transfer.

In addition to releasing a great quantity of heat (for productivity), man had to learn how to prevent it from getting away from him unused (energy conservation, safety). To transfer the heat from the flame to the place where it was needed (quality control), furnaces or combustion chambers had to be built to make efficient use of the heat from the fire. It was necessary to design the fire and the furnace to the job. But good design was not enough. Careful supervision and control of the fire by someone with a knowledge of combustion was necessary to assure the most efficient use of the fuel at all times.

The lessons learned through the years *still apply*. Continuing education is essential.

As a tiny spark can set a great forest on fire,
the tongue is a small thing, but what damage it can do!
And the tongue is a flame of fire . . .

LIVING BIBLE, James 3:5,6

As the wind blows out a candle
and makes a fire blaze,
Absence lessens ordinary passions
and augments great ones.

LA ROUCHEFOUCALD

The fire that's closest kept
burns best of all.

SHAKESPEARE, 1595

Part 6.   FUEL-BURNING EQUIPMENT

INTRODUCTION

Purposes of a fuel-burning system are: 1) to position flames at areas of useful heat release, 2) to initiate and maintain ignition, 3) to mix the fuel and air, 4) to volatilize solid and liquid fuels, 5) to proportion the fuel to air, and 6) to supply fuel and air at the proper rates and pressures to facilitate all five previous functions with safety at any rate required for the process.

Primary functions of a burner (sometimes called a tip, nozzle, fire-bed) are flame positioning and shaping, plus continuous ignition maintenance (without a pilot). Many "burners" also perform a variety of other functions among those listed above for the fuel-burning system.

Solid fuel-burning equipment. Solid fuels, such as coals, lignite, wood, sintering and pelletizing ore-fuel mixes, and solid waste incineration, require additional time for the *char* to burn after all gaseous and liquid compounds have been driven off. The principal method for hurrying this slow char-burning process (mass transfer -- migration of oxygen into the char and of carbon monoxide out) is through exposing much surface area by breaking the fuel into small pieces. (This increases the probability of pollution by particulate carryover.)

The following paragraphs discuss the state of the fast-changing art of solid fuel burning in broad generalities, progressing in size from small particles to large chunks, in heat release rates from fast to slow.

Friable fuels such as coals can be pulverized to 100 micron-sized particles (75% passing through a 200 mesh screen -- about the consistency of talcum powder) so that it can be burned at heat release rates comparable to those often obtained with liquid and gaseous fuels, using burners fundamentally similar to gas or oil burners. In *pulverized coal* burners the conveying air (primary air) is typically 8 to 20% of the stoichiometric requirement and should be less than the lower limit of flammability to prevent flashback.

Finely ground particles of solid fuels can be satisfactorily conveyed pneumatically through the gas tube of conventional *industrial process burners* (such as shown in Figure 6.3g) in some applications such as kilns for structural clay products where (a) the product is not harmed by particles of burning char falling on it, and (b) a system is provided for removing the ash without creating a pollution or dirt problem.

Crushed and ground particles may be burned in a *cyclone combustion chamber* (swirl combustor) or a fluidized bed. The cyclone is used in large steam generators and consists of a refractory-covered, studded, water-cooled, horizontal cylinder. Air and fuel are blown in tangentially onto a fluid mass that includes molten ash, coating the inner walls of the cylinder. Crushed coal (95% passing through a 4-mesh screen) is used, but bark and some waste materials may be added. Pieces of char stick to and burn on the surface of fluid mass. Particulate carryover is minimized and fuels with low melting point ash can be used to advantage. Heat release rates run from 0.5 to 0.9 million Btu/hr ft³. Because they provide recirculation with long residence times, swirl combustors help burn difficult fuels such as low quality coal or waste materials. Stratified or staged entry streams may help reduce noise and emissions of hydrocarbons and nitrogen oxides.

In a *fluidized bed combustor*, combustion air fed through a screen under the solid granules of fuel, causes the whole bed to be fluidized and thus provides intimate contact of air and fuel. Boiler tubes or pieces of material to be heated can be immersed in the fluidized bed and thereby subjected to excellent heat transfer rates. Mixing granular limestone with the fuel bed minimizes SO_2 and SO_3 pollution from high sulfur fuels.

Small chunks can be thrown onto a burning bed by a *spreader stoker** (overfeed) or deposited from a hopper (crossfeed†) on a traveling grate as in a sinter bed or a Herreshoff kiln. Figure 6.1 shows coal size ranges for three kinds of stokers. Larger chunks can be screw-fed or ram-fed (underfeed) into the bottom of a retort stoker‡ (Figure 8.20), or gravity fed into a rotary drum, or simply deposited on a pile to smolder in a big chamber using the starved air principle of incineration with the smoke and gaseous products of incomplete combustion being burned off by an additional air supply in a secondary combustion chamber. Most of these bed-burning arrangements provide underfire air from a windbox under a metal grate.

Some of the information in the first five parts of this handbook can be applied or adapted to solid fuel burning situations. Most of the information in this Part 6 and succeeding Parts 7 and 8 is intended for combustion of gaseous and liquid fuels.

---

* Up to 50% through a ¼ in. screen. Many of these "fines" burn in flight. Grate heat release = 1 000 000 Btu/hr ft².△

† Grate heat release = 300 000 Btu/hr ft².△

‡ Grate heat release = 200 000 Btu/hr ft² for single retort stokers.
                    300 000 Btu/hr ft² for multiple retort stokers.△

△ Rates listed are for comparison. Individual designs vary considerably -- as the Tri-Fuel Boiler, with a single retort underfeed stoker can be fired at 400 000 Btu/hr ft² without exceeding particulate carryover codes when using overfire air jets and at an input rate of 30 000 Btu/hr ft³.

Figure 6.1. Coal size ranges for three types of stokers. The recommended limits apply to the sizes of coal actually delivered to the stoker hopper, not that shipped from the supplier. It is important to maintain uniform sizing across the hopper. Reproduced with permission from Reference 6.a listed at the end of Part 6.

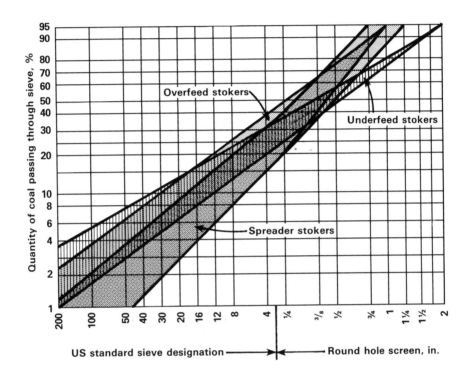

BURNER CHARACTERISTICS

Flame Shape. For a given burner, operating variables such as changes in the mixture pressure or the amount of primary air will affect the flame shape. For most burner types, an increase in mixture pressure will broaden the flame, and an increase in the percent primary air will shorten the flame (input rate remaining the same). Flame thickness is reduced by higher ambient pressure and higher burning velocity.

Burner design, which determines the relative velocities of the fuel and air streams, has much more effect upon flame length and shape than either of the above operating variables. Good mixing, produced by a high degree of turbulence and high velocities, produces a short bushy flame whereas poor mixing (delayed mixing) and low velocities result in long, lazy slender flames. Turbulence and good mixing may be promoted (1) by the use of vanes in the streams to impart

Figure 6.2. Common industrial flame types.

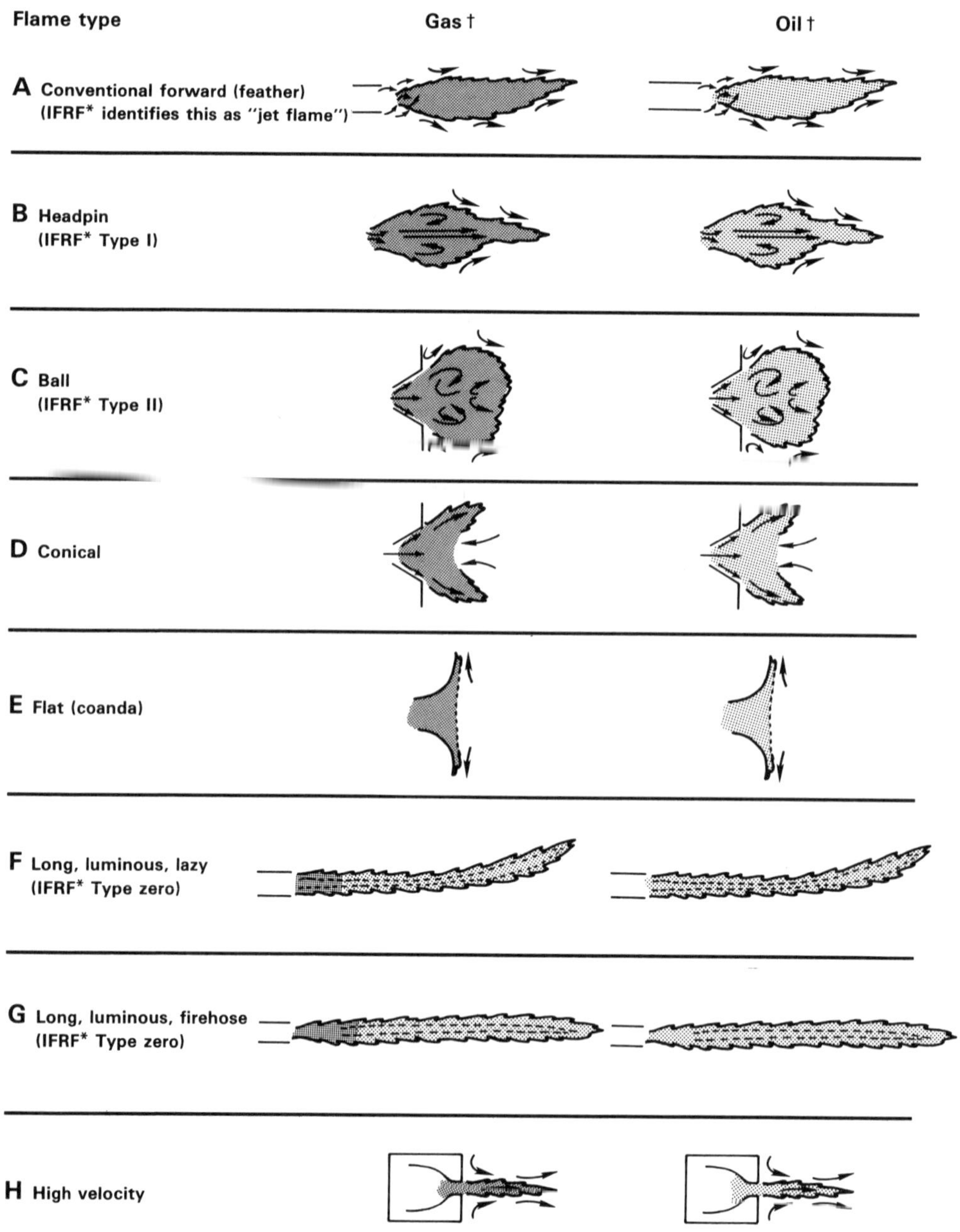

Flame type	Gas †	Oil †
A Conventional forward (feather) (IFRF* identifies this as "jet flame")		
B Headpin (IFRF* Type I)		
C Ball (IFRF* Type II)		
D Conical		
E Flat (coanda)		
F Long, luminous, lazy (IFRF* Type zero)		
G Long, luminous, firehose (IFRF* Type zero)		
H High velocity		

† Dark gray represents blue flame; light gray represents yellow flame.
* International Flame Research Foundation, Idmuiden, The Netherlands.

Figure 6.2. *(concluded)*

Swirl	Recirculation	Jet description	Mixing rate	Uses
none, or weak in center only	little or none	axial	moderate to fast	all purpose combustion chambers
some, swirl number ∼0.3	between primary and secondary jets	primary jet penetrates internal reverse flow, primary jet velocity > secondary jet velocity	moderate	
considerable, swirl number >0.6	hot reverse flow into center, cold forward flow at sides	secondary jet velocity > primary jet velocity	intense	cubical combustion chambers such as stoker-fired boilers, under fluidized beds, in Dutch ovens
high, swirl number >1	cold reverse flow into center	secondary jet velocity >> primary jet velocity	intense	
very high, swirl number ∼2	minimal	radial or swirled flow contained by a refractory shape	fast	to avoid flame impingement; to enhance wall radiation; to focus refractory radiation
none	none	fuel jet velocity and air jet velocity equally low (laminar) -- buoyancy controlled	delayed, slow, diffusion	for uniform coverage in long chambers; to add luminous radiation
none	none	fuel and/or atomizing medium jet velocity >> air jet velocity -- thrust controlled; fuel directed	stretched out	for uniform coverage in long chambers; to add luminous radiation
low	small scale internally, large scale externally	flow contained by refractory shape -- burning inside and outside tile	fast	to drive into loosely-piled loads; to force flow around back sides of loads; to enhance convection; to reach long distances

swirl, (2) fuel and air streams crossing one another, (3) fuel and air streams introduced to the combustion chamber at different velocities, or (4) bluff bodies that form sheltered zones and back-flows. High pressure may tend to throw the fuel farther away from the burner nozzle before it can be heated to its ignition temperature and thus lengthen the flame.

Figure 6.2 illustrates eight common flame types. Types F and G emit more flame radiation than the others, even when burning gas. Type E uses flame convection to heat adjacent refractory; so it heats a furnace load primarily by refractory radiation and is termed a radiant burner, radiation burner, or an infrared burner. (See Figure 6.14.)

Combustion Volume. The space occupied by the fuel and the intermediate products of combustion while burning (flame and invisible combustion) varies considerably with the burner design, the pressures and velocities of the fluid streams, the fuel, and the application. Gas burners with considerable refractory surface and operating with very high mixture pressure and thorough mixing may release as much as forty million gross Btu/hr ft³ of combustion volume. The initial and operating costs are less where less compact combustion is required. The combustion volumes of other types of gas burners range all the way from the above-mentioned figure down to 100 000 Btu/hr ft³. Light oils can be burned at a rate of 100 000 Btu/hr ft³ and heavy oils at 80 000 Btu/hr ft³. (Combustion intensities are discussed on pp 13-14 of Vol. I.)

In some cases, the application itself may limit the rate of heat release. In applications where long luminous flames are required, the delayed mixing type of burner probably will not release more than 40 000 Btu/hr ft³. In boiler furnaces, where continuous operation seldom permits time down for replacement of refractory and boiler tubes, design heat release rates are usually limited to 20 000 to 40 000 Btu/hr ft³.

Stability. This characteristic of burners is very important for safe, reliable operation. A stable burner is one that will maintain ignition, even when cold, throughout the range of pressures, input rates, and air/fuel ratios ordinarily used. (No burner is considered stable merely because it is equipped with a pilot.)

Some burners will function satisfactorily under adverse conditions (particularly, cold surroundings) only if the mixture is rich and if the flame is burning in free air. With such unstable burners, it is necessary to keep the furnace doors open from light-up until a stabilizing temperature develops in the combustion chamber. If the doors are not open, the free air in the furnace will be used up quickly, and an unstable burner flame will be extinguished. It is under these conditions that the presence of a pilot may be a potential source of danger because combustible gases will accumulate rapidly after the flame goes out and they will be ignited explosively by the pilot as soon as a pocket of the mixture in the combustion chamber enters the range between the flammability limits.

A problem that plagued burner users for years was that of burner instability in cold and tight combustion chambers. In many cases, the only way to bring such a furnace up to temperature was to operate at a low firing rate or with a rich air/fuel ratio until the furnace reached about 1600 F above which stability was attainable at full firing rate. In tight combustion chambers, containing no oxygen other than that entering through the burner air connection, rich operation soon uses up the available oxygen but a stable flame must continue burning regardless of the surrounding atmosphere.

Tile-stable burners were developed that could maintain ignition in *cold chambers* with the aid of nothing more than their own refractory tiles; but a problem persisted in *cold and tight chambers* even when burners were operated at stoichiometric or leaner air/fuel ratios. Cold products of complete combustion were recirculating back into the burner tile and surrounding the root of the flame with a cold atmosphere that was low in oxygen. By redesigning burners to eliminate this peripheral recirculation by completely filling the tile with air and fuel, combustion engineers were able to develop burners that were not only tile stable in cold combustion chambers, but also *atmosphere-stable* in tight furnaces. The burners of Figures 6.3d and 6.23 exemplify this ultra-stable feature.

Stability or flame holding may be enhanced by bluff bodies (diffuser plate, step, or ledge), jet tubes, swirl*, or staged air entry. All of these create interfaces between streams of different velocities, producing small scale (fine-grained) turbulence. At every point in a flame front, there is a balance between the incoming gas velocity and the flame propagation rate. A refractory tile surrounding the root of a flame helps by re-radiating heat to the uncombined fuel and oxygen. Recirculation of hot products of incomplete combustion back into the flame center (induced by swirl) also helps provide ignition temperature and added chemical activity.

On the other hand, cooling may be required with premix burners to prevent flashback (a form of instability -- see Flame Speeds, Part 1, Vol. I). A premix flame may be positioned by quenching or cooling with a screen across the mixture stream. To some extent the heat conducted away by the mass of metal in a premix nozzle performs the same function.

---

* Swirl improves flame stability by forming toroidal recirculation zones that recirculate heat and active chemical species to the base of the flame and thereby broaden the range of velocities in which flame stabilization is possible. These toroidal zones form when the swirl number is greater than about 0.6 Swirl number, S, is defined as the ratio

$$\frac{\text{(axial flux of the angular momentum)}}{\text{(axial flux of the linear momentum)} \times \text{(radius)}} = \frac{(M_a \times V_a \times R_a)}{(M_l \times V_l) \times (R_l)},$$

which is dimensionless, where M is mass, V is velocity, and R is the radius of the exit port. Swirl has two added benefits: it shortens combustion time by causing high rates of entrainment of ambient fluids, and it provides quick mixing in the vicinity of the burner nose and along the boundaries of recirculation zones.

Stability is achieved by meeting the requirements for combustion, at least on a localized basis, over a wide range of firing rates and air/fuel ratios regardless of changing temperatures in the surroundings, changing atmospheres, and cross drafts, thus holding the root of the flame in close proximity to the burner nose.

Flame stability depends on many variables, but the following layman's summary explains some of them more specifically. The point of flame initiation must be sheltered so that: 1) air and fuel can be mixed to a ratio that is within the flammability limits (Part 1, Vol. I), 2) the air-fuel mix can be raised to the minimum ignition temperature (Part 1, Vol. I), and 3) the feed speed of the mixture equals its flame speed (Part 1, Vol. I). Figures 6.3a through 6.3h illustrate a number of flame holding arrangements.

Drive. This property of burners relates to the *velocity* and *thrust* of the jet stream of hot gases that they throw into a furnace. Type H of Figure 6.2 is a burner designed to produce high drive.

When fuel was cheap, excess air was used to aid temperature uniformity within a furnace load by (1) reducing the hot mix temperature, (2) preventing stratification, and (3) enhancing convection heat transfer. With high velocity burners that induce recirculation of furnace gases, the recirculating gases produce the above three benefits that formerly required excess air and therefore wasted fuel. The high velocity burners can push their hot gases into a loosely-piled load (such as castings, or a hack of bricks) with greater velocities than were possible with most of the older excess air burners; so forced convection is improved to the interior of the load. Words such as *pierce, punch, scrub,* and *stir* provide good mental images of the advantageous convection and agitation from burners with "drive."

Another use for burners with drive is to reach and wrap around parts of a load located at a distance from the burners. This reduces the heating time for loads in furnaces with long dimensions parallel to the burner centerlines, and for large pieces, the back side of which cannot be "seen" well by radiation or "reached" well by other types of convection burners.

Turndown. The range of input rates within which a burner will operate is specified by the burner turndown ratio. This is the ratio of the maximum to minimum heat input rates with which the burner will operate satisfactorily. For any burner with fixed air orifices, the turndown ratio is also the square root of the ratio of maximum to minimum pressure drops across the orifice. For example, if the maximum supply pressure is 13 in. of water and the minimum is 0.25 in. of water, then the turndown ratio is $\sqrt{13/0.25} = 7.22$ (to 1). Limitations on the fuel supply pressure may limit the fuel flow before the maximum air capacity of the burner is reached.

Figure 6.3. **Flame holding arrangements.** Cases (a) through (f) are various forms of bluff bodies creating fine-grained turbulence in their wake. Cases (c) through (g) constitute air jets blasting through a relatively quiescent volume of raw gas. Case (f) may be cylindrical, like a jet engine burner, or trough-like.

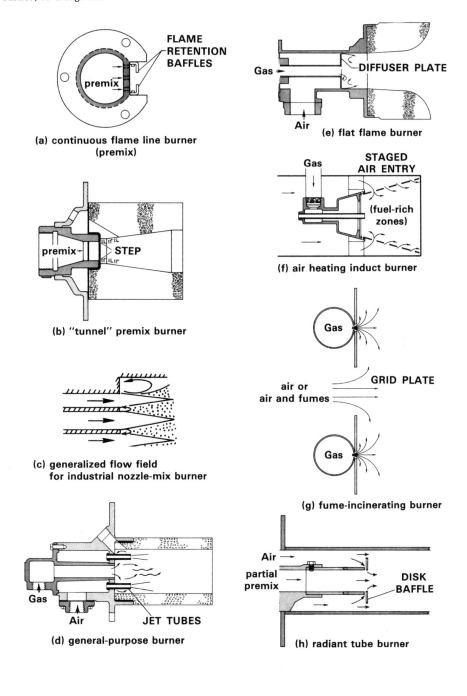

(a) continuous flame line burner (premix)

(b) "tunnel" premix burner

(c) generalized flow field for industrial nozzle-mix burner

(d) general-purpose burner

(e) flat flame burner

(f) air heating induct burner

(g) fume-incinerating burner

(h) radiant tube burner

The maximum input rate is limited by a form of instability known as flame *blow-off* (which results from mixture velocity exceeding the flame velocity) and by the cost of equipment for developing higher pressures. The minimum input rate is limited by the phenomenon known as *flashback* (which results from the flame velocity exceeding the mixture velocity) and by the minimum flow with which the ratio control equipment and its transducers will function. The former limitation applies to premixing, but not nozzle-mixing burners. For low pressure air atomizing oil burners, the minimum firing rate is equivalent to the atomizing air supply rate. This limits the turndown range of these burners. For example, if the atomizing air flow for an oil burner is 1260 cfh and the required turndown ratio is 5 to 1, then the maximum firing rate must be 5 × 1260 = 6300 cfh of air (combustion and atomizing air combined).

A high turndown ratio is particularly desirable in batch-type furnaces where a high input rate is needed during the initial heat-up of the furnace or immediately after charging, but where this high input rate cannot be used during the entire heating cycle. Considerably less turndown is needed for continuous furnaces which are seldom started from cold. The cost of an occasional long starting period may be less than the cost of the larger equipment required for a high turndown ratio.

In some instances where temperature distribution is not too critical, it is possible to shut off some burners when on low fire and thus simulate a high turndown ratio.

BURNER COMPONENTS

The burner *nozzle nose*, or port may consist of single or multiple openings. Their varieties are infinite, and will be discussed in more detail in connection with specific burner types. The nose should be well cooled (a) to protect it from thermal destruction by radiation from the flame and furnace, and (b) to prevent flashback in the case of premix burners. Heat-resistant cast iron of generous cross section assures sufficient cooling by conduction and natural convection in some cases. Others require high temperature alloys, refractories, and/or artificial cooling. The mechanical stress on a burner nose is usually low, but metal scaling due to high temperatures can lead to serious problems.

The *burner tile* (quarl combustion block, burner refractory) is a refractory shape with a conical or cylindrical hole (flame tunnel) through its center. The tile is sold and shipped as part of the burner except for very large burners where it is either handled separately or cast or rammed in the furnace wall. In addition to serving as the insulating separator between the hot furnace and the cool burner parts, the tile radiates heat into the incoming fuel and air, thus helping to maintain ignition. Flames tend to be more stable over a wide range of inputs if a burner tile is used. Proper installation of burner tiles is discussed in Part 8.

The major elements of the backworks of a burner are: a) the *body* or *air plenum*, which may incorporate the air or mixture connection, swirl vanes, and the nose; b) the *mounting plate*, which holds the body, tile, pilot, and flame-monitoring device to the furnace shell in proper alignment; c) the *gas connection*, gas tube, and gas nozzle often with a flame stabilizer; d) the *atomizer*, which includes oil and atomizing medium connections, tubes, vanes, and nozzles.

Burners of the *sealed-in* or *closed* type usually supply all of the air for combustion through the burner, whereas the *open* type may induce air flow into the combustion space through the opening around the burner. Sealed-in type burners permit accurate control of the air/fuel ratio, a wide range of furnace operating pressures, and a wide range of input rates (good turndown). Open burners, however, permit greater capacities by virtue of their induced air. This increased capacity is attainable only if a correspondingly larger fuel capacity is also available. An attempt is sometimes made using an air register shutter to control the amount of induced air, but these usually track poorly if turndown is required.

A *burner pilot* is a small burner used to light the larger main flame. Figure 6.4 illustrates a typical gas pilot with spark ignition for ease of lighting. The functions of the various pilot accessories are explained later in the section on gas burners. An *interrupted pilot* (sometimes called *ignition pilot*) burns during the flame-establishing period and/or trial-for-ignition period, and is cut off (interrupted) at the end of that period. An *intermittent pilot* burns during "light-off" and the entire period that the main burner is firing, and is shut off with the main burner. A *continuous pilot* (sometimes called *constant, standby, or standing pilot*) burns throughout the entire period that the furnace or oven is in service whether or not the main burner is firing.

Figure 6.4.   Pilot arrangement designed to operate with low pressure air and gas. Spark ignition permits its use as an interrupted pilot. The swivel fitting simplifies piping to the pilot hole in the burner mounting, which is usually at an odd angle. The pilot tip is a flame retention nozzle.

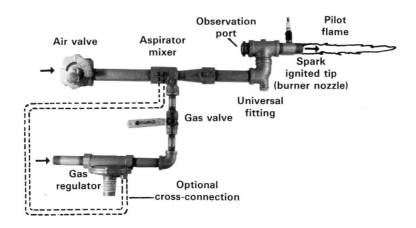

Most pilots are small premix gas burners with flame retention tips, designed for natural gas or LP gas; but direct electric ignition or spark-ignited light oil pilots are also used. Pilot systems are discussed in Parts 7 and 8.

Every burner should have an *observation port* for safety's sake; so that the main flame, pilot flame, and flame rod (if any) can be seen by the operators. A tiny jet of air connected to the side of the port will help keep the glass clean. For safety, and to facilitate burner adjustments, an additional observation port through the furnace wall, allowing a side profile view of the flame, is strongly recommended.

GAS BURNERS

Industrial gas burners may be classified as premix, nozzle-mix, or delayed mix burners, according to the position and manner in which the gas and primary air are brought together.

Premix Gas Burners. In premix gas systems, the primary air and gas are mixed at some point upstream from the burner ports by an inspirator mixer, an aspirator mixer, or a mechanical mixer. The burner proper ("nozzle") serves only as a flame holder, maintaining the flame in the desired location. Theoretically, if mixture velocity equals flame velocity, a flame will stand stationary at any point at which ignition is applied. Actually, a relatively cool burner nozzle (or port) is needed to serve as a flame stabilizer. If the flame advances too far into the port as a result of a momentary reduction in mixture velocity, the cool nozzle tends to quench it to prevent flashback.

Small Port or Ported Manifold Burners. A great many types of burners may be used in conjunction with premixers. One of the most common of these consists of a manifold containing a series of small ports. Most domestic gas burners are of this type. They are often called *atmospheric burners* because of the very low pressure at which they operate and because they rely on a high percentage of secondary air. Small port or ported manifold burners are usually quiet. If the flame on one part of the burner is blown out, the flame from the other part may act as a pilot to re-ignite it if the ports are close together. This type of burner design purposely spreads the heat input over a wide area, which is ideal for low temperature processes. Industrial use of this type is limited to low temperature applications such as make-up air heating, varnish kettle heating, drying ovens, baking ovens, food roasters, and deep fat vats. Figure 6.5 shows examples of small port premix burners.

Large port or pressure type gas burners (blast burners, pressure burners) permit a high rate of heat release within a relatively small space. This type includes a multitude of designs for special applications, but is generally characterized by a single mixture port (or nozzle) that produces a short, intense flame (Figure 6.6). They are often installed with a separate mixer for each port, although several burners may be supplied by a single mixer with proper manifolding.

Large port burners may be of the open or sealed-in type. A rich mixture (only partial aeration) is often supplied to open burners and nozzles, and induced secondary air provides the balance required for stoichiometric combustion. This practice saves on required blower capacity, but usually wastes fuel because the air/fuel ratio cannot be controlled.

When used as a nozzle or open burner without a refractory tile, a large-port premix burner often includes a flame retaining feature consisting of a number of small bypass ports feeding into a recessed piloting ring encircling the main nozzle. See Figure 6.7. These bypass ports have a greater resistance to flow than the main port; so the velocity in their exit ring is lower, and consequently, the tendency for the flame to blow off is greatly reduced. If any irregularity should cause the main flame to be blown off, the ring of flame, fed by the bypass ports, serves as a pilot to relight the main flame. Burners with bypass ports are therefore referred to as *flame retention burners* or *self-piloting burners*. The mixture jet issuing from the main port entrains some of the surrounding air. The incoming air bends the pilot flames toward the base of the main mixture jet. This allows richer primary mixtures, thereby making possible a greater turndown ratio. Most pilot tips are flame retaining nozzles (Figure 6.4).

Figure 6.5. Small port premix burners. a is a continuous flame burner, which is more stable than a drilled pipe or ribbon burner. A second and third row of ports are located under the stainless steel baffle strips. **a** is installed in a duct, as for heating make-up air or on a recirculating oven or dryer. The burners on **b** and **c** are blast tips.

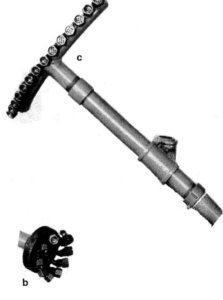

When sealed-in with a refractory tile, large port burners are often referred to as *tunnel* burners. See Figures 6.6 and 6.8. Large port premix burners have been used in a wide variety of industrial applications, including kilns for porcelain, tile, and brick, and furnaces for heat treating, forging, and melting. However, most have been displaced by nozzle-mix burners because (1) premix burners usually do not have wide range stability characteristics, (2) premix burners cannot be adapted to dual-fuel configurations as well as nozzle-mix burners, (3) nozzle-mix burners can be designed for a greater variety of flame shapes, and (4) large manifolds of premixed air and gas can be dangerous.

Types of Premixers. Figure 6.7 shows an *inspirator* or gas-jet venturi mixer, which utilizes the energy in the gas to induce primary air in proportion to the gas flow. This is the only type of mixer with which no air blower is required. A similar mixer is used in the atmospheric burners on most domestic gas burning appliances. Their percentage of primary air required in these applications is small enough to be induced by the low gas pressures available in domestic lines. Industrial applications usually require greater turndown and greater heat release per unit volume, so inspirators are used in industry only where high pressure gas is available. Good practice dictates that industrial inspirators for manufactured gas need at least 5 psi gas pressure, and for natural gas, at least 10 psi gas pressure. Inspirators can rarely be used with propane or butane gas in industries because 25 to 30 volumes of air must be induced by one volume of gas, requiring an oversized nozzle and undersized spud.

Figure 6.6. Typical large port premix burner flame such as from the burner of Figure 6.8. The 12 in. long blue natural gas flame is releasing 296 000 Btu/hr at 8¼ ″wc mixture pressure.

To obtain proportional air inspiration at high mixture pressures, the venturi throat must be smoothly machined and carefully aligned with respect to the gas orifice. The spud size chosen depends upon the desired capacity, air/fuel ratio, gas pressure, and gas gravity. For a given throat size, however, the range of workable spud orifices is limited. Part 5, Vol. I, discusses flow calculations for gas spuds.

Inspirator mixers may be used on a furnace or oven having a steady combustion chamber pressure, but capacity is materially affected by the installation and pressure conditions within the combustion space. Negative pressures increase inspirator capacity and retard flashback. Positive pressures reduce

Figure 6.7. Inspirator (gas-jet) mixer feeding a large port premix nozzle (open burner). High velocity gas from the spud entrains and mixes with air induced in proportion to the gas flow. Gradual enlargement converts mixture velocity to static pressure. The spring-loaded plunger is for cleaning the spud orifice (for dirty gases). The lower half of the flame-retaining nozzle is sectioned to show the bypass ports that relight the main flame if it is blown out. An aspirator (air-jet) mixer (Figure 6.8) could also be used with this burner.

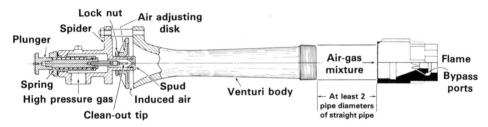

Figure 6.8. Aspirator (air-jet) mixer feeding a sealed-in large port premix tunnel burner. The step between the nozzle and the refractory tile provides bluff body turbulence for flame stabilization. See Figure 6.3b. Blower air enters at lower left. Gas from an atmospheric regulator is pulled into the air stream from the annular space around the venturi throat in proportion to the air flow. A V-port adjustable gas orifice, at top, is for initial setting of air/gas ratio.

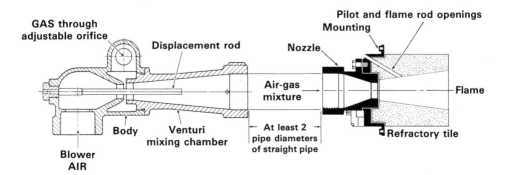

inspirator capacity and increase the probability of flashback. Accurate air/fuel ratio control is difficult because of (1) changing combustion chamber pressure, (2) changing furnace room pressure, and (3) drafts. The amount of piping between the inspirator and burner(s) should be kept to a minimum. An inspirator must induce a large volume of air with a small volume of gas; so the range of turndown is limited. As a rule of thumb, it may be assumed that flashback occurs when the mixture pressure drops below 0.25"wc for natural gas and other slow burning gases, or below 0.40"wc for fast burning gases.

Figure 6.8 is a cut-away view of an *aspirator* (air-jet) *mixer*, a mixing and proportioning device for low pressure air (3 to 24 osi) and "zero gas" (gas at atmospheric pressure). Air is pushed through a venturi so that the low pressure in its throat induces gas into the air stream in proportion to the air flow. Controlling the air flow thus controls the gas flow, giving proper proportioning with single (air) valve control. The diffuser (downstream section of the venturi) gradually reconverts the velocity into mixture pressure (about 30% of the air supply pressure). An adjustable gas port permits manual setting of air/gas ratio. The pilot set of Figure 6.4 includes an aspirator mixer.

For an aspirator type mixer to function properly, about *a 2.5:1 ratio must exist between burner orifice area and mixer throat area.* If the burner orifice is too large, less mixture pressure will be developed, thus reducing available pressure for turndown. If the burner orifice is too small more mixture pressure will be developed but less suction will be available for inducing gas, thus reducing the capacity. This may be critical for low Btu gases. Since the flow resistance in the piping beyond the mixer has to be considered, it is not always possible to establish the best area ratio until after installation of the equipment. The effective area of the mixer throat may be adjusted after installation by inserting a displacement rod of different diameter through the mixer throat.

In *mechanical mixers*, gas is admitted to the air inlet of a compressor, blower, or fan. Such units may include controls for proportioning the air and gas, often using zero gas. Any mixer is susceptible to flashback, but it is more detrimental in a *fan mixer* system because the blower housing is filled with a combustible mixture.

Nozzle Mixing Gas Burners. As the name implies, the gas and combustion air do not mix until they leave the ports of this type of burner. The two fluids are kept separate within the burner itself, but the nozzle orifices are designed to provide mixing of the fluids as they leave. The principal advantages of nozzle-mix burners over premix burners are:

1) The flame cannot flash back upstream of the nozzle, because fuel and air are not premixed. This not only adds to stability, but reduces explosion hazard with larger burners. Premix delivery pipes and manifolds of schedule 40 pipe or thinner may rupture in 4" and larger sizes.

2) A wider range of air/fuel ratios is possible. Premix burner must operate within the flammability limits of the fuel. Only the initial mixing area of a nozzle-mix burner needs to be within these limits; so excess air or fuel can be "staged" into the flame downstream, resulting in apparent air/fuel ratios beyond the flammability limits.

3) Greater flexibility in burner/flame design is possible with nozzle-mix burners, which lend themselves to combination (Dual-Fuel™) designs, and to developing a variety of flame shapes as shown in Figure 6.2.

Figures 6.3d-h show some of the great variety of nozzle-mix configurations in current use. The burners of Figures 6.3d, 6.9, and 6.10 might be classed as *conventional forward flame burners* They simply inject hot combustion gases in a single jet into a large furnace space, where the heat is transferred to the load by a combination of convection, hot gas radiation (CO_2 and H_2O), flame radiation, and refractory re-radiation. If the loading is high relative to the heat input so that the average temperature remains low (below about 1200 F), convection heating will probably predominate. At higher temperatures, radiation is usually dominant.

Figure 6.9.   Flame of conventional forward nozzle mixing burner (type shown in Figure 6.3d) operating on correct air/gas ratio at 8 osi air pressure at a 250 000 Btu/hr rate. Numerals mark inches from the face of the refractory tile. The fuel is natural gas; the flame is blue.

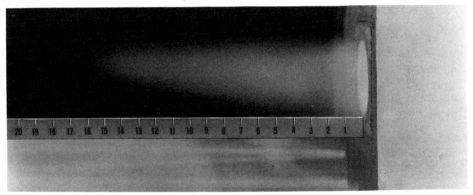

Convection Burners. No burner transmits heat entirely by convection to the exclusion of radiation. Similarly, any "radiation burner" provides some convection. These names simply describe the greater part of the burners' heat transfer.

Figures 6.11a and b illustrate nozzle-mix convection burners that have hot gas velocities at their tile exits on the order of 400 miles per hour. The air wipes the inside of the tile, keeping it cool (no color) despite the intense high speed combustion reaction. The high velocity jet not only aids convection heating, but induces much recirculation of combustion gases within the furnace. This recirculation aids temperature uniformity and fuel economy by lowering the hot mix temperature while re-using the gases with several extra passes over the load. It also prevents stratification (hot tops, cold bottoms).

Figure 6.10.   Forward flame nozzle-mix burner on a fire tube steam generator. A choice of mixing arrangements permits tailoring the flame length to the immersion-tube-like combustion chamber.

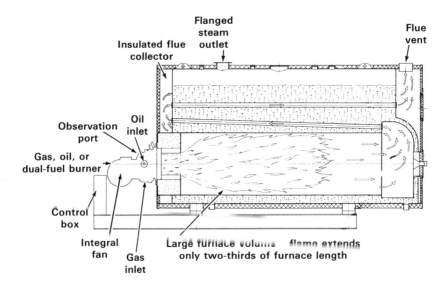

High velocity burners are designed to increase convection heat transfer, regardless of the temperature level or loading in the furnace. This convection heat transfer may incidently heat the refractory walls as well as the load, resulting in considerable radiation heating.

Figure 6.11b shows the flame of a high velocity burner such as that of Figure 6.11a, but with the refractory tile replaced by an alloy chamber having the same internal shape, for use in high temperature air heating or thin-wall furnaces, where a conventional refractory tile would be cumbersome. In Figure 6.11a or b, the refractory or alloy chamber serves as both mixing and pre-combustion chamber.

Figure 6.11a.   Partial cross section of a high velocity nozzle-mix burner designed to produce a jet of hot gases that improves convection heating and induces strong recirculation within a furnace. The flame is the same as in Figure 6.11b.

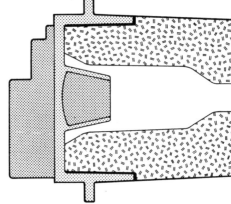

Figure 6.11b.   Flame of a high velocity burner. The internal design is the same as for the burner of Figure 6.11a. Use of this alloy mixing chamber instead of a refractory tile is more convenient for applications such as high temperature air heating and for furnaces with thin-wall ceramic fiber linings.

Figures 6.11c and 6.11d show larger sizes of high velocity burners, such as used in aluminum melting furnaces and for drying/preheating large ladles. As with the burner of Figure 6.11a, 10 to 60% of the combustion may occur inside the nozzle, expanding the poc and as-yet-unburned fuel and air, thereby producing a very high exit velocity. Heat release rates are as high as 418 000 gross Btu/hr for each square inch of nozzle opening.

Figure 6.11c.   High Velocity Dual-Fuel™ Burner. Two stage nozzle mixing permits a stable initial reaction between the eight internal nozzles and the single final nozzle, with the balance of combustion occurring in an external needle-like flame of intense penetrating capability. See Figure 6.11d. An all metal version, without the refractory tile, is used for some lower temperature drying operations.

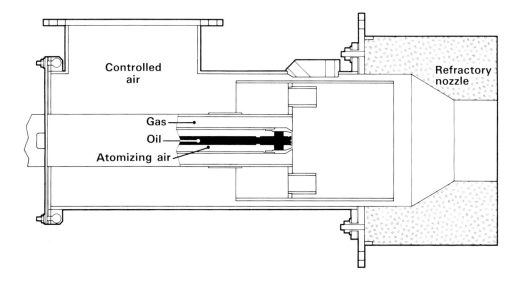

Figure 6.11d.   Flames of the burner of Figure 6.11c. Scale marks are one foot spaces.

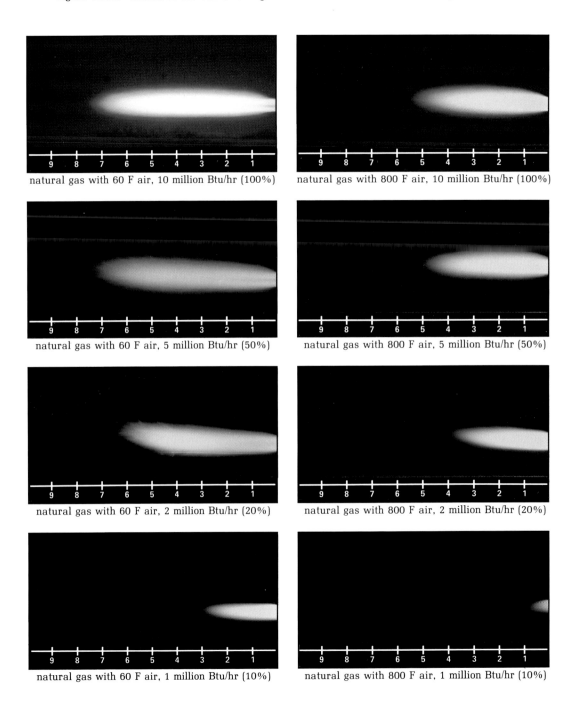

natural gas with 60 F air, 10 million Btu/hr (100%)   natural gas with 800 F air, 10 million Btu/hr (100%)

natural gas with 60 F air, 5 million Btu/hr (50%)   natural gas with 800 F air, 5 million Btu/hr (50%)

natural gas with 60 F air, 2 million Btu/hr (20%)   natural gas with 800 F air, 2 million Btu/hr (20%)

natural gas with 60 F air, 1 million Btu/hr (10%)   natural gas with 800 F air, 1 million Btu/hr (10%)

Figure 6.11d. *(concluded)*

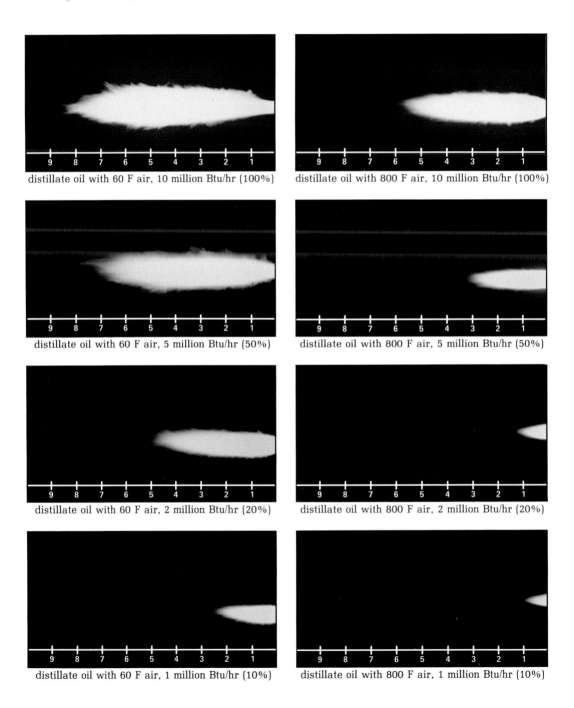

distillate oil with 60 F air, 10 million Btu/hr (100%)

distillate oil with 800 F air, 10 million Btu/hr (100%)

distillate oil with 60 F air, 5 million Btu/hr (50%)

distillate oil with 800 F air, 5 million Btu/hr (50%)

distillate oil with 60 F air, 2 million Btu/hr (20%)

distillate oil with 800 F air, 2 million Btu/hr (20%)

distillate oil with 60 F air, 1 million Btu/hr (10%)

distillate oil with 800 F air, 1 million Btu/hr (10%)

Another approach to convection heating is by heating large volumes of air and then circulating that air over the load in *recirculating ovens, dryers, and air heaters*. Most of these operate at lower temperatures (below 1200 F), often because of potential damage to the load if conveyor stoppage should expose it to higher temperatures. Figure 6.3f illustrates a nozzle-mix convection burner that can be installed as an in-duct air or fume heater (incinerator) or as a through-the-wall burner on an oven or dryer (recirculating or once-through type).

Still another approach to convection heating utilizes the same internal construction that resulted in the Coanda effect in Figure 6.2, type E, causing a flame and hot gases to swirl along the wall of an *immersion tube*, producing a high velocity scrubbing action that results in a high convection heat transfer rate. See Figure 6.12. Unlike radiant tubes, discussed in the following section, immersion tubes can withstand very high heat release rates because their outside surfaces are well cooled by surrounding liquid. For best fuel efficiency, sealed-in burner mountings are preferred because they prevent the tube from acting like a chimney that could pull in heat-absorbing excess air. This combination of the cold walls, confined space, and sealed-in burner presents what is termed a *cold and tight combustion chamber*, a most difficult test of a burner's stability.

Figure 6.12. Immersion Tube Burner. Efficient convection heat transfer and very high heat release rates are made possible because the swirling flame scrubs off the stagnant (insulating) gas film on the inside of the tube wall.

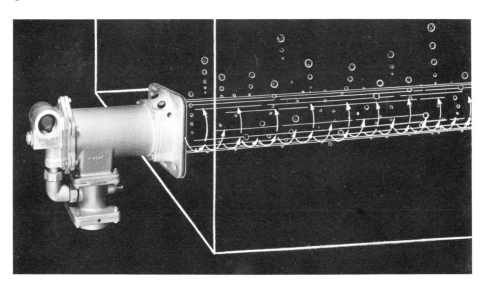

Radiation Burners. Where penetration or forward drive by the flame and hot combustion gases is to be avoided*, or when radiation heat transfer is to be enhanced†, radiation burners should be used. Figure 6.13 shows the flame of the radiation type nozzle-mix burner of Figure 6.2 type E and Figure 6.14c. (A few premix burners are also designed to produce similar flame shapes.) This radiation type of burner actually heats its own refractory tile and the refractory surface of a surrounding furnace wall or roof by convection from the high velocity combustion gases thrown sideways from the burner. These hot refractory surfaces then radiate heat to the furnace load. Figure 6.14 sketches a variety of commercially available refractory shapes for radiation burners. All but "e" utilize the Coanda effect. Where true radiant heating is desired, it is important that the hot gases have no final velocity in a direction toward the work to be heated. Most of the radiation burners of Figure 6.14 also provide some convection heating. In fact, the spreading feature of the flat flame type (Figures 6.13 and 6.14c) has been used to fill more completely a very wide plenum feeding hot convection gases through multi-tubular loads.

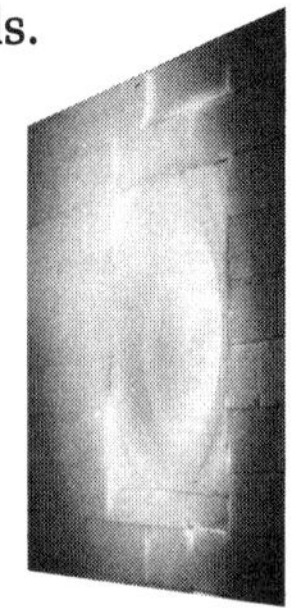

Figure 6.13. Flame of a radiation type nozzle-mix burner. The flame follows the contour of the refractory tile, turning 90° from the burner centerline. The burner is used in both sidewalls and roofs of furnaces. Instead of impinging on a load, the Coanda type flame heats the surrounding refractory by convection, so that it can radiate to the load.

Figure 6.14. Seven styles of radiation burners. Styles **a, c, f,** and **g** are nozzle-mix; **b, d,** and **e** are premix. Burners **b** and **c** heat surrounding refractory as well as their own tiles, and have almost no forward gas motion. Style **e** utilizes a porous refractory "membrane." Style **g** is designed to focus radiation in an intentional hot spot.

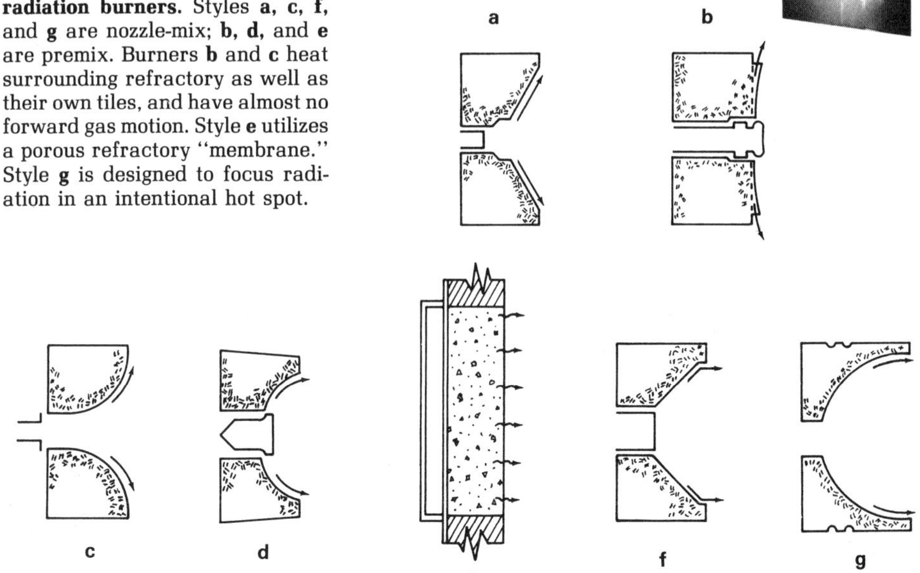

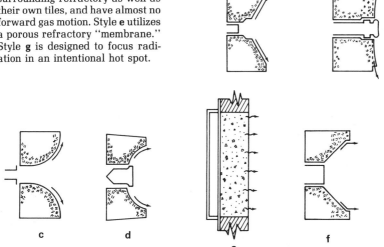

---

* As when firing in close proximity to a valuable container or a bank of tubes.

† Thin, flat, or rotatable loads; dry-hearth melting; top-fired liquid holding.

Delayed Mixing Gas Burners. Long luminous flames are created by this special form of nozzle-mix burner. In some operations, direct flame radiation over a large area is desirable. This is frequently the case in wide or extremely long furnaces where poor heat distribution (that is, hot spots and cold spots) would be obtained from conventional short clear flames, or where all burners must be located at one end. The ideal heating arrangement would probably consist of thousands of tiny clear-flame burners positioned all around the load, but that would not be practical. Luminous flames have a considerable length and they can fill a large volume of combustion space with flames having little temperature variation along their length. The uniform temperature throughout the combustion space permits more effective use of the hearth area.

Long flames can be produced if the rate of mixing of the gas and air is very low so that the two fluids travel a considerable distance from the burner before complete mixing and burning. The flame is often termed a *diffusion flame* because the mixing occurs as the parallel laminar air and gas streams diffuse into one another. See Figure 6.15 and flame type F on Figure 6.2.

Strong heating of the gas in the absense of air causes thermal cracking (polymerization) of the fuel molecules into light and heavier molecules. The latter become micron-sized soot particles. These opaque particles absorb heat from the flame, become luminous, and emit radiant energy. Increasing concentrations of particles radiate more energy.

It is difficult to produce a long gas flame that is not luminous or a luminous gas flame that is not long. The air/fuel ratio of delayed mix burners cannot be judged by the appearance of the flame. It is possible to supply considerable excess air, producing an undesirably oxidizing atmosphere while maintaining what appears to be a rich reducing flame.

Figure 6.15.   Delayed mixing flame, cross section. Burning is at the air-gas interface. See Figures 6.16 and 6.24.

Another way to create a delayed mix flame is with a fire-hose-like jet of fuel that streaks far across the furnace before the viscous friction of the surrounding low velocity air sluffs off its boundary layers, mixing with and burning all the fuel. See Figure 6.15 and type G on Figure 6.2. This is a form of fuel directed burner, discussed below.

Radiant Tube Burners. Radiant tubes are made of expensive alloy or ceramic material; so it is important that no part of them be damaged by overheating, but also that every inch of their length be utilized to fullest advantage. This requires that the flame within the tube must release its heat at a uniformly high rate throughout the tube length. A delayed mixing flame accomplishes most of this requirement, except that it is rather slow in getting started. To avoid a wasteful cool section at the burner end of the tube, a partial premix is incorporated into the burner. This produces a blue flame for about the first foot of tube length until the luminous flame develops. (See Figure 6.16.)

Figure 6.16.   Flame of a radiant tube burner. A pyrex radiant tube used for radiant tube burner development illustrates the premix-then-luminous flame used to obtain ideal heat distribution for efficient radiant tube operation. The burner is at the left. The flame is first blue, then yellow.

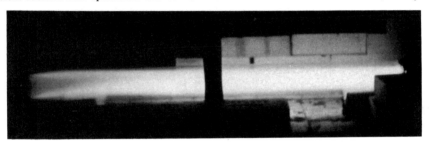

Radiant tube burners constitute one of the most difficult design assignments for combustion engineers. They not only crowd the flame into a very confining cross section and need a long strung-out heat release, but they are tight chambers and, on start-up, they are cold chambers. As a result of all these requirements, their use, operation, and maintenance are subject to more severe limitations than with many burners. Doing all this with oil is doubly difficult.

Fuel Directed® Gas Burners. The nozzle-mix gas burners discussed previously have been "air-directed"; i.e. the energy from the air supply produced the mixing, stability, and flame shape and character. Very low gas pressure was required.

A new generation of industrial gas burners utilizes the energy from the pressure of the fuel gas (3 to 20 psig) to do the mixing, stabilizing, and flame shaping; so less air pressure is required. No vanes or turbulating discs are required; so the burner throat is relatively open, the cantilevered gas tube has fewer weighty objects to support, and there is less exposure of critical components to damage from flame and furnace heat.

Figure 6.17.   Fuel Directed® burner flames.

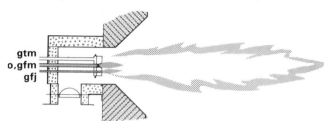

gtm
o,gfm
gfj

gtm = gas, tangential, modulated
grm = gas, radial, modulated
  o   = oil
gfm = gas, forward, modulated
gfj = gas, forward, jet

a) Type G flame from gfm and gfj.

b) Type G long luminous flame.

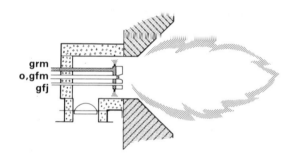

grm
o,gfm
gfj

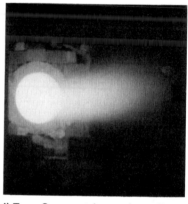

c) Type C flame from grm.

d) Type C prompt heat release flame.

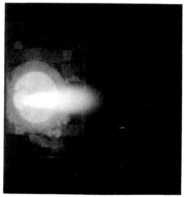

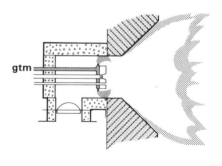

gtm

e) Burner with grm modulated to a low fire
rate, and with gfj.

f) Type D flame, from gtm.

On new installations, the first cost of the combustion air blower is decreased because lower air pressure is required. On retrofit and new jobs, the lower power needed for low pressure air is a saving. The replacement energy comes from the pressure in the fuel supply line, which is usually free and already available with natural gas and with oil atomizers. Some other fuels may require a pressure booster. In areas where there is to be a mix of furnaces using high and low pressure fuel, added pressure regulators may be required.

The flame character of Fuel Directed® burners can be changed by adjusting the fuel pressure, and by supplying the fuel through any of several sets of fuel ports, such as forward, radial, tangential. This allows a start-up setting attuned to the combustion chamber's needs; or programmed changing of the flame configuration. The latter may be in response to timers, temperature sensors, loading swings, changes in load configuration (as a scrap pile melting down), or other parameters.

Figures 6.17a through f show some of many available designs of Fuel Directed burners. A small jet of un-modulated fuel helps maintain the flame momentum even when the burner is turned down to a low input rate. Manipulating solenoids in the separate fuel supply lines can change flame while operating.

GAS BURNERS WITH INTEGRAL HEAT RECOVERY DEVICES

Part 9 discusses heat recovery principles and systems; plus the reasons for using integral burner-recuperators and integral burner-regenerators.

Burners for integral burner-recuperators must have (1) internal materials capable of withstanding the high temperature air that flows through them and (2) hot face materials that will not be damaged by the higher temperature flame. High velocity burners are preferred to prevent short circuiting of the flue gases to the nearby flue/recuperator. It is best to avoid having a flue (recuperator entry) directly above a burner because the low fire flame will short circuit to the flue without transferring much heat in the furnace. In this condition, a recuperator might be damaged by contact with products of combustion near flame temperature.

Integral burner-regenerators are best used with Fuel Directed type burners. (Air-directed types would have their vanes or other stabilizing devices overheated by high temperature air, and they add to the weight that must be supported in the high temperature air stream by the cantilevered gas tube.) It is desirable to keep the pressure drop across the burner air passage to a minimum, taking most of the drop through the heat reclaiming matrix.

A compact burner-regenerator is shown in Figures 6.18a and b and discussed in more detail in Part 9.

Figure 6.18a.   Sectional view of an integral gas-burner-regenerator (Heat Reclaimer). Refer to Figure 6.18b to see how air and flue gas alternately flow through this unit. A bottom-dump and top hatch arrangement permits convenient removal of the bed material for cleaning or replacement. The high surface area of the bed material results in higher heat transfer effectiveness than normally practical in recuperators.

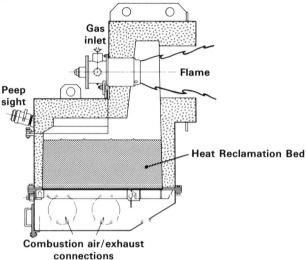

Figure 6.18b.   Schematic piping for a pair of direct-fired integral gas-burner-regenerators (TwinBed®). Regenerative systems always involve moving parts -- in this case, air, fuel, and flue gas valves -- but are capable of developing very high air preheat temperatures without thermal stress/rupture problems.

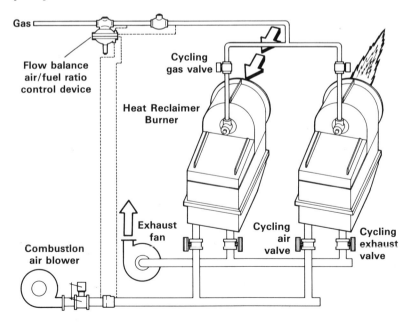

BURNERS FOR LIQUID FUELS

Oil and other liquid fuels must be vaporized before they can be burned. Some small capacity burners accomplish this vaporization in a single step by direct heating of the liquid. Such burners are called *vaporizing burners*. Typical examples of these are the blowtorch, gasoline stoves, wick-type kerosene burners, and early domestic burners wherein oil was vaporized from a flame-heated plate. Most large capacity industrial oil burners use two steps to get the oil into combustible form -- atomization plus vaporization. By first atomizing the oil and thus exposing the large surface area of millions of droplets (10 to 1000 microns diameter) to air and to heat, *atomizing burners* are able to vaporize oil at very high rates.

Requirements for good vaporization after atomization are: 1) a large volume of air must be intimately mixed with the oil particles, 2) the air must be turbulent and at high velocity to produce a scrubbing action for rapid mass transfer from the surfaces of the oil particles, and 3) heat from the flame should be transferred into the incoming spray. This last requirement is the function of the burner tile and recirculation. Functions (1) and (2) may be accomplished by blowing air through the oil, the air velocity being high relative to the oil, or by throwing the oil through calm air at a high velocity relative to the air.

Low pressure air atomizing oil burners are 2-fluid atomizers that utilize air at 1 to 2 psi as the oil atomizing medium. A well designed atomizing unit

---

Figure 6.19. Low pressure air atomizing oil burner. The principle of operation is similar to that of the gas burner of Figure 6.3d except that light oil is sprayed by the atomizing air in a cone of vapor so that it is intercepted by eight air jets. Some designs of low pressure air atomizers can vaporize heavy oils if their viscosity has been reduced to 100 SSU.

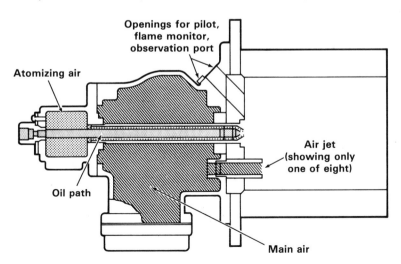

may use as little as 10% of the total air requirement for atomization. This is equivalent to about 150 ft³ of atomizing air per gallon of oil. These figures are only approximate and may be considerably influenced by the design of the atomizer and the viscosity of the oil. Low pressure air atomizers are usually designed to handle oil of 100 SSU viscosity. The oil pressure at the burner is usually 1 to 5 psi -- just enough for positive delivery and flow control. An oil pump is necessary only to deliver the oil to the burner.

Figure 6.19 illustrates a low pressure air atomizing oil burner. Input through this type of burner is controlled by a throttling valve in the air line. The atomizing air must not be throttled, so it is delivered to the burner through a separate atomizing air connection that by-passes the control valve in the "main" (or "combustion") air line. The main air flow may be shut off completely for low fire and for starting. The minimum firing rate and the turndown ratio are determined by the atomizing air flow rate. The atomizing air pressure must remain constant at all firing rates.

Figure 6.20 is a photograph of a typical flame of a low pressure air atomizing oil burner. Low pressure air atomizing type oil burners are applicable to a greater variety of uses than any other single type. Some of their advantages are: no high pressures involved, relatively large air and oil orifices that minimize maintenance, no intricate or delicate parts, simplicity of operation and control, flexibility to changes in loading or fuel, simplicity of installation, no moving parts, and economical operation.

High pressure air or steam atomizing oil burners are 2-fluid atomizers that use steam or compressed air to tear droplets from the oil stream and propel them into the combustion space. The high velocity of the oil particles relative to the air produces the scrubbing action required for quick vaporization. These burners can atomize light to very heavy oils, sludges, pitch, and some tars. They are often used for incineration of liquid wastes. See Figure 6.21a-d.

Figure 6.20. Oil flame of a low pressure air atomizing burner firing at 1 785 000 Btu/hr with 1 psi atomizing air and 1 psi main air. The white lines on the pipe above the flame indicate 1 foot intervals. This forward flame (type A, Figure 6.2) is produced by the burner of Figure 6.23.

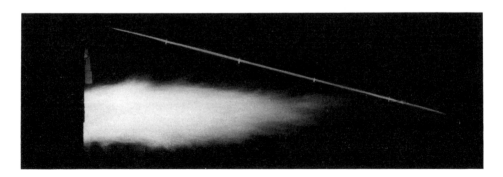

Figure 6.21a.   External mixing oil atomizer uses steam or compressed air. Straight-through oil tube is easily cleaned. It needs only low oil pressure controlled with a conventional oil regulator.

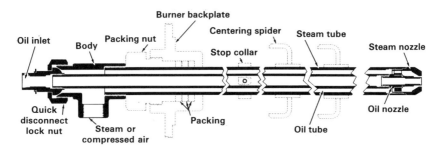

Figure 6.21b.   Internal mixing or emulsion atomizer has large openings (requires less clean steam) and holds a backpressure on the oil, preventing vapor lock. It is limited to high firing rates, requires high steam and oil pressure, and consumes more steam per gallon of oil.

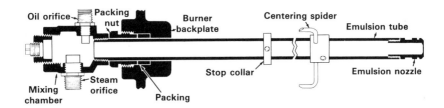

Figure 6.21c.   Tip emulsion atomizer can be designed for a desired spray angle; it provides better atomizing with lower steam pressure, and its steam consumption is low. This model produces a narrow spray angle for a conventional forward flame -- (type A, Figure 6.2).

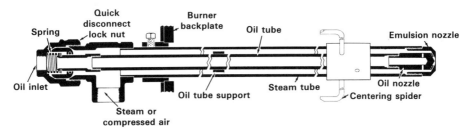

Figure 6.21d.   Tip emulsion atomizer for a wide spray angle (type C or type D flame of Figure 6.2).

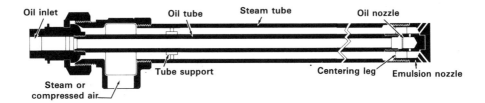

Steam or compressed air at pressures ranging from 5 to 150 psi is used. The steam consumption may vary from 1 to 5 pounds per gallon of oil and air consumption from 22 to 100 ft³ of stp air per gallon, depending on the design and size of the atomizer and the oil viscosity. For the external mixing type, oil pressures of 5 to 10 psi are recommended at the burner, but for emulsion types (oil and steam or air mix inside the unit), the oil pressure must be about the same as the steam or air pressure. The slim compact nature of these atomizers makes them readily adaptable for converting gas burners to combination gas-oil burners.

Oil pressure atomizing burners are 1-fluid atomizers, also referred to as *mechanical pressure atomizing burners*. When oil is permitted to expand through a small orifice, it tends to break into a spray of fine droplets. Atomizers utilizing this principle are usually designed to operate with 100 psi oil at viscosities less than 50 SSU. In some instances they will operate with pressures as low as 75 psi or with viscosities as high as 100 SSU. See Figure 6.22.

Figure 6.22. Oil Pressure Atomizing Nozzle This type is used in mechanical atomizing oil burners. It is similar in principle to a garden hose nozzle. Arrows show the path of oil.

Turndown is poor on this type of nozzle, usually limiting its use to on-off control -- as in domestic furnaces. This type probably has the lowest initial and operating costs but relatively high maintenance problems and lack of flexibility restrict its use. Large boilers utilize a number of oil "guns" that work on this principle, but with hundreds of psi oil pressure. Their steady load, ability to turn off some of the burners, and very high pressure, obviate the turndown problem.

Centrifugal atomizing (rotary) burners use centrifugal force to throw oil from the lip of a rotating cup in the form of a conical sheet of liquid which quickly breaks into a spray. Low pressure air is admitted through an annular space around the rotating cup. If the air velocity is high, it tends to blow the spray into a narrow cone, but if the speed of rotation of the cup is high, this tends to overcome the effect of the air stream, producing a wide angle spray. Shape of the spray is also determined by the relative positions of the cup and air orifice. Horizontal rotary cup burners are still used for boilers but not for high temperature furnaces, because they have electric motors and other moving parts immediately adjacent to the burner.

Sonic and ultrasonic atomizers usually use an arrangement similar to a vibrating reed to produce a fog of minute oil droplets. Burners must be specially designed to transport the fog to the combustion space and mix a high volume air stream with it.

COMBINATION GAS AND OIL BURNERS

Periodic changes in the fuel supply and price picture may sometimes necessitate changing fuels. Burners that are capable of burning either gas or oil are called combination burners.* There are probably as many types of these burners as there are combinations of the different types of gas and oil burners, but only the most common ones will be described here.

When a necessity for switching fuels arose in the past, attempts were made at (a) adding some sort of atomizer to an existing gas burner, or (b) adding a gas ring around the nose of an existing oil burner. Either retrofit had to insure entrainment, mixing, and positive delivery of the fuel to the combustion chamber by the main air stream.

As flame monitoring controls have been mandated, most engineers have concluded that such do-it-yourself efforts should be abandoned in favor of burners designed and tested for multiple fuel use. Some combination burners can be purchased as stripped-down models that burn only one fuel; and in such cases an after-fit for a second fuel is practical by purchasing the omitted components.

General-Purpose Dual-Fuel™ Burners. The burner of Figure 6.23 is a low pressure air atomizing dual-fuel burner that produces a type A flame (Figure 6.2) with gas, light oil, or heavy oil.

Figure 6.23. General-purpose combination burner capable of burning heavy oil or any gas. This burner can be operated over a wide range of air/fuel ratios and firing rates in cold and tight combustion chambers. The gas tube and gas connection can be omitted and a shorter oil tube substituted to make a less expensive straight oil burner. The low pressure air atomizer can be omitted and a blank backplate substituted on the gas connection to make a less expensive straight gas burner. The metal tile support shown should be used only when the burner is installed in a thin metal wall -- not when installed in a refractory wall.

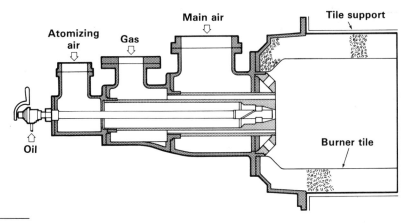

---

* All North American combination burners are identified by the name "Dual-Fuel."

Figure 6.24 shows an integral fan Dual-Fuel™ burner which includes burner, blower, blower motor, control valve, control valve motor, pilot, and atomizer engineered together as a package. (Gas-only and oil-only models are possible by omitting some components.) This burner's fuel-directed type internal design allows its flame shape to be adjusted to suit combustion chamber requirements. Such pre-packaged assemblies are popular for boilers (Figure 6.10), dryers, ovens, air heaters, incinerators, and process heaters, and other applications where the purchaser may not have the manpower nor experience to select and assemble burner system components. Installation of such units requires only attaching the burner to a refractory tunnel in the combustion chamber wall, connecting fuel, electric power, and control lines.

Smaller packaged automatic burners are designed for on-off control using gas or light (distillate) oil. Larger sizes have modulating control (responding to temperature or steam pressure sensors), and can burn gas, light (distillate) oil, or heavy (residual) oil. Such packages often include a control panel programmed for light-up, shut-down, and monitoring procedures to meet the customer's insurance requirements.

Figure 6.24. Integral Fan Dual-Fuel™ Burner. In addition to the 25 million gross Btu/hr unit shown (12 to 50 foot similar assemblies range from 30 to 200 million Btu/hr flame length), fuel-directed internals permit a choice of flame characteristics.

Modern Special Purpose Burners. The days of the "universal," all purpose furnace are almost gone. The necessity for saving fuel while doing a fast uniform heating job has resulted in the development of tailor-made furnaces, heating systems and heating machines. To facilitate these, burner manufacturers have developed many special purpose burners.

Burners for Use with Preheated Air. Preheating the combustion air broadens the flammability limits, increases flame velocity, elevates flame temperature, and raises the available heat. These effects combine to make burner flames more stable, and to reduce fuel consumption markedly. Flames are shorter and brighter.

Depending on the materials of construction, burners intended for use with cold air can usually be used with air preheated to 500 or 600 F without damage. With added cooling arrangements, a few have been used with 800 F air.

Rising fuel costs have justified smaller sizes of recuperators, regenerators, and hot air burners (Figure 6.25). A traditional larger size refractory-lined burner is pictured in Figure 6.26. Most of the flame types from Figure 6.2 are available in hot air burners of either the fuel-directed or air-directed type. With very high air preheat, all flame types tend toward higher luminosity because the hot air polymerizes the fuel molecules before they burn.

Figure 6.25.   Small Dual-Fuel™ hot-air burner for use with air preheated as high as 1100 F. Eight sizes of this burner release from 0.16 to 2.16 million Btu/hr with 8 osi air, producing a forward flame (type A, Figure 6.2). The heat-resistant cast iron body is lined with vacuum-formed refractory fiber. In keeping with modern safety requirements, it has pilot, flame monitoring, and observation ports in positions engineered to provide coverage over a wide range of operating conditions.

Figure 6.26.   Combination burner for use with air preheated up to 1100 F. The burner body is lined with refractory. The flame-stabilizing disk that forms the annular air orifice may be refractory-faced, or of very high alloy to protect it from furnace and flame radiation. By use of a variety of orifice, disk, and atomizer configurations, it is possible to produce several of the flame types of Figure 6.2. The model shown uses low pressure air (2 psig) as an atomizing medium, but others have atomizers designed to use compressed air or steam.

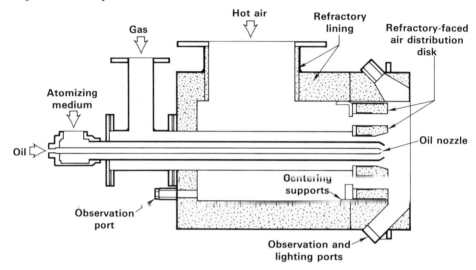

With high temperature air passing through a burner, it needs (a) better materials to withstand the higher air and flame temperature, (b) more capacity to pass enough weight of oxygen to release available heat equal to that with cold air, and (c) temperature compensating air/fuel ratio control to correct for the ever-changing air stream temperature produced by most air heating equipment. Reasons (a) and (b) result in hot air burners being much larger than cold air burners. The hot air piping to burners should be oversized and insulated.

It is not normal practice to preheat atomizing air, natural gas, nor low viscosity liquid fuels. Compared with the main (modulated) air, they constitute much smaller quantities; so the fuel saving from heating them would be small compared to the added piping and equipment costs. However, preheating of the fluid immediately surrounding a burner's heavy oil tube helps keep the oil hotter, and therefore may aid atomization.

Adjustable heat pattern burners are used to set the heat release pattern in a furnace upon initial light-up, or to automatically change flame shapes upon the command of differential pyrometers, as in Figure 6.27, or as directed by a programmer.

Fuel Directed® burners, mentioned earlier, are particularly adaptable to such requirements. Other forms of adjustable heat pattern burners have either a large number of manually adjustable vanes in the air passage, or a flip/flop valve that deflects the air through alternate passages.

Figure 6.27. Adjustable flames. Temperature uniformity can be maintained at both ends of a long one-way-fired soaking pit by using Fuel Directed® burners to achieve an adjustable heat pattern, changing the flame shape from the long luminous type G flame of the top photograph to the short wide type C flame of the lower photograph.

Radiation Burners. Most oil burners emit considerably more radiant heat than their straight gas counterparts. The conventional forward flame types heat by both radiation and convection. Coanda and centrifugal effects used in many radiation type burners tend to be counterproductive to thorough atomization, making such burners more prone to sooting if not well tuned. Dual-fuel versions of radiation burners are generally available for gas and light oil only.

Figure 6.28 illustrates a light oil or gas-fired radiation burner designed to thaw hopper cars. It transfers heat to the undersides of the car hoppers by radiation from the refractory trough, radiation from the glowing alloy covers, flame radiation, and convection.

Figure 6.28.   Radiation type hopper car thawing burner utilizing the principle of most so-called infrared burners -- venting combustion gases through a grid so that the grid becomes radiant.

Convection burners for dual-fuel operations fall into two broad classes -- high velocity burners and large air-heating burners. Dual-fuel forms of both are available for gas or distillate oils only.

High velocity (type H) flames can (a) penetrate into loosely-piled loads, such as a latticework of bricks in a hack in a kiln, stacks of parts in a heat treat furnace or oven, a load of scrap to be preheated or melted, (b) reach farther than type A flames and drive hot gases around the back sides of loads or pillars, and (c) stir furnace gases to provide more uniform heating, better convection, and more passes for improved efficiency.

Figure 6.29a.   High velocity Dual-Fuel™ burner with a spark-ignited pilot (right side) and an air-cooled ultraviolet flame monitor. The dial-type rotary plug V-port oil valve on the back of the burner serves dual functions -- as a shutoff and a limiting orifice.

Figure 6.29b. Gas and oil flames of a Dual-Fuel™ high velocity burner. Although the combination burner versions cannot generate as high a flame momentum as the gas-only models (Figures 6.11b), their velocities are several times higher than those of conventional (type A) burners.

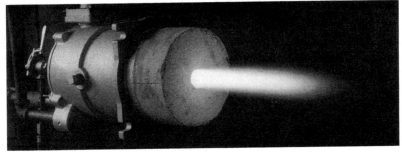

For drying of glass tanks and newly-lined refractory chambers, operators prefer to use one big burner because it involves preparing only one opening. This means, however, that the drying burner must have a considerable "throw" to be effective in reaching all corners of the chamber. High velocity is also helpful in inducing recirculation and it can be used to inspirate fresh air. The same applies to drying or preheating of large ladles for molten metals and for rotary drum dryers for fertilizers and aggregates. All of these are really air heater applications, requiring very large volumes of air with a very lean ratio for low hot mix temperature. Fuel only control is used.*

Waste Incinerating Burners -- Liquid. Standard Dual-Fuel burners are often used for this purpose with the liquid waste delivered through the oil tube and gas burned simultaneously as an auxiliary fuel (for start-up or to sustain combustion if the waste alone cannot do so). Great care must be exercised to be sure (a) the waste will burn, (b) the liquid waste can be atomized, and (c) the products of combustion will not be toxic or polluting. The user must set up a pilot test. The composition and properties of the waste may change from time to time. This calls for constant vigilance by the operators.

---

* Theoretically, such lean, low temperature use of fuel is inefficient. The ideal way to heat such jobs is with waste gases from higher temperature operations.

Waste Incinerating Burners -- Fume. Conventional forward flame burners have been used for this purpose, but a grid type fume burner (Figure 6.30) assures burning of all of the fume with less auxiliary fuel, lower peak temperature, and shorter burning length in the duct (which must be of high temperature materials). It is impractical to handle oil in the many tiny fuel jets required for this configuration. Oil has been successfully atomized into the fume stream upstream of the grid, using the grid as a flame holder only, but such an arrangement calls for special engineering and precautions against flashback. All except warning (b) above apply to planning a fume incinerating operation.

Figure 6.30.   Fume incinerating flame grids for use with gas as the auxiliary fuel. Sizes to span two duct sizes are shown, the larger with two small sections set differently to show the adjustability of the oblong fume orifices.

Oil Vaporizing Burner Arrangements. The use of heat as well as atomization to vaporize distillate oils permits combustion of the oil through existing gas burner systems with a gas-like blue flame. Some burner designs recirculate hot combustion gases within the burner to transfer heat to the incoming oil.

Oil-to-gas converter systems use a direct-fired air heater, burning a portion of the fuel oil in that heater, to provide 800 F air through the mixture manifolds of a premix burner system. See Figures 6.31a and b. The fuel burned in the air heater is not wasted. It is just released in a different location, but the hot air lines should be insulated and kept as short as possible.

Oil-to-gas converter systems are more expensive to install than conventional dual-fuel burners, but they are the only solution when oil standby is required for little burners. (It is not practical to try to build or operate industrial-type oil burners at less than about 1.9 gph or ¼ million Btu/hr.) Although the oil-to-gas vaporizing systems require attentive operators and maintenance, they have been used successfully world-wide for many years.

Figure 6.31a. Oil-to-gas converter vaporizing system for burning light distillate oil through existing small premix gas burners. The resulting flames have essentially the same characteristics as when burning gas.

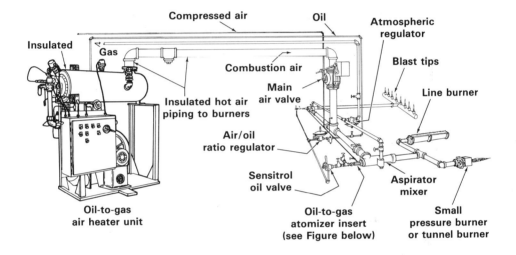

Figure 6.31b. Close-up view showing compressed-air-oil atomizer inserted into the aspirator mixer of a premix gas burner system for oil standby when gas is unavailable. Hot air aids vaporization so that the hot-air-and-vapor-mix burns like a gas at the burner ports.

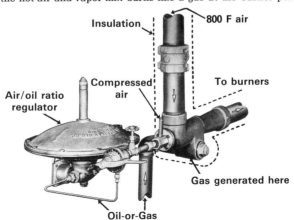

Burners for Oxygen Enrichment. Oxygen enrichment of the air for combustion elevates the flame temperature and reduces the percentage of nitrogen passing through the furnace, thereby lowering the stack loss and raising the thermal efficiency. Whether oxygen enrichment produces any overall energy saving depends on the cost of the oxygen.

Oxygen may be premixed with the combustion air or nozzle-mixed through the burner but the higher heat release rates and resultant high temperatures may not be compatible with existing burner materials. If the burner does not have a connection designed for oxygen, the burner manufacturer should be consulted to see if it is practical to admit oxygen through the atomizing air connection, or unused oil, observation, or flame supervisory connections. Oxygen must not be premixed with fuel. There must be no possibility for oxygen back-flowing into either fuel or air piping.

REFERENCES

6.a   **Shore, D. E. and McElron, M. W.:** "Tuning Industrial Boilers, II -- Establish Excess Air Levels", vol. 121, no. 5, pp. 76-79, POWER, May, 1977.

6.b   **Beer, J. M. and Chigier, N.A.:** "Combustion Aerodynamics", pp. 125-127, Applied Science Publishers Ltd., London, 1972.

ADDITIONAL SOURCES

Babcock and Wilcox: "Steam -- Its Generation and Use", 39th ed., Babcock and Wilcox Co., Barberton, OH, 1978.

Chiogioji, M. H.: "Industrial Energy Conservation", Marcel Dekker, Inc., New York, 1979.

IHEA: "Combustion Technology Manual", 4th ed., Industrial Heating Equipment Association, Arlington, VA, 1988.

Lupton, H. P.: "Industrial Gas Engineering", vols. 1, 2, 3, North Western Gas Board, London, 1960.

Palmer, H. B. and Beer, J. M.: "Combustion Technology -- Some Modern Developments", Academic Press, New York, NY, 1974.

Singer, J. G. (ed.): "Combustion -- Fossil Power Systems", 3rd ed., Combustion Engineering, Inc., Windsor, CT, 1981.

Stambuleanu, A.: "Flame Combustion Processes in Industry", Abacus Press, Tunbridge Wells, Kent, England, 1976.

Part 7.   COMBUSTION CONTROL

The field of combustion control includes a great many specialized subjects -- more than can possibly be treated in this handbook. Combustion engineers are mainly interested in valves and regulators used in fuel systems, air/fuel ratio controls (atmosphere control), fuel pressure control, furnace pressure control, and safety controls, which are covered in detail. Automatic control of furnace input as a function of temperature, steam pressure, or steam flow is a field in itself and will be mentioned only briefly here.

CONTROL VALVES

Valve capacities and pressure drops are discussed in Part 5, Vol. I.

Two important specifications of control valves, particularly those used for modulating control, are turndown range and characteristic. The *characteristic* is the manner in which the flow rate changes with changes in valve handle position and is described by a plot of flow rate vs. valve handle position. (See Figure 7.1.) This is often assumed to be the same as the variation of valve opening area with valve handle position which is termed the *area characteristic*. The flow may not be directly proportional to the valve opening area, because of the aerodynamic characteristics of the upstream and downstream piping, and because the pressure drop across the valve proper is reduced as the pressure drop across the other resistances in the system increases. Because the area and flow characteristics are not always the same and it is difficult to predict their divergence accurately, the only sure way to achieve a desired characteristic is to adjust the characteristic after the system is installed. It is desirable however, to have some basis for the initial selection; so the following discussion of valve types describes the area characteristics of each.

To assure uniform control loop stability throughout the anticipated valve operating ranges, it is important that the valve characteristics be specified so that the valve response will match the dynamics of the process. A *quick opening* valve characteristic is desirable for on-off control. A *linear* characteristic (flow rate directly proportional to valve travel) is good for most flow control and liquid level control systems. *V-port* and *modified parabolic* type valves give fine throttling action in the early stages of valve opening, followed by a linear characteristic for the balance of the valve travel. An *equal percentage* characteristic (each equal step in valve movement results in the same percent gain in flow rate above that at the end of the last step) is best suited to pressure control or flow control systems with wide-swinging pressure drops. Figure 7.1 shows a range of characteristics that are possible with valves having adjustable ports.

An effect similar to partially closing the port of an adjustable port valve is achieved with reduced port butterfly valves or by down-rating any valve so that its 100% rated flow rate corresponds to less than its wide-open position. A butterfly valve rated for 100% of flow at 90° open might have a characteristic similar to the quick opening or fixed port curves; but when rated for 100% flow rate at 60° open, its control characteristic will be similar to the lowest curve on Figure 7.1. This underscores the folly of always selecting valves on the basis of C_V factor or flow per dollar.

Sliding Plug Valves. Sliding plug valves adjust the valve orifice size by moving a plug axially along a valve stem as in an ordinary globe valve. The sliding action may be produced by direct axial motion of a control motor or pneumatic

Figure 7.1.   Characteristics of various control valve types. The solid lines on either side of the linear line suggest the possible range of an adjustable port valve. Curves are based on a constant pressure drop through each valve. Valve characteristics that match the process on which they will be used give best control.

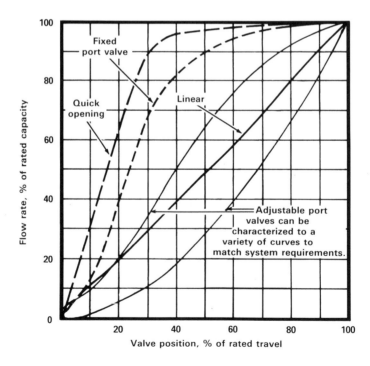

operator, or it may be produced by rotating a threaded valve stem. The characteristic is determined by the shape of the plug, which may be disk type, V-port, parabolic, or any modification of these three basic types. The disk type plug, such as in an ordinary globe valve, gives equal increments of open area for equal increments of stem movement: that is, the area characteristic is a straight line. Although this straight line characteristic is a desirable feature, the disk type plug has the disadvantage that its entire turndown range is expired within a very small range of axial movement. The V-port plugs permit a more gradual change of valve opening. Figure 7.2 shows a typical V-port valve used for limiting gas flow. The characteristic of V-port and parabolic plugs approaches a semi-logarithmic curve; equal increments of valve movement give equal % increases in open area.*

Rotary plug valves consist of a ported sleeve or plug which is rotated past an opening in the body. The characteristics of such valves depend upon the shape of the port and the opening in the body. Figure 7.3 shows a rotary plug valve with a V-shaped opening, which gives a characteristic similar to that of a V-port sliding plug valve. Figure 7.4 shows another rotary plug with a cylindrical plug having a straight-through rectangular port, which approaches a straight line characteristic.

The valves of Figure 7.4 each contain a curtain that can be moved axially, providing an adjustable port. These valves therefore have an adjustable characteristic, or can be *characterized*, as one can change the valve resistance relative to the system resistance to match the valve to the system after it is

---

Figure 7.2 V-port sliding plug gas valve used as the limiting orifice that comprises the constant resistance in gas lines in pressure type air/fuel ratio control systems. The valve of Figure 7.3 is used for the same purpose with oil and a motorized version is for direct oil input control.

Figure 7.3. Rotary plug V-port oil valve.† The parts on the left are the precision-ground rotary shear plates, the right-hand one of which has a V-port that exposes varying areas of the off-center port in the disk. Internally electrically heated versions of this valve are used for heavy oil.

---

* It is sometimes possible to simulate a linear characteristic even though it is not inherent in the valve by using a linkage with a characteristic such that equal increments of movement of the control motor will produce equal increments of flow.

† See Sensitrol™ in the Glossary.

installed. This convenience results in some sacrifice of turndown range, but it is minor in properly sized systems. The straight-through flow minimizes pressure drop and the consequent undesirable reduction of turndown ratio.

Butterfly Valves. A butterfly valve is a section of straight pipe containing a rotating vane or disk such as a stovepipe damper. (See Figure 7.5.) The pressure drop across a wide open butterfly valve is small. (See Table 5.24b.) The amount of leakage depends upon the tolerances of the disk and valve casing (which often serves as the valve seat). The characteristic is similar to that of a V-port valve. Butterfly valves are used principally for low pressure air. For gas, they must have seals around the shafts.

Reduced port butterfly valves are intended to accomplish the same characterizing effect as an adjustable port valve with its curtain partially closed, but their fixed ports preclude any final fitting to the system.

Figure 7.4. Adjustable port rotary plug valves for low pressure air (left) and gas or oil (right).
An adjustable curtain is used to vary the width of the rectangular opening to produce a desirable system characteristic after a valve is installed. (See Figure 7.1.) Closing the curtain permits an adjustable port valve to approach a linear or an equal percentage characteristic.

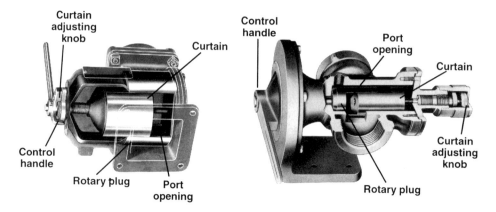

AIR/FUEL RATIO CONTROLS*

The fuel savings from tighter air/fuel ratio control can be evaluated by using the data from Part 3, Vol. I. In addition to fuel economy, other benefits of better air/fuel ratio control are less pollution, greater safety, improved temperature uniformity and product quality.

Area control of air/fuel ratio, also termed valve control of air/fuel ratio or a proportioning valve (pv) system, is achieved by use of constant pressure

---

* For a discussion of air/fuel ratio controls for blower mixers and carburetor-type mechanical mixers, see Reference 7.a listed at the end of Part 7.

Figure 7.5a. Threaded manual butterfly valve. The locking handle can be replaced with one of a variety of linkages and 2-position or modulating control motors.

Figure 7.5b. Butterfly valve with integral electric drive for 2-position control. This direct-connected valve with motor has lower first cost and is easier to install and adjust than a separate valve and motor connected by a bracket and linkage. It has a shaded pole motor, actuated by a conventional 3-wire controller. Adjustable stops set the end points of the disk travel.

Figure 7.5c. Wafer type butterfly valve sandwiched between flanges on the connecting pipes, only two bolts being required for alignment. This one is equipped with a modulating electric motor drive.

Figure 7.5d. A 42 in. wafer type butterfly air valve. Pneumatic, hydraulic, or electric operators can be used to turn some sizes of these valves, but large valves are often operated by cylinders having one end attached to a rigid structural member.

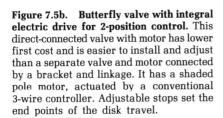

and variable areas. A simple mechanism can be used to cause the opening areas of fuel and air valves to vary in proportion to one another. This requires that the two valves have identical characteristics and that the mechanical connection between them produce directly proportional movement. In other words, if one valve is caused to rotate through a 45° angle, then the other must also rotate through a 45° angle, and if this movement causes a 25% change in flow rate of one fluid, it must also produce a 25% change in flow rate of the other fluid. If the valve characteristics are not the same, the fuel and air flows will match at only a few points throughout the range. If the movement is not directly proportional, the mixture may be lean at some firing rates and rich at others.

Two rotary valves on a common shaft may be used for area control, but a side-by-side arrangement with a parallel arm or Z-type linkage is preferred because it is sometimes possible to adjust the linkage to correct for different characteristics in the two valves. One or preferably both of the valves should embody manual adjustment of its opening (in addition to handle adjustment -- see Figure 7.4) for setting air/fuel ratio.

The valve control system requires an air blower with a constant pressure characteristic and constant oil or gas pressure regulators ahead of the control valve because the upstream pressures for both air and fuel must be constant. In the case of fuel oil, the oil temperature must be constant at the valve because variations in the oil viscosity would affect the flow rate.

Figure 7.6. Area control of air/fuel ratio on a packaged automatic Dual-Fuel™ burner. The pneumatic operator (bottom center) actuates the air shutters (top center), inlet butterfly air valve, oil valve, and gas valve (lower left) via a factory-calibrated linkage.

Even though a pair of proportioning valves may have the same area characteristics, it is unlikely that they will have the same flow characteristics because differences in the density, phase, viscosity, and surface tensions of the fluids make them aerodynamically dissimilar, causing the air/fuel ratio to veer off to rich or lean, or some of each, in various parts of the operating range. Some more expensive arrangements have manually adjusted cams or electronically adjusted schemes to attempt to correct for these effects. Adjustable linkages, such as shown in Figure 7.6, also facilitate correction for such effects, but require much experience on the part of the adjuster.

Area control is usually less expensive to purchase, but more expensive to operate because it provides less precise ratio control and therefore uses more fuel. As energy costs rise, better quality air/fuel ratio control will be justified on smaller and smaller installations.

Pressure control of air/fuel ratio works on the assumption that the resistance to flow downstream from the control valves is a constant in both the fuel and air lines of any burner system. From the principles outlined in Part 5, Vol. I, it is known that the flow through a constant resistance is proportional to the square root of the pressure differential across that resistance. Therefore, if the fuel and air pressure are kept equal (or proportional) then the fuel and air flow rates should be proportional throughout the entire range of firing rates. Whereas the previously discussed valve control system worked on the principle of constant pressures and variable areas, the pressure control system works with constant areas and variable pressures. This is more accurate and adaptable to a wide variety of arrangements.

In air primary systems (common for <10 000 000 Btu/hr), (Figure 7.7), the main input control valve is in the air line and a ratio regulator is used in the fuel line, the regulator being *cross-connected* to the air line. When burning *gas*, the ratio regulator is a device that duplicates the air pressure in the fuel line. It consists of a globe-type valve in which the plug is attached to and moved by a diaphragm. The pressure on one side of the diaphragm is that of the air line, conveyed to the space below the diaphragm by an *impulse line* (small pipe). The pressure on the other side of the diaphragm is the feedback pressure in the fuel line downstream from the regulator. Thus, if the fuel and air pressures are not the same, an unbalanced force exists on the diaphragm causing it to move. This in turn moves the valve plug, adjusting the fuel flow, and thereby correcting the downstream fuel pressure until the difference between the fuel and air pressures is zero.

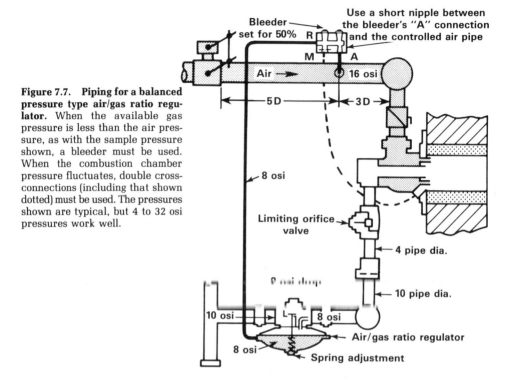

Figure 7.7.   Piping for a balanced pressure type air/gas ratio regulator. When the available gas pressure is less than the air pressure, as with the sample pressure shown, a bleeder must be used. When the combustion chamber pressure fluctuates, double cross-connections (including that shown dotted) must be used. The pressures shown are typical, but 4 to 32 osi pressures work well.

Figure 7.8.   An air/gas ratio regulator (pressure type a/f ratio control) produces an outlet gas pressure equal to whatever pressure is exerted under its diaphragm. If the air line pressure is piped to the lower chamber, gas and air pressures will be equal, and flows proportional. If the lower chamber is vented to atmosphere, the outlet gas pressure will always be zero, regardless of flow demand.

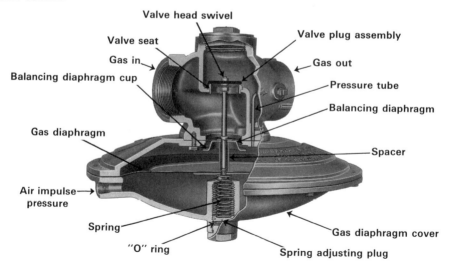

Figure 7.8 shows a cutaway view of a ratio regulator used to control gas flow. Note the small pipe connection below the diaphragm for the impulse line and the pressure tube leading from the upper chamber to the downstream side of the valve. If the impulse connection is left open to the atmosphere, the regulator will produce zero gas for suction type mixers and premixing burners. The spring below the shaft counterbalances the weight of the shaft, plug, and diaphragm assembly so that the diaphragm floats freely. The upstream gas pressure acting on the underside of the valve plug might lift the plug; so a balancing diaphragm is attached to the shaft between the valve body and the diaphragm case.

In applications where the maximum available gas pressure is less than the maximum combustion air pressure, the above system would fail to maintain proportional fuel and air pressures at higher air flow rates; so it is necessary to produce an impulse pressure less than but proportional to the pressure in the combustion air line. This is accomplished by use of a *bleeder*, which permits a certain amount of leakage from the impulse connection (Figure 7.7). The opening of the limiting orifice valve may be set to give the desired air/fuel ratio when the system is first put into operation, and thereafter becomes the constant area resistance in the gas line.

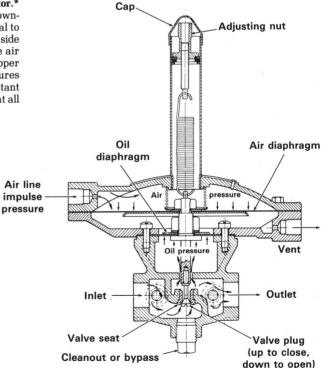

Figure 7.9. Air/oil ratio regulator.*
This regulator produces a downstream oil pressure proportional to the pressure exerted on the top side of its diaphragm. By piping the air supply line pressure to this upper chamber, the oil and air pressures will be proportional and a constant air/oil ratio will be maintained at all firing rates.

Cap

Adjusting nut

Oil diaphragm

Air diaphragm

Air line impulse pressure

Air   pressure

Oil pressure

Vent

Inlet

Outlet

Valve seat

Cleanout or bypass

Valve plug (up to close, down to open)

* See Ratiotrol™ in the Glossary.

When burning oil in low pressure air atomizing burners, it is desirable that the oil pressure be several times greater than the combustion air pressure, rather than equal to it. This ratio is maintained by an air/oil ratio regulator* such as shown in Figure 7.9 in which the upper diaphragm has several times the area of the lower diaphragm. The impulse air pressure pushes down on the upper diaphragm, tending to move the valve shaft assembly downward (opening the valve). The downstream oil pressure (feedback) pushes up on the lower diaphragm (via the clearance space between the shaft and the body) tending to raise the valve shaft assembly (closing the valve). The space between the diaphragms is vented to the atmosphere. If the area ratio were 12:1, the oil pressure would have to be twelve times greater than the impulse air pressure for the upward and downward thrusts to be balanced. Any unbalanced force on the diaphragm moves the oil valve until the proper pressure ratio is attained.

An adjustable tension spring balances the weight of the diaphragms, shaft, and valve plug. The spring can be adjusted for less tension to maintain a leakage rate when no impulse is applied. This leakage should be equivalent to the minimum air flow rate to the burner (usually the atomizing air rate).

Figure 7.10 shows a typical piping arrangement using such a ratio regulator. The oil valve at the back end of the burner can be used for manual individual-burner adjustment to set the air/fuel ratio for the desired furnace atmosphere. Thereafter it becomes the constant area resistance in the oil line.

Figure 7.10. Schematic piping for an air/oil ratio regulator*. When cross-connected to the controlled air line, air/oil ratio regulators produce an oil outlet pressure that is some fixed multiple of the impulse pressure (such as 10 or 12 to 1). The manual oil valve on the back of each burner and air orifices built into each burner constitute the constant resistances in each line that make it possible to control air/fuel ratio by control of pressure in the air and fuel lines. If the combustion chamber pressure is not steady, double cross-connections (including the one shown dotted) must be used.

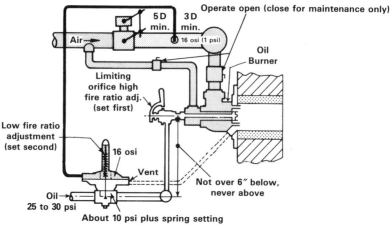

---

* See Ratiotrol™ in the glossary.

There are limitations on the maximum upstream pressure that should be applied to any ratio regulator. These are the result of a compromise between ruggedness and accuracy. It is desirable to have diaphragms of high rupture strength, but this makes them less sensitive. Use of an upstream *pressure-reducing regulator* is recommended to protect the ratio regulator and provide a constant inlet pressure at a level that assures its optimum performance.

Flow control of air/fuel ratio actually measures the air flow and fuel flow and adjusts the flow of one of these fluids accordingly. The flow rates are measured by pressure taps detecting the pressure differentials across resistances in the air and fuel lines. These pressure differentials are transmitted to some controlling device that automatically adjusts the flow of the air (or of the fuel) to maintain the desired ratio.

Pressure impulses from upstream and downstream of an orifice in the air line act on opposite sides of a diaphragm connected to a shaft. Pressure impulses from upstream and downstream of a limiting orifice valve in the fuel line act on another diaphragm connected to the same shaft so as to oppose the action of the air diaphragm. The resulting movement of the shaft opens or closes a valve in the regulator until a balance is reached, providing proportional control of gas flow.

Figure 7.11. Pneumatic fully-metered or flow-balanced air/gas ratio control system. This 3rd generation system balances pressure differentials with a 4-signal regulator whereas the 2nd generation system balanced pressures with a 2-signal regulator. The upper diaphragm spaces sense a pressure differential corresponding to the fuel flow; the lowest two diaphragm spaces, the air flow. This is an *air primary* arrangement. The limiting orifice valve serves dual functions: as the gas flow metering device, and as the air/gas ratio adjusting device.

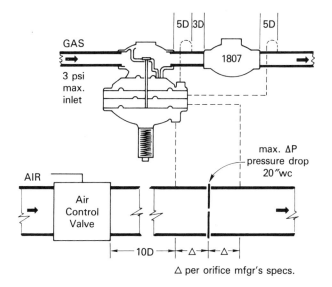

The preceding paragraph described an *air primary* system of air/fuel ratio control. In an air primary system, the input controller (usually temperature, except pressure in the case of a boiler) adjusts air input, and the air/fuel ratio control causes fuel input to follow. In a *fuel primary* system, the input controller adjusts fuel flow, and the ratio control causes air input to follow.

A basic schematic diagram of a digital electronic version of flow control of air/fuel ratio is shown in Figure 7.12. Advantages of this are: compactness, electric lines are substituted for long impulse lines, controller can be remote in a preferred environment, more actuating power than a regulator type, adaptability to additions of sophisticated control cascades, usable with systems that are air primary or fuel primary. (See Glossary.)

Figure 7.12.   Schematic diagram of a fully-metered, or flow-balanced, electronic air/fuel ratio control. (Fuel primary shown.)

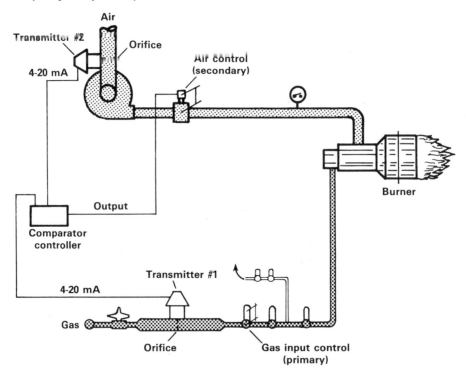

Fully-metered electronic controller systems include a manual ratio adjuster, usually at the control panel, for setting a desired air/fuel ratio and for precise trimming. On installations consuming large quantities of fuel, this feature is used for frequent correction for weather changes and fuel variables, such as shown in Figure 7.13. This is often accomplished with an oxygen trim system cascaded to automatically modify the a/f ratio setpoint. Similarly, with preheated air systems, a temperature signal can automatically adjust the a/f ratio. Addition of such a temperature compensating system makes a fully-metered system a *mass flow control* system.

The same principles are used for large gas mixing stations and for propane-air mixing controls.

Figure 7.13. Effects of changes in combustion air temperature and humidity and in fuel gas gravity on combustion stoichiometry.*

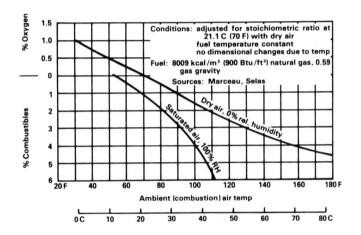

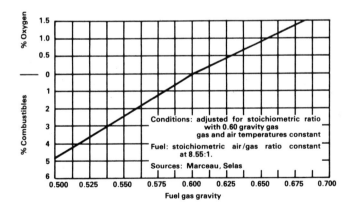

---

* From Reference 7.b listed at the end of Part 7. Other graphs therein show the effects of high temperature air (recuperator or regenerator outlet) changes, fuel temperature, and fuel calorific value.

Substitution of alternate fuels simply requires more selector switches and valves. Addition of multiple fuels (simultaneously) requires the addition of square root extractors so that pressure differential signals can be added. (From formula 5/37, Vol. I, flow is proportional to the square root of pressure differential.) This is applicable when a waste fuel stream is added to purchased fuel to fill out the heat demand, or when air supply to a process is enriched with oxygen.

Figure 7.14.   Digital electronic flow type air/fuel ratio controller. This contains the comparator that determines whether the ratio of air flow to fuel flow is off balance from the setpoint ratio (e.g. 10:1 for typical natural gas), and if so, sends a correction command to a control motor in the air line if fuel primary, or in the fuel line if air primary. The setpoint can be changed by keying it in on this console. This unit has several other capabilities, which will be described on later pages.

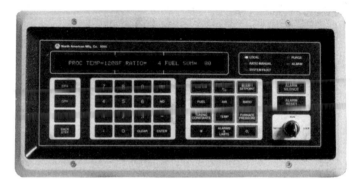

The inputs to an electronic a/f ratio controller can be any variables that can be converted and transmitted as 4-20 milliamp signals; so flow measurements are no longer limited to orifice plates which may consume considerable pumping power because of the required pressure drops. They are not even limited to other inferential (differential pressure) techniques such as annular orifices and annubars, but can respond to a variety of flow measuring methods, including anemometers, Doppler principle devices, magnetic counter systems, positive displacement flowmeters, sonic and ultrasonic schemes, or vortex flowmeters.

The output of an electronic a/f ratio controller can actuate electrically, pneumatically, or hydraulically operated valves, dampers, eductors, or fan speed controls.

More accurate and reliable sensors and controls are being developed every year. As fuel cost and supply problems increase, better air/fuel ratio controls are needed for smaller sizes of fuel-consuming installations. The use of preheated air or oxygen enrichment can rarely be justified without first installing accurate air/fuel ratio control. Then, the addition of preheated air or oxygen enrichment necessitates compensating corrections in the automatic air/fuel ratio control.

Justification for better air/fuel ratio controls can be calculated using formula 3/37, Vol. I, and available heat data from Part 3; or by Figure 7.15.

See also the segment on "Complete Combustion Controls" later in this Part 7.

Figure 7.15. Required fuel for excess air and preheated air operation. Excess air should be minimized by automatic air/fuel ratio controls or heat recovery equipment will have to be oversized to handle the extra air.

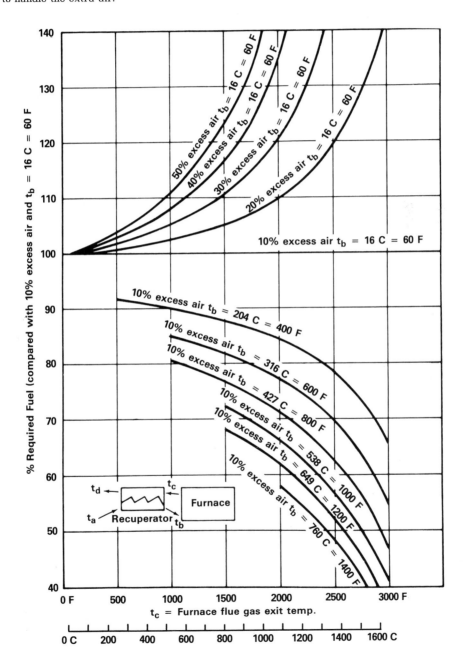

PRESSURE CONTROLS

Pressure Regulators. *Pressure-reducing regulators* or *line pressure regulators* should be used upstream of most air/fuel ratio controls to drop the supply pressure to a practical level and to maintain a constant inlet pressure at all flow rates. One such regulator may supply a number of ratio regulators.

Figure 7.16a shows a gas pressure-reducing regulator and an oil pressure-reducing regulator. In both, the downstream (regulated) pressure acts on one side of a diaphragm while a preset-spring is balanced against it on the other side of the diaphragm. The valve will remain open as long as the downstream pressure is too low to balance the constant thrust of the spring. The upper chamber of a gas pressure regulator should be vented to outdoors for safety in the event of diaphragm rupture. A relief valve vented to outside should be located downstream from a gas pressure regulator to avoid an excessive pressure build-up against valves and ratio regulators when the burners are off.

Pressure-relieving regulators or *relief valves* are similar to a pressure-reducing regulator except that the regulated pressure is the upstream pressure. They are used to bleed off excess pressure.

Pressure controllers can perform either of the above functions, but use an external source of power (hydraulic, pneumatic, or electric). Controllers are used instead of regulators for dirty gases and for larger line sizes. The furnace pressure control systems discussed next are examples of pressure-relieving controllers.

Automatic furnace pressure controls aid accurate air/fuel ratio control by providing a constant downstream pressure. Uncontrolled furnace pressure may affect the air/fuel ratio if the furnace (1) uses natural draft to induce air, (2) has open burner mountings, (3) uses any but flow-type or double cross-connected pressure-type air/fuel ratio control, (4) is exposed to strong winds, or (5) has leaks in its lining or around doors and the hearth. (See also pp 218-227 in Vol. I.)

Maintaining a positive pressure in all parts of the furnace or kiln at all times prevents cold air infiltration through leaks in the structure. This is important in any operation where the burners are frequently on low fire while the furnace is hot, creating a strong chimney effect. Preventing cold air infiltration not only maintains the desired furnace atmosphere, but also improves temperature uniformity by avoiding chilling of parts of the load, and saves fuel by not having to heat infiltrated air.

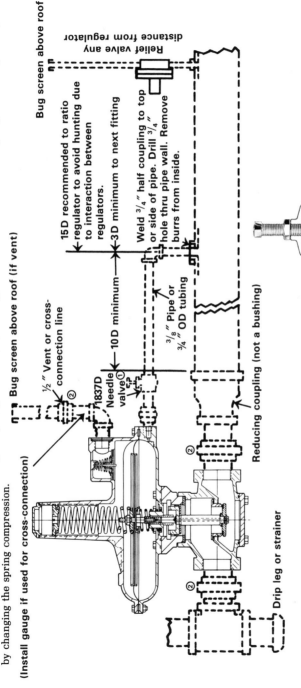

Figure 7.16a. A gas pressure-reducing regulator, used to reduce high pressure gas to a safe constant pressure for control equipment. The outlet pressure pushes up on the large diaphragm and is balanced against the constant force of a spring, which may be adjusted by changing the spring compression.

(Install gauge if used for cross-connection)

Bug screen above roof (if vent)

Bug screen above roof

Relief valve any distance from regulator

15D recommended to ratio regulator to avoid hunting due to interaction between regulators.

3D minimum to next fitting

Weld ³/₄″ half coupling to top or side of pipe. Drill ³/₄″ hole thru pipe wall. Remove burrs from inside.

½″ Vent or cross-connection line

1837D Needle valve①

10D minimum

③/₈″ Pipe or ³/₄″ OD tubing

Reducing coupling (not a bushing)

Drip leg or strainer

②

Figure 7.16b. An oil pressure-reducing regulator functions in the same manner as the gas pressure-reducing regulator of Figure 7.16a. It reduces oil pressure to a safe constant value at the inlet of air/oil ratio regulators or valve control systems.

① When necessary to dampen pulsation, use a fixed orifice, or a needle valve with internal bypass, which allows on-line ''tuning out'' of outlet pressure pulsation.

② Use pipe unions to simplify regulator repairs or replacement.

The section on "flow of flue gases" at the end of Part 5, Vol. I, gives information on evaluating pressures in furnaces, ducts, and stacks, and discusses the effects of firing rate and configurations on furnace pressure. The general conclusion is that a very slight positive pressure (such as + 0.02"wc) is usually the better compromise between (a) and (b):

(a) high positive pressure, which...causes stingers or fluing at doors and cracks, blows sand or water from seals, adversely affects furnace maintenance, and produces operator discomfort.

(b) negative pressure, which..........induces cold air, adversely affecting furnace atmosphere, chills the bottom of the load, causing poor product quality, and costs fuel to heat the cold air to restore desired uniformity and production rate.

In high temperature furnaces, the cost of cold air infiltration caused by negative furnace pressure (heating the air and chilling the load) is usually greater than the cost of hot gas escaping as a result of a slightly positive furnace pressure. See Figure 7.17.

Figure 7.17.   **Cost of additional fuel required for non-neutral furnace pressures.** Based on: 60 F (16 C) ambient air, 1900 F (1038 C) exit gases, $4 per million gross Btu fuel cost. Courtesy of G. Rimsky, Loftus Div., Thermal Electron Corp.

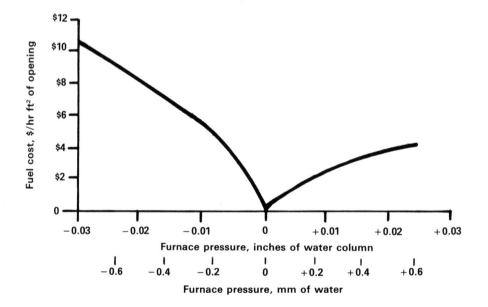

Figures 7.18 and 7.19 are helpful in evaluating the extent of the negative pressure developed within a hot furnace, the air infiltration, and the extra fuel input required to heat the infiltrated air.

Example 7-1. Find the cost of a ½ in. wide gap all around the charging door of a 1700 F aluminum reverberatory furnace (Figures 7.20a and b) that is 6 ft × 10 ft × 5 ft above the metal line, operating 6000 hours per year. Fuel cost is $1.75 per million Btu.

Figure 7.18. Negative pressure developed in a hot furnace, and fuel input required to heat infiltrated air. "Elevation difference" (left vertical scale) means the height between a cold air inlet at the bottom (crack, door opening) and a hot gas outlet at the top (flue, stack top, top of door opening).

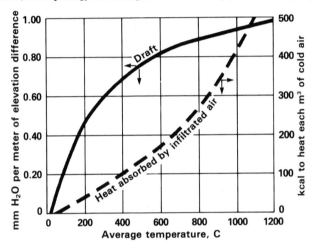

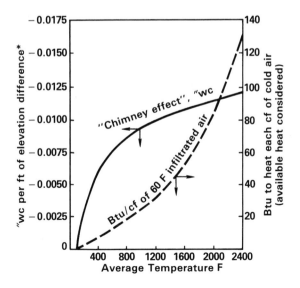

Figure 7.19.   Cold air infiltration through furnace openings. Use this graph in conjunction with Figure 7.18 as in Examples 7-1 and 7-2.

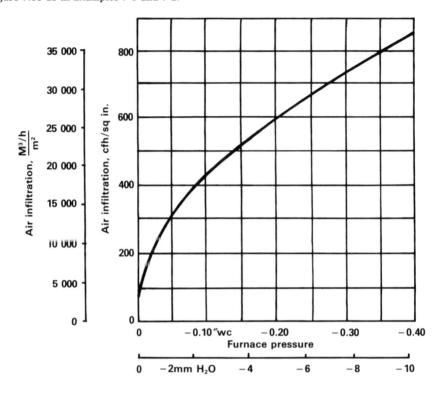

The bottom gap has 48 in. × ½ in. = 24 in.² with a 3 ft height for stack effect. From the top curve of Figure 7.18 at 1700 F, read 0.011″wc per ft of furnace height. This creates a furnace pressure (stack effect suction) at the base of the door of − 0.011″wc/ft × 3 ft = − 0.033″wc. From Figure 7.19 at − 0.033 in., read 250 cfh of air infiltration per in.². This × 24 in.² = 6000 cfh.

The side gaps are 2 sides × 36 in. high × ½ in. gap = 36 in.² with a mean height of 1½ ft. This produces a furnace pressure of − 0.011″wc × 1½ = − 0.0165″wc. From Figure 7.19, the cold air infiltration at the sides will be about 175 cfh/in.², × 36 in.² = 6300 cfh.

The total infiltrated air is therefore 6000 + 6300 = 12 300 cfh. As the furnace ages, more and wider cracks will develop because continued fluing through the cracks aggravates the situation. Considerable judgment is required to evaluate this effect; but for this sample problem, estimate that the *average* cold air infiltration over the remaining life of the furnace lining will be 150% of the above calculated figure for the present situation. Therefore, 1.50 × 12 300 = 18 450 cfh. Much of the infiltrated air will go up a flue without reaching furnace temperature. In this case, estimate that the equivalent of ½ of the 18 450 cfh is heated to 1700 F and the other half exits with no temperature rise.

From the bottom curve of Figure 7.18, the unnecessary heat for infiltrated air is 60 Btu/ft³ × 18 450 cfh × ½ = 554 000 Btu/hr = 554 cfh gas or 4.1 gph of light oil. If the flues were properly sized for a slightly positive pressure while the burners operated at high fire rate, the negative pressure condition calculated above probably would exist only during about ⅓ of the 6000 operating hours; so 554 000 Btu/hr × 6000 hr/yr × ⅓ × $1.75/1 000 000 Btu = $1939.00 possible fuel waste per year. Most of this money could be saved by installation of automatic furnace pressure control to minimize air infiltration.

It is conceivable that air to the burners could be cut back to compensate for air in-leakage. If burners could be adjusted to just the right degree of richness to utilize all infiltrated air, zero money would be wasted. However, it would be difficult to achieve such thorough mixing of infiltrated air with excess fuel as to produce a flue gas analysis of 0% O_2 and 0% combustibles at all flues; but use of automatic furnace pressure control to minimize infiltration is a sure way to reduce the waste of fuel.

Figure 7.20a.   Aluminum reverberatory furnace of Example 7-1.

Figure 7.20b.   Flow pattern through the ½ in. gap all around the door of Example 7-1.

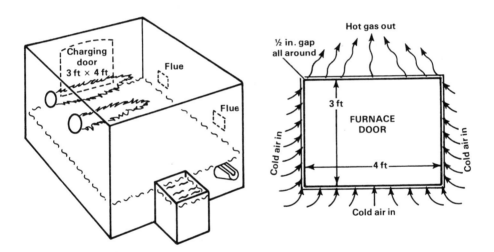

Example 7-2. A **car hearth forge furnace** with dimensions as shown in Figure 7.21, operates at 2300 F about 80 hours per week; and at an average temperature of 1200 F (bring-up, cooling, idling) for about 60 hours per week. Fuel cost is 35¢/gallon, or $2.33/million Btu. Estimate the fuel saving possible by adding furnace pressure control.

The high fire input will keep the furnace pressurized, but when at 1200 F, Figure 7.18 shows the furnace draft will be − 0.010″wc/ft × 7 ft = − 0.070″wc which, from Figure 7.19, induces 360 cfh/in.² For this furnace, 360 cfh/in.² × 288 in.² = 103 700 cfh. In this configuration, it is estimated that an equivalent of ⅔ of the infiltrated air "short circuits" to the flues without being heated. At 1200 F on the bottom curve of Figure 7.18, the fuel input required to raise the infiltrated air to flue temperature is 30 Btu/ft³ × ⅓ × 103 700 cfh × 60 hr/wk = 62.2 million Btu/wk. There will probably be more air in-leakage around the door, as in Example 7-1.

In addition to the bad effect on product uniformity and the longer bring-up time, the cold air infiltration could waste up to 62.2 million Btu/wk × $2.33/million Btu = $145.00 per week.

Figure 7.21.   Car hearth furnace of Example 7-2.

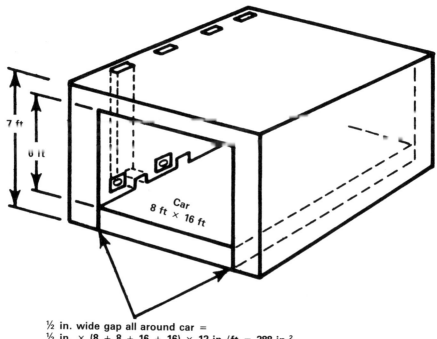

7 ft

0 it

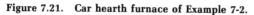

Car
8 ft × 16 ft

½ in. wide gap all around car =
½ in. × (8 + 8 + 16 + 16) × 12 in./ft = 288 in.²

Control Equipment. A downdraft flue arrangement (Figure 5.44c, Vol. I) is an attempt to minimize the chimney effect, but the hearth-level flues and double wall construction are very expensive. Updraft fluing with an automatic furnace pressure controller involves lower first cost *and* provides constant positive furnace pressure down to the hearth at all times.

Figure 7.22a shows a typical automatic system with a furnace pressure setpoint adjustable from − 0.05 to + 0.15"wc. A compensating line at the same level as the pressure tap, but immediately outside the furnace, serves as a standard for comparison. The pressure is usually controlled -- that is, the tap is located -- at the hearth level. In cases where hearth scale or splashing metal might plug a low level sensor, the tap is located at an alternate position and the setpoint adjusted to maintain the desired hearth-level pressure.

Figure 7.22a.   Furnace pressure control system. Better uniformity and constancy of temperature, atmosphere, and input result from automatic control of furnace pressure. Placement of the pressure tap for most effective control at all firing rates is largely a matter of experience. A hydraulic relay can be used to amplify the signal to a magnitude capable of moving heavy dampers.

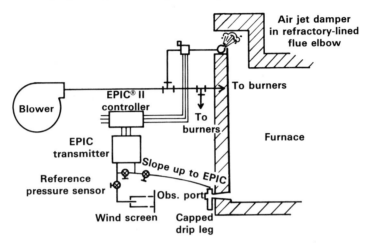

Dampers. Furnace pressure is usually controlled by adjusting flue dampers of which there are endless variations (Figures 7.22b through i). For heavy dampers, a hydraulic system is desirable.

Bell-crank dampers can be counterweighted; enabling use of a minimum size operator. They are often used on small furnaces or with a common shaft operating dampers on a row of flues.

Barometric dampers admit sufficient cold air to reduce stack effect. They require careful proportioning and locating of the cold air inlet. Furnace pressure response is slower than with a damper that throttles flue gas. Stability-enhancing devices are sometimes required.

Butterfly dampers present no particular control problems, but may require considerable maintenance. Power required to operate them may be underestimated through not allowing for warpage and heat damage to the bearings.

Inclined dampers need cylinders sized to overcome the friction between the damper and slide frame. These and other cable operated types permit more convenient location for the operator than with a direct connection. Counterweights can be used to reduce operator size. It is best to provide for alternate manual damper operation over its normal range of travel. The sheave on the piston rod in Figure 7.22g works in the bight of the cable; so speed and travel of the damper are twice that of the piston. A double-ended piston rod with a counterweight conserves space and minimizes the size of cylinder required. In Figures 7.22f and g, the counterweight size and cylinder area must be twice those of counterweights and cylinders traveling the same speed and distance as the damper.

Figures 7.22b-i.   Dampers, eductors, and variable speed induced draft fans can be used as final control elements in automatic furnace pressure control.

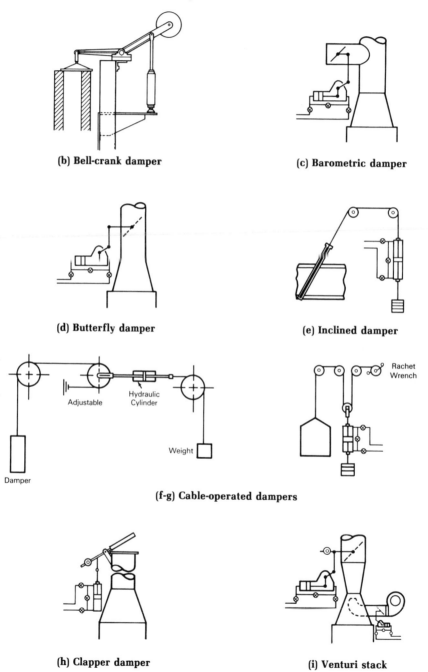

(b) Bell-crank damper

(c) Barometric damper

(d) Butterfly damper

(e) Inclined damper

(f-g) Cable-operated dampers

(h) Clapper damper

(i) Venturi stack

Clapper dampers are often used at the tops of tall stacks. All but one of the moving parts is out of the hot gas.

Venturi stacks control furnace pressure by modulating flow of a cold air jet, thereby avoiding moving parts in the hot gas stream. Control characteristics are similar to those with a variable speed induced draft fan.

Air jet dampers eliminate the maintenance problems of moving mechanisms in the hot flue gas stream for flues no wider than 18 inches (0.45 metres). The operating cost of the blower air is minimal, and the first cost of the blower can sometimes also be negligible if the required capacity can be "piggybacked" onto a combustion system blower. Care must be exercised in designing the air jet manifold and flue arrangement so as to avoid air injection (back-feeding) into the furnace, which may produce an excessive oxidizing atmosphere, particularly when the burners are on low fire. See Figure 7.22a. The automatic furnace pressure control needs to move only a small butterfly valve in the air supply line to an air jet damper; so a light-duty pneumatic or electric operator can be used.

Electronic Equipment. A problem in the past with furnace pressure controls has been the need for very large (16 in.) diameter sensing diaphragms necessary to detect the very small furnace pressure changes. These, in turn necessitated large pressure tap lines and incurred considerable time delay if the lines were long. A spinoff from the space program has made available highly sensitive diaphragm arrangements less than one-fourth the previous size with electronic detection and amplifying circuitry. See Figure 7.23.

Figure 7.23. Furnace pressure control system -- electronically amplified. The transducer (1) senses differentials from − 0.05 to + 0.15″wc through taps (2) and (3), and the demodulator and preamplifier produce a dc current signal that is linearly proportional to the input ΔP. This is compared with the adjustable control point to determine the output control signal.

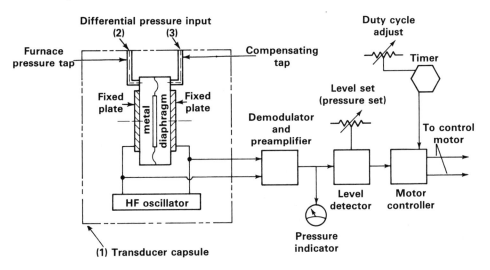

Figure 7.24. A batch-type heat treating furnace equipped with an electronically amplified furnace pressure controller. A pressure signal (2) and compensating signal (3) are connected to the transducer capsule (1)*, the output of which actuates a cylinder (4) connected to a counterweighted damper arm (5).

* See EPIC® in the Glossary.

The output of an electronic pressure indicating controller such as Figure 7.23 is fed to a pair of 110 V ac solid state relays that serve as a bidirectional motor controller, or they may be used to operate external motor starters, relays, or electro-hydraulic valves. The electronic controllers are more compact, easier to install, need only small pressure tap lines, and respond more quickly. Figure 7.24 shows such an installation. In some cases, there may be a delayed response in system pressure. A manually adjustable cycle timer is used to achieve control stability despite this system time lag, by setting the fraction of each minute during which the motor controller is actuated (from 3 to 60 seconds).

See also the next section.

COMPLETE COMBUSTION CONTROLS

Complete combustion system controls automatically control the 3 major aspects of combustion, and provide for later addition of many sophisticated control features.

Integration of several combustion control functions into a single controller can simplify operation and maintenance for industrial heat processing systems.

Figures 7.14, 7.25a, and 7.25b illustrate such controllers, which can be configured to control the following: 1) air/fuel ratio, with options for lead-lag, air preheat compensation (mass flow control), O_2 trim, 3 fuels simultaneously, air plus oxygen simultaneously (enrichment); 2) furnace pressure; 3) input (as sensed by temperature or steam pressure); and 4) communication with higher level control systems.

For any furnace, process heater, kiln, incinerator, boiler, or oven that will use (a) more than about 5 million Btu/hr of purchased fuel, (b) two or more fuels simultaneously, (c) preheated air, or (d) oxygen enrichment, a complete combustion control system will save fuel costs and simplify control. If any of situations a, b, c, d, or an appreciable rise in fuel cost, are anticipated in the next few years, a complete combustion control system (such as shown in Figures 7.25a and b) will be a good investment because it will accommodate growing into more sophisticated control schemes.

Modern combustion control systems incorporate scaling in a variety of engineering units, and they are adaptable to use with distributed control systems. They facilitate remote supervision and provide management with recorded information for review.

Figure 7.25a. **Furnace with a typical complete combustion system controller.** This type of control is ready for future addition of air preheat, more fuels, or oxygen enrichment.

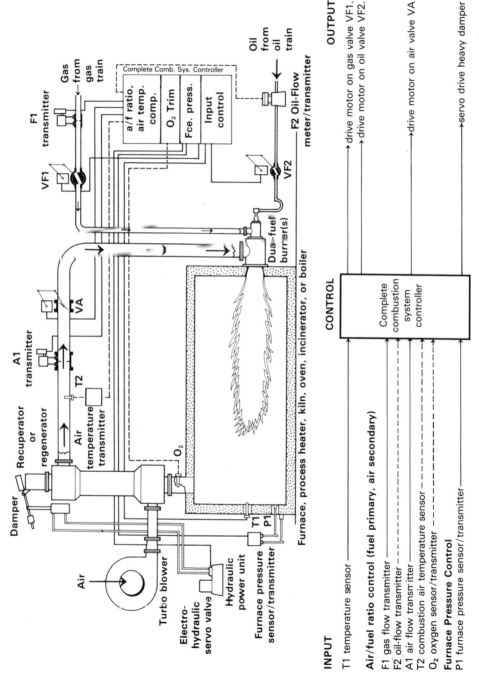

Figure 7.25b. Full scope of possible arrangements with a complete combustion control system such as shown in Figures 7.14 and 7.25a.

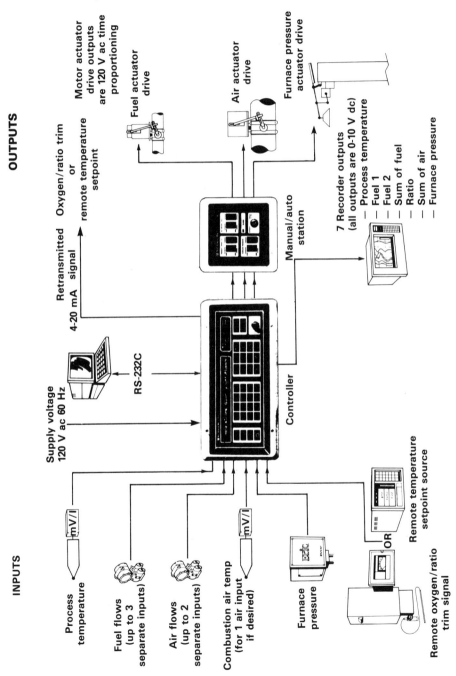

COMBUSTION SUPERVISING CONTROLS

The objective of a combustion supervising system is to stop the flow of fuel if the flame should happen to be extinguished. If the fuel flow is not stopped, the combustion chamber (or an entire building) may be filled with an explosive mixture of fuel and air.

A pilot is not enough protection. It may go out or become inadequate to relight an extinguished main flame promptly; or a pilot may be unable to relight the main flame if the air/fuel ratio is too rich or too lean, if the feed rate is too fast or too slow, or if atomization is poor.

The old idea that a *constant* pilot was helpful because it was always there to relight an extinguished flame has fallen into disrepute. Too many pilot flames have been unable to light a main flame when needed but have later served to ignite an explosive accumulation of air-fuel mixture. An *interrupted pilot* with its programmed trial-for-ignition period is the best way to avoid a pilot-ignited explosion. To prevent accumulation of unburned fuel in a combustion chamber, flame monitoring devices should be used to govern automatic fuel shutoff valves.

Automatic Fuel Shutoff Valves. Any automatic fuel shutoff valve (manual reset or automatic reset) should be wired to interlocks so as to close upon failure of (1) combustion or atomizing air blower, (2) any element of the input control system (such as temperature limits, steam pressure limits), (3) air pressure at the burner, (4) fuel pressure at the burner, (5) current from the flame detector, or (6) current from other safety devices such as low water cutoff and high pressure cutoff. In special cases there may be additional items connected to the shutoff valve. The fuel shutoff valve must be connected in series with all of these elements as shown in Figure 7.26a. Fuel, air, and steam pressure can be converted to an electrical signal by means of a bellows, Bourdon tube, or diaphragm operated switch.

The prime requisites of any fuel shutoff valve are that it cannot be manually locked open, that it shut tightly, and that it be sensitive to any possible failure in the system. In addition it is desirable to have a manual shutoff arrangement, high mechanical advantage for easy opening, and an auxiliary switch.

In a typical manual reset fuel shutoff valve installation, interruption of any interlock circuit cuts off power to an electromagnet that has been holding the valve open, and a strong spring snaps the valve shut so that fuel flow is stopped quickly. When the trouble has been eliminated so that the circuit is again closed, the valve can be opened by the action of a hand lever. If, however, the trouble has not been satisfactorily corrected, the circuit remains open and the valve cannot be opened by moving the hand lever because the valve stem remains disengaged from the handle. This is termed a *manual reset fuel shutoff valve* as differentiated from an *automatic reset fuel shutoff valve* that automatically "resets" and reopens when power is restored. The manual reset type is used wherever the presence of an operator is required to assure a safe, low-fire relighting of the burners. Spring action closes the fuel valve when the electro magnet is de-energized. Actuating the handle will not reopen a manual reset valve until the electro magnet circuit is energized. When the valve is open, the handle may be used to close it.

Automatic reset fuel shutoff valves are used only with programmed relighting cycles that include automatic pre-purge and post-purge and purging with a timed trial for ignition on a monitored interrupted pilot followed by a monitored main flame. Except in very small sizes, such as pilot solenoid valves, automatic reset type valves are designed to open slowly so as not to damage regulating equipment or blow out pilot flames, and so as to provide a smooth main flame light-up.

Figure 7.26a.   Wiring for an automatic fuel shutoff valve. The circuit can be arranged as shown so that the valve will close upon failure of any of several interlocks of a combustion system. Other possible interlocks might include such items as a circulating air flow switch, a conveyor stoppage switch, and emergency pushbuttons at exits.

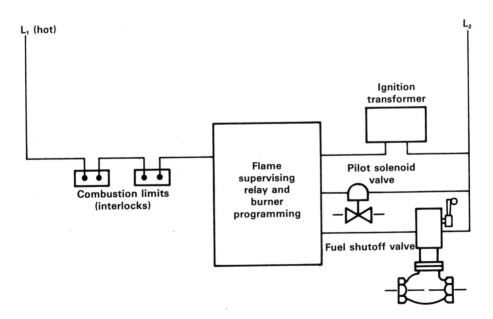

Momentary power interruptions can cause nuisance shutdowns where manual reset fuel shutoff valves are used. If this is a frequent problem, some people use an electronic device known as a *valve trip delay* to maintain power to the fuel shutoff valve for 1 second, supposedly long enough to span a typical interruption but not long enough to allow a hazardous fuel accumulation. With modern flame supervising systems, it is better to use a total battery-backed power outage system.

Because a fuel shutoff valve affects the final action commanded by a trouble-detecting system, a leak or failure of such a valve could be extremely dangerous. For this reason, most insuring authorities insist that in gas systems there be two fuel shutoff valves in series (redundancy -- main and blocking valves), plus provision for a periodic leak test of both valves. See Figure 7.26b.

All of the leak test methods consist of some sort of shutoff valve immediately downstream from the fuel shutoff valves with provision for detecting pressure build-up between the valves (as would occur if a fuel shutoff valve were leaking). The simplest form of leak test utilizes a petcock with a rubber hose, the open end of which is immersed in a beaker of water so that one can observe bubbles of fuel forming in the water if a leak exists. The most reliable leak check system is a programmed automatic check device such as a "Double Checker." It makes an automatic check after each closure of the fuel shutoff valve, using a pressure switch and alarm.

Figure 7.26b. Gas train component arrangement should be in this order: 1) manual shutoff valve, 2) step-down gas pressure regulator, 3) low gas pressure switch, 4) main automatic shutoff valve, A) leak test cock, 5) vent valve if required, 6) high gas pressure switch, 7) blocking valve, B) leak test cock, 8) manual shutoff valve, and 9) air/gas ratio regulator if used.

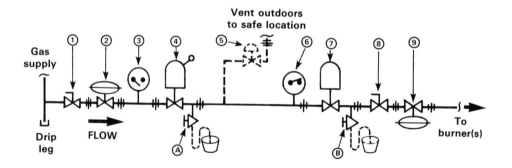

Current national standards do not require the use of a vent valve between the main and blocking valves, but a few authorities insist that a vent valve be used. Vent valves have occasionally stuck open, allowing large volumes of fuel to vent to atmosphere unnoticed. This could be extremely costly and may create a hazard. Leak testing devices are not permitted on vent lines; so a more expensive vent valve that contains a "proof-of-closure" switch might be justified.

A vent line pipe must connect the vent valve discharge to the outdoors, preferably above a roof. The vent pipe cross-sectional area must be equal to or greater than the vent valve port area. Vent lines must not be manifolded, nor exceed 40′ in length.

Supervising Valve System. When gas furnace explosions occur they often happen during light-up because someone carelessly left one of the gas burner valves open after the previous shutdown, thus allowing the furnace to be filled with unburned fuel as soon as the main fuel valve was opened. The supervising valve system reduces the probability of such light-up explosions.

A special supervising gas valve at each burner is drilled with a small secondary passage located so that it is open only when the main passage is closed. A checking pressure line is connected through the secondary passages of all the gas valves so that a pressure switch at the end of the line cannot close a circuit to the fuel shutoff valve unless all of the valves are shut. If the operator finds that he cannot open the fuel shutoff valve because a valve has been left open, he must not only close the valve but also he must understand that the furnace is to be purged before he lights any of the pilots or burners. Operator intelligence is necessary. The supervising valve system is a start-permissive concept; whereas flame supervision is a run-permissive concept.

A system equivalent to the supervising valve system can be obtained by use of end switches on the gas valves. The checking pressure line is thereby replaced by an electrical circuit that is closed when all fuel valves are turned to the off position. This method could easily be adapted to oil valves as well as gas valves. For protection equal to that provided by the checking pressure system, the checking circuit must be wired through pressure switches on the fuel line and air lines.

A supervising valve system is *not* a substitute for a flame supervising system. Burner management systems with programmed purge cycles will become more universally used, obviating the need for supervising valve systems.

Flame monitoring devices are sometimes termed flame detectors, scanners, sensors, eyes. *Thermopiles* and *bimetal warping devices* are limited to domestic and low-input heating applications where fast cooling is assured upon flame failure. *Photocells* (cadmium sulfide) are rarely used because they are too easily "fooled" by light sources other than flames. *Lead sulfide* (infrared) cells are used only with circuitry to detect the flicker of a flame.

Flame electrode or flame rod type detectors rely upon the ability of ionized gases in the flame to rectify an ac current. The point where the rod intersects the pilot and main flames must be a point where the pilot can assuredly ignite the main flame. Either flame alone can then rectify the current between the rod and a ground. The ground, usually on the burner body, should have at least four times as much area as the immersed portion of the flame rod. Flame electrodes are used for gas flames only, and usually on smaller installations. Care must be exercised to keep the electrode and its porcelain insulator clean and dry to avoid nuisance shutdowns. Excessive temperature on the electrode may result in drooping and system shutdown.

Ultraviolet detector tubes are sensitive to the ultraviolet (UV) radiation emitted by all flames in small amounts. They are completely insensitive to infrared and visible radiation; so they cannot be fooled by radiation from hot refractory. When UV radiation impinges on the cathode within the detector tube, electrons start to flow toward the anode, ionizing the gas within so that it can conduct a d-c current as long as a voltage is applied to the terminals. To make such a system fail-safe, an interruptor, such as a shutter, ensures that the tube will conduct a current only while receiving the UV radiation that repeatedly re-establishes its conductivity.

Combustion products and steam are opaque to ultraviolet radiation; so the scanning device must be aimed near the root end of a flame, and steam atomization may require special positioning. However, this assures positive discrimination from adjacent and opposing burners. The arc of a spark igniter emits much ultraviolet radiation; so the scanner must not view such an igniter nor its reflection unless it is interrupted. If a UV detector is to sight through a closed port, the lens must be of an ultraviolet-transmitting material such as fused quartz.

Pilots and direct electric igniters should be rigidly positioned so that the pilot flame or igniting arc intersects the main burner flame and can light it at any rate or air/fuel ratio expected to exist whenever light-up is necessary. See Figure 7.27a. Similarly, a flame monitoring device should be positioned so that it can detect the presence of a pilot* and/or the main flame at any safe rate or air/fuel ratio expected to be used during the entire period of burner operation. See Figure 7.27b.

It is usually safer and more practical to program burners (particularly bigger burners) to be ignited at low fire. Most modern burners are designed with built-in pilot and main flame monitor holders that have been proven to meet all of the above requirements. See Figures 7.27c and 7.27d.

---

* In a programmed flame supervising system.

Figure 7.27a. Pilot location relative to the main flame. A pilot must be capable of igniting the main flame whether the feed stream through the burner is a high rate, low rate, rich ratio, or lean ratio.

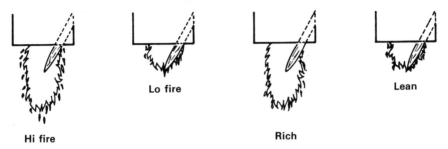

Hi fire   Lo fire   Rich   Lean

Figure 7.27b. Monitor location relative to the main flame. A flame monitoring device must be positioned so that it will be capable of detecting the main flame regardless of whether the feed stream issuing from the burner is at high rate, low rate, rich ratio, or lean ratio.

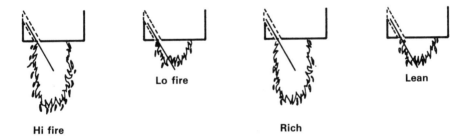

Hi fire   Lo fire   Rich   Lean

Figure 7.27c. Burner-pilot-monitor geometry. The desirable geometry is with the centerline of the pilot* or electric igniter and the flame monitoring device intersecting the flame at low firing rate; but it is essential that the arrangement be tested for smooth lighting characteristics and adequate monitoring signal at all anticipated light-up and operating conditions. See also Figure 8.7a.

Figure 7.27d. High velocity dual-fuel burner with spark-ignited pilot and flame monitor positioned by the burner manufacturer as a result of extensive testing.

Pilot   Monitor

Main flame

---

* In a programmed flame supervising system.

Interrupted pilots are turned off after a programmed trial-for-ignition period whether the main flame is established or not. They are required* instead of constant, standing, or intermittent pilots (which continue to burn after the main flame is established). The logic behind this requirement is, if something goes wrong causing the main flame to go out, the flame monitor would still detect a constant pilot, and therefore hold the fuel valve open, causing the chamber to be filled with an air-fuel mixture. This situation is most likely to happen when something goes wrong that causes the air/fuel ratio to go rich, and the main flame goes out because it is too rich to burn. If someone then either opens a furnace door, turns off the fuel, or does anything that brings the accumulated mixture back within the limits of flammability (explosive limits), the standing pilot may ignite the whole accumulated air-fuel mixture.

IF YOU EVER COME UPON THIS SITUATION -- pilot on, fuel on, main flame out -- TURN OFF ANY SPARK IGNITER, PILOT AIR, and pilot fuel IMMEDIATELY.† Many persons' first impulse may be to shut off the fuel, but that might allow the mix in the chamber to become lean enough to explode. It is usually better to permit the rich mix of fuel and air to cool everything well below the ignition temperature. Then, when you are sure there are absolutely no hot spots or other sources of ignition in the chamber, ventilate the room, open the furnace doors, and shut off the fuel.

The above-described explosion hazard can occur on a furnace without flame monitoring equipment if the human monitor fails to turn off the pilot when he sees or hears the main flame go out. *Observation ports* should be kept clean and used for frequent flame surveillance on *every* furnace.

Burner management systems start and stop the burner(s) upon signals from the flame monitor, limit control interlocks, and the input controller; assure safe starts by proper sequencing and by checking their own components before each start; check for the presence of an adequate ignition source and monitor the main flame during burner operation; shut down the burner in proper sequence if an unsafe condition develops. Some burner management systems are interfaced with other controllers in a process system.

Nonprogramming flame supervising systems consist of one or more flame monitoring devices and control circuits which when properly applied provide the following protection. During light-up they sense the presence of a pilot flame at a location where the pilot will reliably ignite the main burner before permitting the main fuel shutoff valve to be energized and opened. Following main flame ignition it is recommended that the system shut off the pilot flame and monitor only the main flame. In the event of flame failure, the circuitry causes all fuel

---

* Per NFPA 86 code, with some exceptions. (Reference 7.c.)

† If at a filling station you saw a stream of leaking gasoline, wouldn't you put out your cigarette before attempting to stop the leak?

to be shut off. An operator's attention is necessary before the next pilot flame establishing period or trial-for-ignition can start. Because of this characteristic, these systems are also termed nonrecycling flame supervising systems. These systems must be used with other items to provide minimum standards of operation.

Programming flame supervising systems are similar to nonprogramming systems but they can also provide additional functions such as pre-purge, post-purge, and automatic relight without operator assistance.

Approved flame supervising systems, either programming or nonprogramming, have nominal flame-failure response times of 2 seconds, maximum of 4 seconds. They may have flame-failure contacts that can be wired into alarm and signal circuits. They usually have a built-in safe-start check to prevent light-up if the flame-sensing relay is in the flame-present position due to component failure within the flame supervising system, or due to the presence of actual or simulated flame.

A typical programmed flame supervising system common to both pilot and main flame and using spark-ignited pilots, might include the following light-up sequence, as illustrated in Figure 7.28.

1) Prove air flow; check that air pressure is above low limit; check that gas pressure is above low limit; check that gas pressure is below high limit.

Figure 7.28. Burner lighting sequence. This generalization is for a common flame monitor supervising either pilot flame or main flame. Different insuring and governing authorities may require alternate trial-for-ignition times, air changes, and sequences.

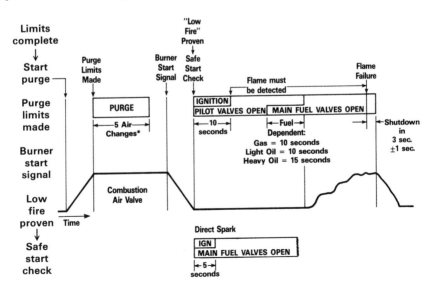

2) Time delay for a minimum purge of 4* standard cubic feet of fresh air or inert gas per cubic foot of heating chamber volume and flue passage. The minimum required purge flow must be proven.
3) Energize (open) pilot solenoid(s) and energize electric ignition source.
4) Prove pilot flame and time out electric ignition source, allowing pilot flame to continue burning. (If pilot is not proven within 10† seconds, de-energize pilot solenoid and require manual reset before recycling is possible.)
5) Energize circuit to allow manual or automatic opening of the fuel shutoff valve(s)† and automatic closing of the vent valve (if any) in a block and vent arrangement.
6) Within 10‡ seconds after main fuel flow begins, the pilot valve(s) should be de-energized to interrupt the pilots.
7) Prove the main flame.
8) Release control to the input control system.

Use of flame supervising systems is required on each operating burner on any furnace, boiler, kiln, heater, or oven when below 1400 F (760 C). Even when such combustion chambers are normally operated above 1400 F (700 C), it must be remembered that they are not above this temperature level during start-up and warmup, which is a particularly hazardous time. The above-mentioned 1400 F furnace temperature level is a generally agreed upon minimum temperature at which an industrial burner flame will be reliably ignited from the hot furnace interior. For installations that run continuously above 1400 F for long periods, it may be desirable to remove the flame supervising equipment to protected storage, or to electrically bypass it after the furnace is above 1400 F. See Reference 7.c at the end of Part 7.

The minimum temperature at which accidentally accumulated flammable gas or vapor input might be ignited by the hot furnace interior, or by a standing pilot or spark, and thereby cause an explosion, is the ignition temperature of the fuel (Table 1.10 in Volume I) or of flammable volatiles or combustible material being heated (Table 5-2 in Reference 7.c at the end of Part 7).

There are innumerable standard and custom-designed wiring systems for single- and multiple-burner and single- and multiple-zone flame supervising systems. These simplify installation for the user. For installations having many burners, where having more than one burner shut down at a time would jeopardize a costly load, independent burner operation may be required. Such a system provides separate flame supervision and fuel shutoff valves for each burner. Thus failure of one burner will not shut down other burners in the zone.

* Some insuring authorities may require different timing or different numbers of air changes.
† Some program sequences automatically turn the input down to low fire rate for the main flame lighting. If not automatic, it is advisable, and in many cases necessary, to manually turn the input rate to low fire in order to satisfy a low fire interlock switch before the lighting sequence can be started.
‡ 10 seconds for gas or light oil; 15 seconds for heavy oil.

Nuisance shutdowns are a frequent complaint about flame supervising systems. An *annunciator* simplifies trouble-shooting.

Recent and continuing developments are minimizing many of the objections to flame supervising systems, such as the time and money required for wiring, and the time and money involved in trouble-shooting shutdowns. An honest appraisal of the cost of an explosion or fuel-fed fire will almost always justify the cost of a flame supervising system. The Industrial Heating Equipment Association's *Combustion Technology Manual* (Reference 7.a listed at the end of Part 7) states the rationale for flame supervising systems very well:

> "*The need for flame supervisory and sequence controls has been firmly established by experience. There is a continuing reluctance to use protective controls because of potential nuisance problems. Properly installed and maintained, modern systems are reliable and simple to operate. An explosion may cause extensive property damage, personal injury and business interruptions. Many companies cannot recover from business interruption as a result of an explosion. It is, therefore, imperative that all precautions available be taken to prevent potential hazards.*"

The *irony of safety controls* is the fact that they sometimes are used so rarely that they easily fall into neglect and become unsafe. An essential corollary of a safety equipment investment is a safety maintenance commitment -- regular checking of safety circuits and mechanisms to make sure that there has been no tampering, jumpering, clogging, galling, wearing, corroding, or other irregularity. If an interlock or valve has not been actuated for months, there is a chance that it never will be -- even when needed!

REFERENCES

7.a   **I.H.E.A.:** "Combustion Technology Manual", 4th ed., pp. 93-102, 202, Industrial Heating Equipment Assn., 1901 N. Moore St., Arlington, VA 22209, 1988.

7.b   **Marceau, W. D.:** "Combustion Systems and Combustion Control", vol. 36, pp. XLVIII-LI, Canadian Ceramic Society Journal, 1967. (Some of the data in this paper is through the courtesy of Selas Corporation and the Glass Container Industry Research Corp. Graphs from this, plus additional data, are available in North American HBS 273.)

7.c   **N.F.P.A. 86:** "Ovens and Furnaces", pp. 47 and 13, National Fire Protection Association, Batterymarch Park, Quincy, MA 02269, 1995.

ADDITIONAL SOURCES of information relative to combustion control

Bodurtha, F. T.: "Industrial Explosion Prevention and Protection", McGraw-Hill Book Co., New York, NY, 1980.

Factory Mutual Engineering Corporation: "Handbook of Industrial Loss Prevention", 2nd ed., McGraw-Hill Book Co., New York, NY, 1967.

Fisher Controls Company: "Control Valve Handbook", 2nd ed., 4th printing, Fisher Controls Company, Marshalltown, IA, 1977.

N.F.P.A.: "Prevention of Furnace Explosions in...
Fuel Oil & Nat'l Gas-Fired Sgl Burner Blr/Fces",   NFPA 85A, '87;
Natural Gas-Fired Multiple Burner Boiler/Fces",   NFPA 85B, '84;
Fuel Oil-Fired Multiple Burner Boiler/Furnaces",   NFPA 85D, '84;
Pulverized Coal-Fired Multiple Burner Blr/Fces",   NFPA 85E, '85;
National Fire Protection Association, Batterymarch Park, Quincy, MA 02269

Vervalin, C. H. (ed.): "Fire Protection Manual for Hydrocarbon Processing Plants", vol. 1, 3rd ed., 1985; vol. 2, 1st ed., 1981; Gulf Publishing Co., Houston, TX.

Part 8.   COMBUSTION SYSTEMS

BURNER INSTALLATION

Refractory Tiles. Major causes of burner and furnace failures are refractory expansion and leakage of hot combustion products into the supposedly cool areas around burner mounting and between the furnace shell and refractory. Figures 8.1, 8.2, and 8.6 illustrate typical problems. Expansion joints are particularly important in tall refractory walls with burners near the top. Figures 8.3, 8.4, 8.5, and 8.6 suggest construction details for preventing tile damage. Burner tiles must not be expected to support refractory above them -- use spanner tiles or a cast arch or slab.

Tiles should be supported by the refractory wall. Small lightweight burners may be supported by their tiles, but the burner piping should have independent support. In thin walls or walls of low strength materials, a heat-resistant metal support should surround the tile or high temperature alloy expanded metal may be cast within the tile. The opening for a burner in a ceramic blanket wall should

Figure 8.1.   A tile failure in shear due to unequal expansion of the refractory and shell or unequal expansion between layers of refractory. Expansion joints in refractory brickwork and a matrix around the tile can prevent this.

Figure 8.2.   A tile failure in tension as a result of bowing or buckling of the shell or brickwork. Prevent this by using more rigid shell, buckstay, and bracing construction. Install a high temperature matrix to prevent "fluing" of gases through to the shell.

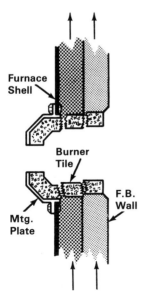

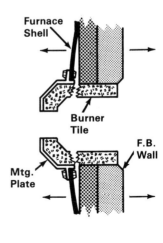

be cut undersize and the excess material tucked in tightly around the supported burner tile. Because of shrinkage after firing, periodic repacking with extra ceramic fiber is advisable. Treat tiles tenderly -- support, contain, avoid strain.

A gap around a burner tile is not the only means whereby hot furnace gases may reach and damage lower grade refractory or insulation or the furnace shell. As shown in Figure 8.6, observation ports and other openings, even doors may provide the passageway. A very tiny crack can "snowball" into a warped or molten mass. One never knows when or where some pressure effect may develop that could cause hot gases to flow within the wall; so it is wise to design walls and ports to prevent the problem. All refractories should be anchored tightly against one another and against the furnace shell.

Figure 8.3. Recommended method for installing a burner tile. Elevation section through a vertical furnace wall.

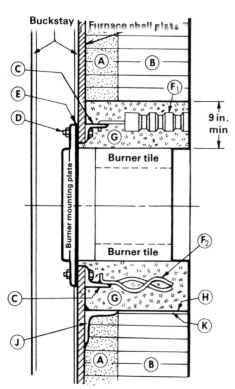

A = Insulating refractory or block insulation.

B = High temperature refractory. All refractory must be tied securely to furnace shell plate. Horizontal and vertical expansion joints must be provided in surrounding refractory to prevent pressure from being exerted on the burner tile.

C = Horizontal angles welded to furnace shell plate between buckstays. Vertical angles should also be welded to the shell plate on each side of the burner.

D = Burner mounting bolts with heads welded in place.

E = Gasket -- may be used to provide better seal between burner mounting plate and furnace shell plate.

F_1 = Anchor tiles tied back to angle (preferred to F_2), or

F_2 = Stainless steel bent-rod anchors, mastic-coated.

G = Castable refractory all around -- minimum thickness = 9 in. or ½ tile OD, whichever is greater. Rammed refractory is an alternative, but anchors then must be refractory type. (Castable is preferred because its entire mass sets up without firing.)

H = Waterproofing (all around) to prevent surrounding refractory from absorbing water from the castable refractory. Plastic sheet is suggested.

J = Shelf support angle. This and expansion joint K prevent vertical pressure from being exerted on burner tile. This construction is especially helpful when burners are located high in a wall.

K = Expansion joint, densely packed with refractory wool.

Figure 8.4.   Suggested refractory installation around large burners for which no tile is supplied.
Expansion joints must be provided in surrounding refractory to prevent strain on the cast or rammed
burner tunnel section.

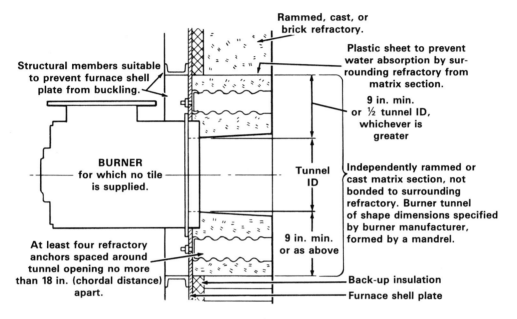

**Figure 8.5.   Section through a fiber-lined vertical furnace wall showing the generally recommended
method for installing a burner's refractory tile (2200 F maximum).** Standard non-supported or non-
jacketed tiles are not suitable for long life in fiber-lined furnace.

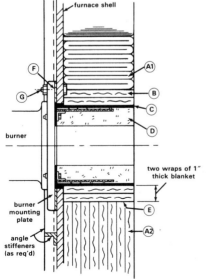

A1 = Ceramic fiber modules. (Does not require
      "B" if modules are tight to burner.)

A2 = Fiber type blanket, block, or board insu-
      lation.

 B = Blanket insulation wrapped around tile. Use
      blanket 1″ thick with at least 8 lb/ft² density
      and temperature rating equal to furnace
      lining.

 C = High temperature, air setting refractory
      cement -- thin wash coat.

 D = Must be burner refractory tile constructed
      specifically for fiber-lined furnaces. (Non-
      supported cast or prefired tiles usually are
      not suitable for fiber-lined furnaces.)

 E = Non-metallic fastening device (tape, rope,
      plastic, cheesecloth, etc.) to compress and
      secure wrapped blanket about tile.

 F = Gasket: to provide a seal between burner
      mounting plate and furnace shell.

 G = Burner mounting bolts with heads welded
      to inside of furnace shell.

Figure 8.6. Observation port design is important near burners with negative tile pressure.

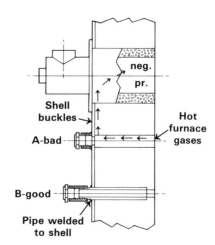

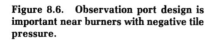

Observation ports on burners and pilots are useful, and should be kept clean, but additional and larger ports should be installed through the furnace wall (a) to view the main flame from a location within reach of the adjusting valve, and (b) to view into the tile.

GENERAL PIPING PRACTICE

Piping must conform to specifications of insuring authorities and to local, state, and national codes. See Reference 8.c. For fuel gas lines, the local gas utility's industrial department may be able to supply information on such codes; and they may have their own regulations. For propane piping, consult the propane supplier and Reference 8.d.

All fuel oil systems must be laid out and installed in accordance with rules of the American Insurance Association (85 John St., NY, NY 10038, tel. 212-669-0400), and Reference 8.b, as well as local and other governmental codes. Consult local fire officials.

Burner Piping. No pipe weight nor stress should be transmitted to a burner. (Remember: People may walk on horizontal pipes.) Flexible connections minimize stress and vibration transmission, but their necessarily thin sections must be checked frequently for rupture if the vibration is prolonged or severe. Figure 8.7a suggests preferred positions for air and fuel supply manifolds and for pilot and flame monitoring systems relative to burners.

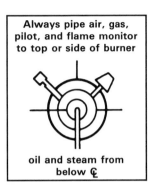

Always pipe air, gas, pilot, and flame monitor to top or side of burner

oil and steam from below ℄

Figure 8.7a. Preferred piping positions for horizontally-fired burners. Neither steam nor oil should drip into an air line (where it might burn) nor a gas line (where it could damage regulator diaphragms, and where its weight on a diaphragm would upset air/gas ratio control). Pilots and flame monitoring devices have very small openings; so they should be located above the burner centerline where neither refractory crumbs, dirt, nor drips can interfere with their operation.

Fuel lines should be "black iron pipe" (steel, not coated). The sulfur in fuels will react with the zinc in galvanized pipe to form a sludge. Malleable iron or welding fittings are preferred for strength and to minimize the possibility of leaks developing. Welding type fittings or fabricated connections can usually be designed to offer less flow resistance. Soldered joints should not be used for fuel lines because they could melt in the event of a fire. Drip legs should be installed upstream of regulators, valves, and burners to protect them from dirt and water. (Drip legs are often run all the way to the floor to help support the piping.) See References 8.c and 8.d at the end of Part 8.

Air lines may be of almost any type of pipe, tubing, or duct material if it is clean, airtight, and strong enough for the anticipated pressure and temperature. Workmen often walk on horizontal air pipes; so they may need greater structural strength than dictated by the flowing fluid. Piping must have its own support brackets to avoid strain on burners, blowers, and accessories.

Manifolds. To minimize dripping when burners are not in use, oil and steam headers (manifolds) should be located below the burner centerline. For oil and dual-fuel systems, air and gas headers should be above the burners or piped so that the air and gas connections are on top of the burners so that liquids cannot drip into them. Similarly, for burners in vertical walls, the pilot and flame monitoring device should be located above the centerline of the burner mounting plate.

To avoid large pressure drops, care should be exercised to prevent excessive contraction of fluid streams as they enter pipes of smaller cross section. Uniform flow distribution in the downcomers fed by a manifold is usually enhanced by following the suggestions in Figure 8.7b.

Figure 8.7b.   Manifold design suggestions for promoting uniform flow distribution and minimum pressure loss.

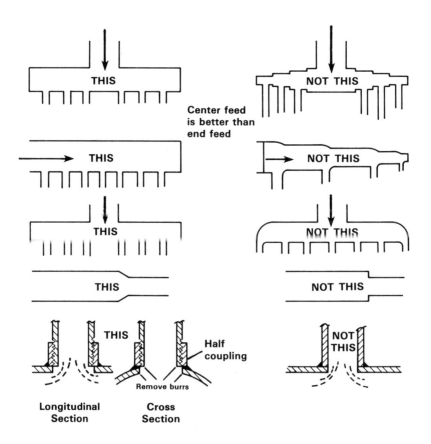

Piping. Many local codes prohibit the use of reducing bushings because they are more prone to leak than reducing couplings. They also cause more pressure loss than reducing couplings. Holding the use of elbows and tees to a minimum will result in less pressure drop. The ends of all pipe should be reamed before fitting. Pipe should be inspected and blown or flushed out to avoid future plugging of regulators, valves, and burners. Every section of pipe should be blown out *before and after* assembly.*

---

* The need for frequent inspection of piping during installation cannot be overstressed. Banana peels, apple cores, pop bottles, cloth, and paper have been found to cause expensive and time-delaying start-up troubles.

Figure 8.8. Many otherwise sound combustion installations have failed to operate correctly because of improper piping. Too often, the only cure has been extensive repiping, causing back charges and lost production time.

Location/*Explanation*	**Minimum run of straight unobstructed pipe, in pipe diameters**
Between **mixer** or mixture piping and premix burner **nozzle**/*to insure thorough mixing and flame stability*	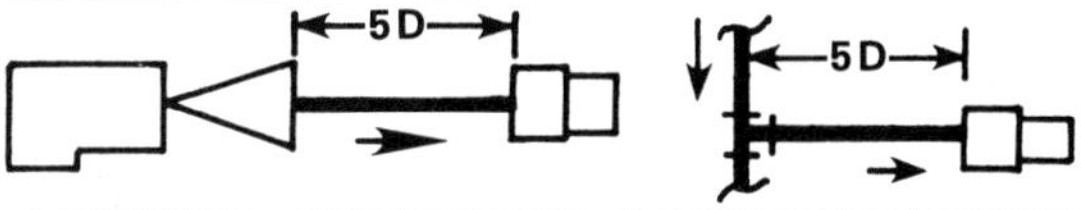
Between **impulse tap** and upstream **control valve** and downstream fitting or valve and on the same centerline as the shaft of the butterfly valve/*to avoid false signal to a ratio regulator or controller.*	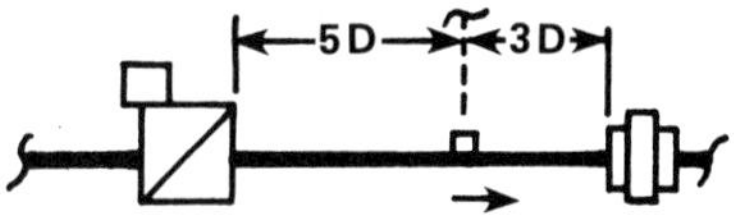
Between **control valve** and first downstream **pipe branch** or manifold downcomer/*to achieve equal flow and pressure distribution to all downstream branches.*	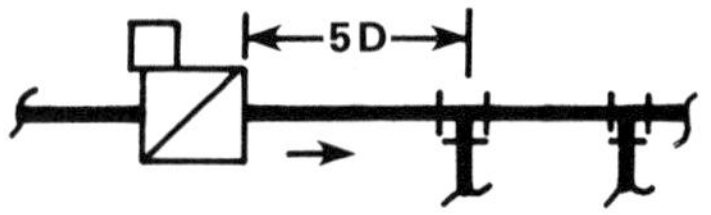
Between supply **elbow or tee** and first downstream **branch** or manifold downcomer/*to achieve equal flow and pressure distribution to all downstream branches.*	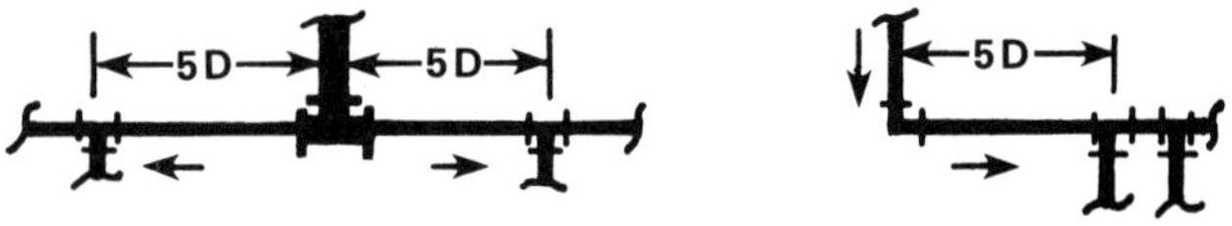
Between pressure gauge or pressure switch **tap** and upstream or downstream **valve or fitting**/*to avoid false readings because of lopsided flow.* (Do **not** use a valve with a pressure switch.)	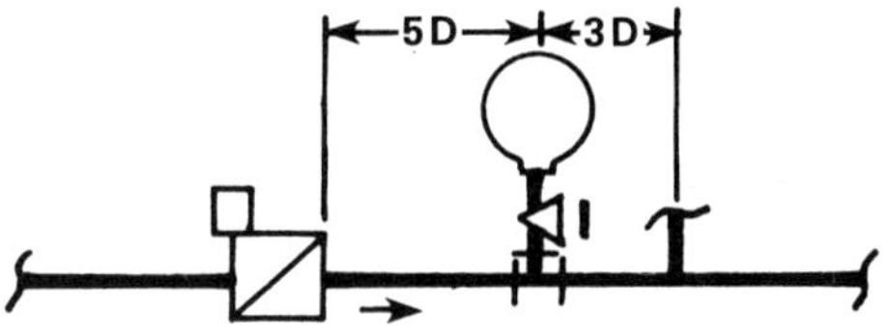
Between gas **metering orifice** and upstream or downstream **valve or fitting**/*because accurate repeatable measurement requires undisturbed flow.*†	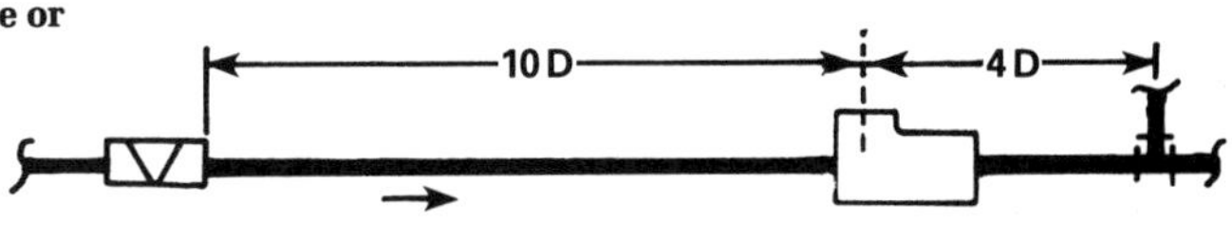
Between air/oil **Ratiotrol** and **burner centerline**/*to avoid an error in the oil pressure and flow at the atomizer due to the weight of a column of oil.*	**12″ maximum regardless of pipe size**

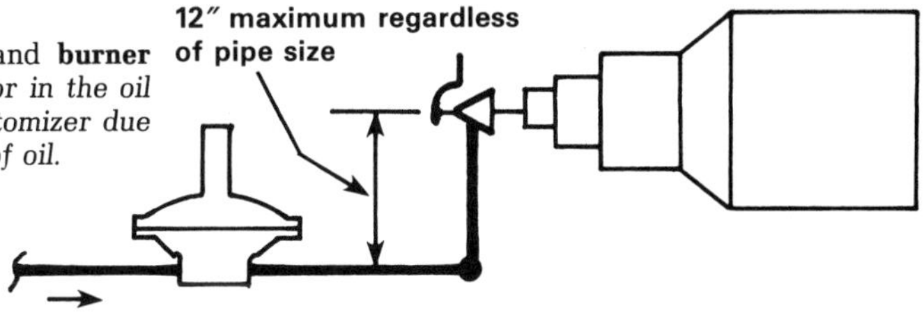

† 10 straight unobstructed pipe diameters upstream and 4 downstream are sufficient for staying within 2% flow error with 8697 Metering Orifices. For straight pipe requirements for installation of orifices (other than 8697) see Reference 8.a (at the end of Part 8).

Failure to follow the straight, unobstructed pipe requirements outlined in Figure 8.8 may necessitate extensive and expensive repiping. Premix burners should not be piped to mixers with elbows or short nipples because such a compact arrangement often results in flame instability at other than closely held air/fuel ratios. It may also result in mixture pressures having to be held lower than desired to avoid flame liftoff; or in flames being poorly developed, frequently resulting in nuisance outages by flame supervisory devices.

Poorly installed impulse lines transmit false pressure signals, defeating the purpose of the air/fuel ratio control system and causing fuel waste, burner instability, incorrect furnace atmosphere, and poor temperature uniformity. Pressure gauges and manometers connected in incorrect locations give false readings and thereby mislead furnace operators when they are attempting to set firing rates or air/fuel ratios.

Use of street elbows should be avoided because they constitute a greater constriction than a standard elbow and nipple of the same nominal size. If corrugated flexible connectors are needed to minimize transfer of vibration or stress to other parts of the piping system, the connector should be installed with little or no bend. Using one size larger helps reduce turbulence and line pressure loss.

Anti-seize compounds should be used on threaded connections in burners and mounting plates, which may get hot. Teflon pipe thread sealing paste, commercial pipe dope compounds, or Teflon tape should be used, but none of these should be used in excess or over the end of the thread where they will be sheared off and plug small downstream orifices. There are many compounds available for the above two purposes. Care should be used in selecting them for the anticipated temperature, and to make sure that they do not dissolve in or react with any fluid that may flow in or around the pipe. Slow drying, non-hardening compounds are usually preferred.

Unions, flanges, or couplings should be installed wherever necessary to allow easy removal of burners, regulators, valves, or accessories for cleaning, maintenance, and inspection. ANSI flanges (formerly American Standard) or flange unions can be used in either fuel or air lines. Lightweight square 4-bolt flanges are ideal for air lines. Morris or Dresser type couplings are convenient in place of unions to allow for expansion and some variation in pipe lengths and alignment.

Accessories. Great caution is necessary in specifying sizes for valves, strainers, regulators, and the pipe itself. Mere matching of the pipe to the size of the pipe connection on the accessory or vice versa will often result in trouble. Every selection should be carefully engineered for flow and pressure requirements in accordance with the procedures outlined in Part 5, Vol. I. The accessory manufacturer's data on pressure loss should be consulted. Even after careful selection of sizes, specifications must be emphatic to avoid the natural tendency of an unknowing pipe fitter to install pipe of the same size as the accessory pipe connection.

Gate valves, rotary plug type valves, and blast gates are to be used as shutoff valves only -- not for flow control. Air filters should be installed on all blower inlets to exclude such foreign matter as lint or dust which might clog burners and control equipment. Strainers should be located upstream from all oil and steam control equipment.

OIL HANDLING SYSTEMS

The properties of commercially available fuel oils have become very changeable because (1) low sulfur requirements have forced blending or different refining methods and (2) shortages and the resulting changes in supply sources have brought about changing mixes of base materials. As a result, the oil user can no longer count on steady pour points and viscosities, and he must expect problems with separation and degradation in his storage tanks. This necessitates that he design or revamp his oil handling system to prepare for the worst oils.

A few problems common to large and small distributing circuits for both light and heavy oil are outlined below. The warnings of the preceding section regarding piping should be re-read.

Air in oil lines may result in poor control, irregular burning, and limited capacity. Removal of air is best accomplished in a system of the recirculating type, that is where oil is continuously pumped from the tank to the furnace and back again. "Dead end" systems do not adequately purge themselves of air. The branch lines from the circulating manifold to the burners should be as short and straight as possible. Air bubbles will not follow the oil if it is forced to flow downward, but will collect at a high point in the line. For this reason, oil lines should not rise over (or drop under) a furnace and should enter the burners from below. Manual air bleed valves should be installed at high points.

Dripping of oil through burners that are shut down results in an accumulation of carbon soot in the burner nozzle or tile because the dripping oil "cracks" upon exposure to the heat remaining in the furnace. This carbon may impair the operation of the burner when next used. Dripping may be minimized by installing the oil lines so that they rise to the burners from a manifold located below the burners, that is, so that less length of pipe can drain through the burner. Most burners are equipped with a solenoid oil shutoff valve at the oil line connection. Dripping will be minimized if this valve is always closed tightly whether other upstream oil valves are closed or not. Most dripping occurs shortly after the oil is shut off; so it is good practice to permit air to blow through the burner for several minutes after the oil has been turned off. This tends to remove dripping oil from the burner nozzle and tile to minimize carbon buildup.

Oil expansion in pipelines may seriously damage gauges, regulators, and other apparatus. After a burner is shut off, heat from the furnace expands the oil more than the pipe. The normal shutdown procedure often leaves a section of pipe with closed valves at both ends. The expansion of oil in this section causes a rise in pressure that may burst diaphragms, Bourdon tubes, or bellows. Accumulators (bellows or bladder type expansion chambers) should be installed to take up the oil expansion between pairs of valves in sections of line subject to heat.

Oil Piping Recommendations. Figures 8.10 through 8.16 are recommended piping diagrams designed to provide the greatest reliability and convenience of operation and maintenance. Sizing of the pipe requires use of the tables in Part 5, Vol. I. Heat loss from pipes may be determined by the use of the tables in Part 4, Vol. I. An oil handling system consists of three parts: 1) the storage tank and associated equipment, 2) the main circulating loop, and 3) the branch (booster) circuits.

Oil Storage Tanks. (See Figure 8.10.) Indoor storage is not recommended. Insurance regulations and local safety codes should be consulted as to the depth of earth and/or concrete required above an underground tank, or the dike arrangement around an aboveground tank. Such codes and regulations also specify minimum distances from storage tanks to property lines, buildings, and other tanks. The various elements of the storage tank and associated equipment are discussed below. Unless otherwise stated, the following applies to underground and aboveground tanks for light or heavy oil.

Table 8.9.   Thermal expansion in 10 ft of pipe for a 10 degree (F) oil temperature rise.

Nominal pipe size, inches	¼	³/₈	½	¾	1	1¼
Cubic inches expansion	0.063	0.115	0.183	0.320	0.519	0.898

In climates where the temperature drops to within 20 degrees Fahrenheit of the oil pour point, the tank should be insulated and heated. The unpredictable properties of blended residual oils negate relying on a self-insulating effect from congealed oil clinging to the interior surfaces of cold tank walls, as was possible in years past.

The *suction line inlet* should be at least two inches above the bottom of the tank for light oil and at least four inches above the tank bottom for heavy oil to avoid drawing water, dirt, and sediment into the piping system. There should be an accessible shutoff valve in the suction line close to the tank. The suction line should pitch toward the tank. Suction lines should be kept short, preferably with no more than 15 feet of lift.

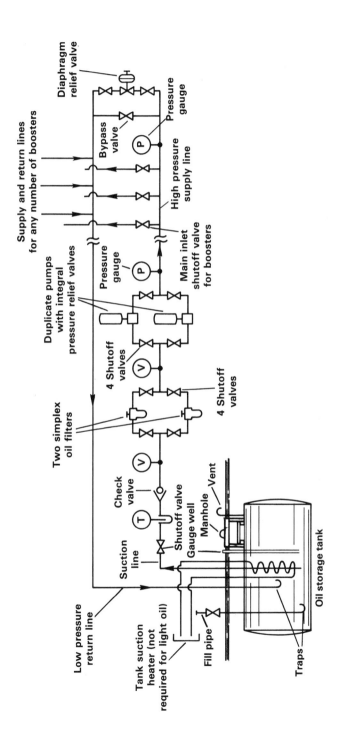

Figure 8.10. Oil storage tank and main circulating loop. A circulating system similar to this is recommended for all installations regardless of size. The accompanying text discusses each element of the system.

The *return line* should contain a trap (liquid seal such as in any sewage drain) so that it cannot act as a vent for oil vapors when the oil level in the tank is low. The return line to an overhead tank should be fitted with a check valve and a manual shutoff valve so that oil will not siphon out of the tank when the line is opened.

The *fill pipe* should either extend below the level of the suction inlet or it should contain a trap so that it cannot act as a vent. Fill terminals should not be located inside buildings. They should be tamperproof, waterproof, and dirtproof. The fill pipe should never be cross-connected to the vent.

Vent pipes should be arranged to drain to the tank. The lower end of the vent pipe should not extend more than one inch below the uppermost point of the tank. Vent pipes should not be cross-connected with fill lines, return lines, or other vent pipes. They should be visible from the filling connection, weatherproof, clogproof, and filtered, with flame arresters. If the tank is to be filled through tight connections by means of a pump, the vent pipe size must be at least as large as the pump discharge.

Manholes. Tanks of more than one thousand gallons capacity should be provided with bolted and gasketed manholes, which should not be used for gauging, return lines, or any purpose other than inspection and repair.

Tank Heaters. Heaters and insulation should be installed to keep all of the oil in a tank at least 20°F above the pour point to avoid separation of blends or sludge formation. Thermostatically controlled heaters are usually located in the suction bell or thimble to assure pumpability. The temperature to which the oil is heated in the tank depends upon the type of oil, the pump, and the length and diameter of the delivery line, but heating to more than about 150 F may result in distillation of lighter ends. See Figures 2.8 (Vol. I), 8.11, and 8.16; Tables 2.9, and 2.10 (Vol. I).

Main Oil Circulating Loop. The main oil circulating loop is the section of the oil handling system that delivers the oil from storage to the branch booster circuit or (in small systems) to the furnace, the boiler, or the point at which it is to be used. (See Figure 8.10.) It may be a large oil header supplying all of the furnaces in an entire plant, or it may exist solely for the purpose of transporting oil to a single boiler or furnace. Regardless of the size of the system, this delivery loop should be a closed circuit, returning to the tank. The use of a closed circuit facilitates quick air elimination, permits future load additions to the line, and minimizes pressure fluctuations in the line due to load changes. The various items of equipment in the main circulating loop are discussed in the following paragraphs.

Figure 8.11. Bell arrangement for oil suction and return through the top of a horizontal cylindrical tank. Small vent holes in the bell permit escape of returned air bubbles or accumulated vapor. Returning the hot oil to a bell speeds warmup, minimizes oil heater energy consumption, and minimizes distillation. Steam heaters are preferred for in-tank heating. If electric heater must be used, redundant safety switches should be used to prevent a sticking contact from causing overheating problems.

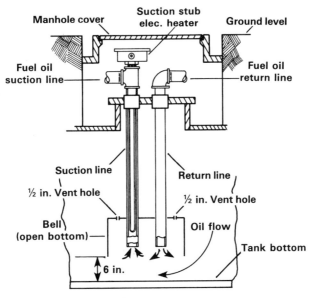

The *oil filter* or *strainer* prevents dirt, scale or other foreign matter from entering the system. This filter usually contains a 20 mesh (about ¹/₃₂ in. perforation) or 30 mesh basket but finer ones may be used for light oil. The basket is easily removed for cleaning. This provides an opportunity for observing the foreign material in the system. For systems in continuous operation, use of two filters in parallel permits one basket to be serviced while using the other. By observing the growing difference in readings of gauges upstream and downstream of the filter, maintenance personnel can develop an experience factor for determining when the baskets need cleaning. Duplex strainers may be more prone to letting air bubbles leak into the oil system than two simplex strainers, each with upstream and downstream shutoff valves.

Piping. Suction line difficulties outnumber all other oil system problems. Air in-leakage can cause burners to sputter and go out; or it can lodge in a down-turning elbow and restrict oil flow capacity. Type K copper tubing may be used for light oil suction lines. It is available in long lengths (minimizing joints) and is less liable to be flattened. To protect it from dents, it is desirable to run it inside protective conduit or tile. Wrought iron, steel, or brass pipe and their standard fittings may be used, but must conform to all codes. Cast iron fittings are not allowed.

Suction piping must be tested for leaks before connection to the pump. Tank suction must be tested with 100 psi air pressure and not drop below 90 psi for 24 hours. (An alternate test consists of subjecting the suction line to a vacuum of not less than 20″ Hg for 24 hours without a noticeable drop.)

The *pumps* should be the positive displacement type. They deliver an approximately constant volume of oil at all times and will develop whatever pressure is necessary to accomplish this delivery. Spring-loaded relief valves are usually built integral with the pumps to prevent the development of excessive pressures in the system if all of the outlets are closed. The relief valves act as short circuits around the pumps. They are for pump protection only and are not suitable for loop pressure control. Duplicate pumps are recommended to prevent shutdown when one pump needs servicing. Shutoff valves should be located on each side of each pump to permit its complete isolation from the system. The pump lift (vertical distance through which the pump must pull the oil on the suction side) should be as short as possible. The vertical distance from the pump to the open end of the suction line plus

$$\left(\frac{\text{pipe and fitting loss in psi} \times 144}{\text{oil specific weight in lb/ft}^3} \right) \text{ should not exceed 15 feet.}$$

Although some pumps may be capable of developing a lift of almost 1 atmosphere, more than the 15′ lift is not recommended because of the possibility of vapor binding, noisy operation, and excessive wear.

The pump capacity may be approximated by one of the following rules of thumb: maximum oil burning rate plus 0.8 to 1.0 gph (depending upon the effectiveness of the tracing and insulation) for each foot of loop length; or 1½ times the maximum burning rate. Use whichever is smaller. The heaters should be sized for the pump capacity and an oil temperature high enough for a pumpable viscosity. The method is illustrated in Example 8-1 and Figure 8.12.

Example 8-1. As a general rule, the oil temperature at a burner should not vary more than 10°F or the viscosity will vary more than 25 SSU, affecting the quality of atomization, and making it impossible to maintain a constant air/fuel ratio at all firing rates. Figure 8.12 shows the burner spacing and piping for feeding 14° API oil to 3 widely separated burners with a 1 in. oil circulating line insulated with 1 in. of calcium silicate insulation. The lowest expected ambient temperature is 50 F and h_o is predicted to be 4 Btu/ft² hr °F. If burner B is to be turned on and off, will it be possible to use a 90 gph circulating pump without causing more than 10°F change at burner C, or will it be necessary to use the next larger (150 gph) pump?

The temperature drop in each section can be calculated by use of formula 4/6e, Vol. I. From Table 2.3, Vol. I, a 14° API oil has 0.973 sp gr and weighs 8.106 lb/gal. Evaluate specific heat from formula 2/12 (Vol. I) using an estimated average oil temperature of 230 F: c = [0.388 + (0.000 45 × 230)]/$\sqrt{0.973}$ = 0.498. From Table 4.22, Vol. I,

$A_f = 0.8668$ and $X = 1.532$. From Figure 4.20b, Vol. I, $k = 0.4$. R_i is assumed to be negligible in this case, and there is no X_2/k_2; so

$$R = \frac{X}{k} + \frac{1}{h_o} = \frac{1.532}{0.4} + \frac{1}{4} = 4.08. \text{ For section AB with burner B on,}$$

$$D = \frac{I - M}{\dfrac{c \times p \times R}{L \times A_f} + \frac{1}{2}} = \frac{240 - 50}{\dfrac{0.498 \times (224 \text{ gph} \times 8.106) \times 4.08}{350 \text{ ft} \times 0.8668} + \frac{1}{2}} = 15.007.$$

Using this procedure, follow the method of solution by reading each horizontal line of the following tabulation consecutively.

pump gph	B on, off	section	L, ft	flow, gph	°F drop, by 4/6e	temperature at end of section
90	on	AB	350	90 + 95 + 39 = 224	15.01	240 − 15.0 = 225.0
90	on	BC	180	90 + 39 = 129	12.40	225 − 12.4 = 212.6
90	off	AB	350	90 + 39 = 129	25.32	240 − 25.32 = 214.7
90	off	BC	180	90 + 39 = 129	11.66	214.7 − 11.66 = 203.0
150	on	AB	350	150 + 95 + 39 = 284	11.93	240 − 11.93 = 228.1
150	on	BC	180	150 + 39 = 189	8.71	228.1 − 8.71 = 219.4
150	off	AB	350	150 + 39 = 189	17.66	240 − 17.66 = 222.3
150	off	BC	180	150 + 39 = 189	8.43	222.3 − 8.43 = 213.9

A 150 gph pump is preferred because it allows only 219.4 − 213.9 = 5.5°F variation as opposed to 212.6 − 203.0 = 9.6°F variation with the 90 gph pump.

Figure 8.12. Schematic piping for Example 8-1.

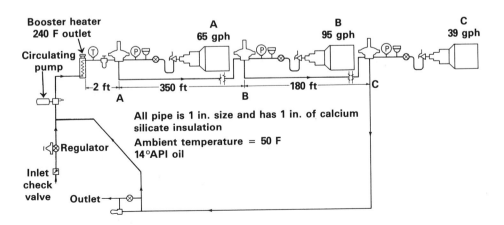

Shutoff Switches. Provision should be made for stopping the flow of fuel in the event of a main oil line break. Even small leaks, which are difficult to detect by pressure switches, may spray fuel and feed a fire. This may best be minimized by locating a number of manual electric switches (in series) at convenient spots about the plant. These switches should close electric fuel shutoff valves and stop pumps. *Pressure gauges* facilitate adjustment of relief valve settings and location of trouble.

The *diaphragm relief valve* at the extreme point of the main circulating loop serves to maintain a constant pressure in the high pressure side of the loop. The setting should be as high as possible without opening the bypass relief valve on the pump. To make this setting, gradually raise the opening pressure of the diaphragm relief valve. The reading of the pressure gauge will gradually rise until the pump bypass relief valve opens. When the pressure gauge reading stops rising, reduce the opening pressure of the diaphragm relief valve until the gauge reading drops slightly below its previous maximum reading. This setting should be made when all of the oil is being recirculated to the tank; that is, when none is being drawn off for use in the branch lines.

Take-off lines to the branch circuits lead from tees in the high pressure side of the main circulating loop. Figures 8.13a and b suggest configurations for these take-off points. Return lines from the branch circuits deliver to the low pressure side of the main circulating loop. These return lines are not necessary in light oil systems.

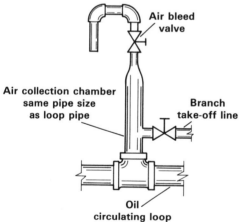

Figure 8.13a. Top take-off from an oil circulating loop minimizes dirt entry into the take-off line, but permits air to enter the take-off line. To minimize air problems, install a vertical standpipe with an air bleed valve and a U-diverter so that regular maintenance can tap off accumulated air and catch the first slug of oil in a bucket.

Figure 8.13b. Bottom take-off (downcomer) from an oil circulating loop minimizes air entry into the take-off line, but permits dirt and water to enter the take-off line. To minimize water and dirt problems, install a vertical dirt-pipe with a drain valve and L-diverter so that regular maintenance can allow flow into a bucket until the oil stream runs clean.

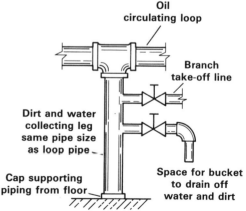

To initiate circulation in the main circulating loop; (1) Turn on the steam or hot water to the tank heater and tracers adjacent to oil lines. (2) Open the shutoff valves on the suction and return lines to the tank. (3) Open the valve to bypass the diaphragm relief valve. (4) Set the valves for one of the two pumps and start that pump. (5) When warm oil is flowing through the return line to the tank, close the valve that bypasses the diaphragm relief valve and set the diaphragm relief valve inlet pressure to the desired value, as described 2 paragraphs previously. This procedure may consume considerable time if the lines are full of air or if the oil is very cold. After oil is flowing through the main circulating loop, the branch circuits may be opened and started.

Much time is required to start up a heavy oil system; so it is suggested that this be done early every autumn and that the system be left running until late spring. This assures that the standby fuel system is truly standing by. Later in this Part 8, schemes are suggested for substituting light oil, steam, or air in heavy oil systems during summer shutdown in order to facilitate faster start-up by eliminating heavy oil that would otherwise congeal in the lines.

Branch Circuits for Light Oil. Figure 8.14 is a schematic piping diagram for a light oil system with two control zones.

A main return line to the storage tank is important to flush dirt and air bubbles back to the tank so they do not go through burners nor accumulate within the oil delivery system.

A shutoff valve should be located at each branch circuit inlet so that it may be isolated from the main circulating loop.

An oil filter should be located immediately upstream of each pressure regulator to prevent clogging of the burner oil valve or excessive leakage pressure due to dirt on the valve seat of the regulator. This oil filter should contain a 40 mesh basket.

A pressure regulator reduces the pressure from that of the circulating loop to the required inlet pressure for the ratio control system (25-30 psi for an Air/Oil Ratiotrol).

A pressure gauge should be installed for checking outlet pressure of the pressure regulator.

A manual reset fuel shutoff valve should be used to prevent further admission of oil to the branch circuit in the event of failure of flame, blower, or controls. (See Part 7.) In the *fuel trains*, the components and their relative positions must be approved by insuring authorities. Figure 8.22 shows a typical dual-fuel combustion system with gas, pilot, and light oil trains.

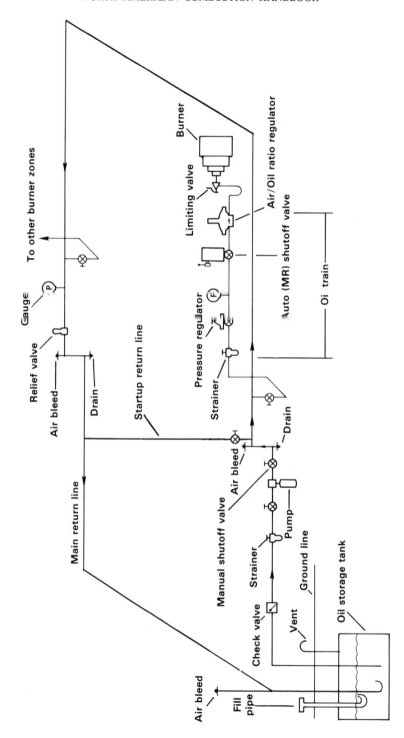

Figure 8.14. **Light oil system piping.** Problems with air bubble elimination will be minimized if the non-recirculating sections (dead ends to oil trains and burners) are kept short.

At least one *ratio regulator* (Air/Oil Ratiotrol) should be used for each control zone.† A control zone may contain one or more burners (and one or more ratio regulators), but the capacity of each regulator must be equal to or greater than the sum of the capacities of all of the burners it supplies. If a control zone contains burners on more than one level (more than one foot difference in elevation), separate ratio regulators must be used for each level. Air/oil ratio regulators must not be above nor more than one foot below the centerline of the burners they serve. The length of piping to the control zones should be kept to a minimum to reduce the total length of dead end piping from which air must be eliminated.

Startup return lines to the low pressure side of the main circulating loop are used to facilitate air removal from long branch circuits. The shutoff valves in such startup return lines should be closed as soon as the air has been eliminated from the circuit. A 30 psi pressure gauge should be located on the downstream side of the ratio regulator for the purpose of checking its adjustment.

An *expansion chamber* or *accumulator* is necessary to protect the ratio regulator against pressure build-up due to thermal expansion of the oil in the dead end of the line.

Each burner should have a *shutoff valve*, and an *oil limiting orifice valve* (for setting air/fuel ratio).

To *start* flow through a light oil system: (1) start the pump and open valves to allow circulation through the main loop and back to the tank; (2) open each air and drain valve, one at a time, until a clean, bubble-free oil stream flows into a bucket, then close the valves; (3) open one branch circuit shutoff valve; (4) provide a downstream outlet by opening the line at a point near its extreme end (or by opening the startup return line, if any) and providing a collecting bucket for the oil; (5) permit oil to run through this outlet until all foreign matter is flushed out, and until the flow is no longer interrupted by air bubbles, then close it; (6) light one burner on low fire, watching it carefully until all air is forced out of the dead end of the line; (7) light the other burners in the same manner, one at a time, and set all burners at the desired rates and ratios; (8) repeat steps 3 through 8 until all branches are operating.

Branch (Booster) Circuits for Heavy Oil. Branch circuits for heavy oil differ from those for light oil in that the oil must be heated and circulated within the branch or booster circuit. For most burners, heavy oil must be heated enough to reduce its viscosity to 100 SSU for atomization. If the oil were not continually circulated past the burner, its temperature at the burner would change with the burner firing rate, and the consequent variation of oil viscosity would have an adverse effect upon the control apparatus and the degree of atomization.

---

† Or see manufacturer's literature for specific cases.

Part 2 of Volume I lists the preheating temperature required to obtain suitably low viscosities for pumping and atomizing. Heating can be by steam, hot water, heat transfer fluid, or electricity, but should be thermostatically controlled. Steam heating is preferred, because oil temperature will not exceed the saturation temperature of the steam; so a steam pressure control avoids overheating. (See the example of steam pressure determination under "Heat tracing and insulation", below.)

If electric heating must be used, redundant cutoff switches in series should be used to prevent a sticking contact from causing overheating problems. A maximum of 10 watts per square inch of electric element surface is recommended because of danger of vapor lock or coke formation, either of which can result in inadequate cooling of the element. Because electric power continues to flow into the element, it burns out quickly. Air accumulation in electric heaters will cause overheating for the same reason. For systems where air is a problem, or for oils containing light ends (such as crudes), even 12 watts per square inch is too much; so a lower watt density electric heater or a steam heater should be used.

Excessive temperature will cause vaporization and coking, leading to quick burnout of an electric heating element, particularly if the oil should stop flowing.

A typical design figure is based on

[8/1]   required watts heater capacity = circulating pump capacity in gph × required temperature rise in degrees F.

This formula is based on an oil density of 7.5 pounds/gallon, a specific heat of 0.455 Btu/lb °F, and a conversion factor of 1 watt = 3.413 Btu/hr. A safety factor of 1.3 is often applied.

Heat tracing and insulation should be applied to all heavy oil lines. Part 4 (Vol. I) deals with insulation. The purpose of heat tracing is to balance heat loss from the line between the heater and the point of use. Heat tracing cannot be expected to "melt out" heavy oil that has been allowed to get cold. Heat tracing may be done electrically, or with steam. Figure 8.15 shows a cross section of a steam-traced oil line. The maximum steam pressure should be the saturation pressure from the steam tables (Table A.4 in the Appendix) corresponding to a saturation temperature equal to the design outlet temperature from the oil heater. **For example** if it is determined from viscosity requirements that 240 F oil is required at an atomizer inlet, the steam pressure regulator for the booster loop heater and the steam trace lines should be set for about 10 psig maximum.

Figure 8.15. Suggested method for steam tracing heavy oil lines. Heavy oil must be traced, insulated, and recirculated to maintain temperature up to the burners and to help in starting up a cold system.

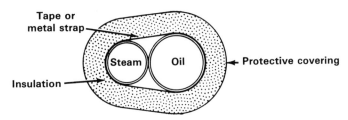

The steam pressure is best controlled by a pressure-reducing regulator. Throttling the steam through such a regulator may produce some temporary superheat. If the pressure drop through the regulator is expected to be more than 10 or 15 psi, it would be advisable to leave a few feet of steam pipe uninsulated before it contacts the oil pipe. This will desuperheat the steam and prevent coking when the oil flow rate happens to be low. (To avoid wasting the heat from this uninsulated section, make sure that it is indoors so that it will help with space heating, but provide protection against persons being burned by the hot pipe.) *Vapor lock*, formation of oil vapor bubbles that may inhibit full flow capacity, can be prevented by maintaining at least 30 psig pressure in all oil supply lines.

Electric heat tracing has been done by induction and by using the pipe itself as an electrical resistor. If electric tracing is to be used, control convenience is achieved by a wrap-around resistor tape, the resistance of which rises as its temperature rises so that it automatically throttles its heat input as the temperature rises.

Heavy oil line pumps, heaters, and tracing should be turned on before cold weather and allowed to run continuously until warm weather. If this is not done, a standby oil system will not be "standing by" ready to go when needed. Start-up in cold weather will be tedious, costly, and sometimes impossible.

At shutdown, it is advisable to pump light oil, steam, or compressed air into an oil system to displace the heavy oil. This greatly facilitates the next start-up. Even with light oil, it is a good policy to blow out atomizers after the oil is shut off. Some installations incorporate a 3-way solenoid valve to shut off the oil and immediately blow steam or compressed air through the atomizer.

The oil volume circulated in any loop depends on the heat loss, but in general, recirculating pumps, piping and heaters may be sized to handle the total capacity of all ratio regulators in the circuit *plus* about 0.8 to 1.0 gph for each foot of circuit length. See Example 8-1 and Figure 8.12. When circulating oil through several regulator bodies, the required pipe size will be greater than the regulator connection. Reducing couplings should be used -- not bushings.

Figure 8.16 shows a heavy oil piping diagram with two possible control schemes. The top left scheme uses individual ratio regulators for each burner. Each regulator could supply several burners, but that would increase the length of dead end line. Heavy oil is actually circulated through the body of each regulator to minimize the length of dead end piping.

To *start* the burners of Figure 8.16, (1) turn on the steam to the tank heater and main loop tracer lines; (2) close the light oil tank and crossover valves, open the heavy oil return valve, start the main loop pump; (3) when heated oil is circulating in the main circulating loop, open the warmup valves and set the 3-way valves for the shortest possible circuit so that warm oil can force out air and cold oil; (4) after a period of time sufficient to eliminate air from the branch circuit, turn on the booster heater; (5) when hot oil starts to flow through the return line, start the branch circulating pump and close the pump bypass valve; (6) set the pressure regulator to produce an outlet pressure of 30 psi on the gauge immediately downstream from the regulator (a pressure relief valve prevents the development of excessive pressure in the branch circuit due to heating of the oil as it circulates in the closed circuit); (7) as the system warms up, gradually close the warmup valves one at a time, and circulate warm oil into the farthest sections of the system; (8) when the thermometer at the return to the booster heater indicates the oil in the branch loop is up to atomizing temperature, open the shutoff valve of the burner nearest the ratio controller and light that burner; (9) light the other burners, and set all of the burners at the desired rates and ratios.

Large oil systems supplied from a single storage area often need intermediate circulating loops between the main and branch loops. A pumping and heating unit at the tank(s) might heat the oil to 125 F and circulate it to a number of buildings. An intermediate loop in each building might boost the oil to 180 F and circulate it past branch circuits at each furnace, each with a recirculating pump and heater for heating the oil to final atomizing temperature.

RATIO CONTROL SYSTEMS

The operation of air/gas and air/oil ratio regulators was described in Part 7. Regulators should always be installed in horizontal lines unless the manufacturer's instructions specifically state that it is permissible to do otherwise. They must not be tipped or the spring may not properly balance the weight of the diaphragm assembly.

Each control zone (that is, each manual or motorized air control valve) requires a separate ratio regulator. One regulator per burner provides greatest flexibility, but several burners may be controlled by a single regulator if the capacities are properly matched. If the sum of the pressure drop across a regulator and the applied impulse pressure exceeds the upstream fuel pressure, the regulator cannot function properly. Use a bleeder -- Figure 8.17d.

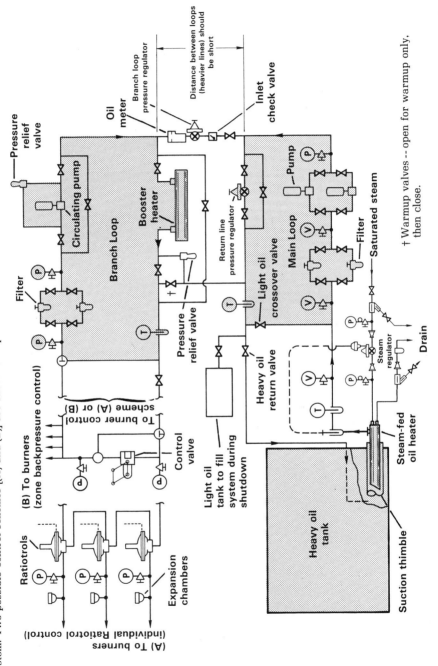

Figure 8.16. Heavy oil system piping. The shaded areas indicate the three major sections--tank, main loop, and branch loop. Innumerable designs are in use and the experienced oil system engineer may have good reasons for adding to or altering this system, but the inexperienced can save himself many headaches by employing an experienced consultant or incorporating all the details of this system. Two possible control schemes [(A) and (B)] are shown at top left.

Ratio regulators should be located as close to the burners as possible without overheating, so as to minimize the variable downstream pipe friction. Each oil ratio regulator should be at the same elevation as the burner atomizer(s) that it serves. A ratio regulator requires a constant downstream resistance.

Impulse line connections to the main combustion air line should be at least five pipe diameters downstream from the control valve and three diameters upstream from the next fitting so that the impulse pressure will not be adversely affected by turbulence (uneven pressure distribution) in the pipe. Impulse taps should be flush with the pipe's internal surface, and parallel to the axis of the control valve. Taps should not be located on the bottom side of the air line, where they may become plugged with dirt, water, or oil. Atmospheric vents should be shielded from wind, water, and dirt.

Air/Gas Ratio Control Systems. Premix gas burner systems usually require "zero" gas; that is, gas at atmospheric pressure. Figure 8.17a shows a ratio regulator used for this purpose. In applications involving a furnace pressure different from atmospheric pressure or a fluctuating furnace pressure, it is necessary to vent the diaphragm chamber to the furnace chamber as in Figure 8.17b.

Nozzle mixing burners (Figure 8.17c) require gas supplied upstream of the ratio regulator at a pressure greater than the sum of (a) the furnace pressure, (b) the maximum modulated air pressure, and (c) the pressure drop expected across the ratio regulator at maximum flow rate. As shown in Figure 8.17c, an impulse line is used to convey the air pressure set by the air control valve to the gas ratio regulator, which then automatically produces an outlet gas pressure equal to the air impulse pressure. Thus the pressure drop between the air control valve and the furnace is the same as the pressure drop between the gas regulator and the furnace. By adjustment of the limiting orifice valve in the gas line it is possible to establish the desired ratio of gas to air flow. This ratio will then be maintained at all rates of firing because the regulator will always maintain equal pressure drops across the orifices in the air passages of the burner and the limiting orifice gas valve. This is true, of course, only if all of the air for combustion is being supplied through the burner (100% primary air). Fluctuations of furnace pressure have equal effect on both air and gas flow; so there is no need for any other connection between this system and the combustion chamber pressure.

For any nozzle mixing burner system wherein the available gas pressure is less than the sum of the air pressure and the regulator pressure drop, it is necessary to use a bleeder as shown in Figure 8.17d. A bleeder is simply a constant impulse line leak that reduces the impulse applied to the regulator to a proportionally lower value. This means that the regulator outlet pressure will be less than, but proportional to, the air line pressure. The limiting orifice valve would necessarily be wider open in Figure 8.17d than in Figure 8.17c.

Figure 8.17. Air/gas ratio regulator hookups.

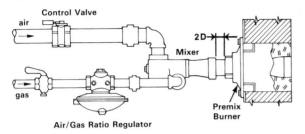

a. Premix burner (gas suction proportional to air flow).
Combustion chamber pressure steady at zero.

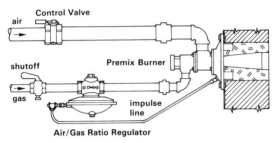

b. Premix burner (gas suction proportional to air flow).
Combustion chamber pressure fluctuating.

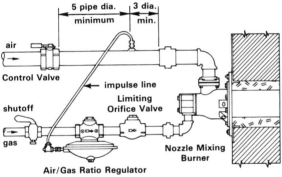

c. Nozzle mixing burner. Gas pressure equal or greater than air pressure.
Combustion chamber pressure steady or fluctuating.

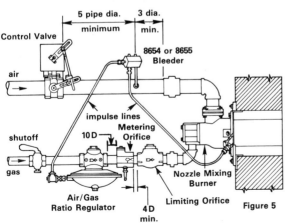

d. Nozzle mixing burner. Gas pressure less than air pressure.
Combustion chamber pressure fluctuating.
Motor-operated air input control valve and gas metering orifice shown here could also be used in a, b, and c. A bleeder could be used in c.

Air/Oil Ratio Control Systems. Elevation is a critical factor in the installation of ratio regulators for oil because a column of oil one foot high may be heavy enough to create a "head" or pressure which would affect the fuel flow. For this reason, regulators should be located on the same level as the burners or no more than 12 in. below the burners. Burners which are more than 12 in. above or below one another require a separate regulator for each level. A strainer (40 mesh or equivalent) should be located immediately upstream from every ratio regulator to prevent foreign material, such as pipe scale, from lodging on the valve seat. In dead end systems, an expansion chamber or accumulator should be installed immediately downstream from each regulator to prevent rupture of the diaphragm as a result of oil expansion after shutdown. Figure 7.12 illustrates a combination air/oil and air/gas ratio control system utilizing flow-type ratio control.

Ratio control systems for use with preheated air are many and diverse.† If there is only one zone of control for each air preheater, and if there is no hot air bleed or leak, the air metering orifice can be located upstream of the air heater and any flow-type ratio controller can be used. Otherwise, the control systems become quite complex as shown by the examples in Figures 8.18 and 8.19. Other aspects using preheated combustion air for heat recovery are discussed on pages 69-76 of Volume I, and in Part 6 of this volume.

Ratio control system for use with oxygen enrichment. The oxygen, air, and fuel should be mixed and controlled in such a way that the oxygen could never back up into either the air or fuel pipes if there should be a loss of pressure in either of those lines. Oxygen and fuel should never be on opposing sides of diaphragms because a leak or break in a diaphragm would result in dangerous premixing. See also pages 76-78 in Volume I and Parts 7 and 13 in this volume.

When using oxygen to achieve high temperatures, it is costly to turn on the oxygen in the lower temperature range where air alone will do. It has therefore been common practice to use air to bring the furnace to as high a temperature as possible, or until the rate of temperature climb has slowed to an impractical rate. At that point, when the air control valve is wide open and there is still a demand for higher temperatures that is not being met, the temperature controller should be switched over so that it actuates an oxygen control valve, with the air valve still at its maximum rate, an "oxygen additive system". This procedure also prolongs burner life. If oxygen is used to reduce NOx pollution (oxy-fuel firing) or to lower poc volume, this oxygen additive system will probably not be satisfactory.

For larger furnaces and for continuous furnaces with oxygen enrichment, it is generally best to use an electronic air/fuel ratio control system with square root extractors that permit addition of the air and oxygen flow rate signals. This was described in Part 7.

---

† Any preheated air installation should incorporate provision for protecting the recuperator or regenerator from overheating.

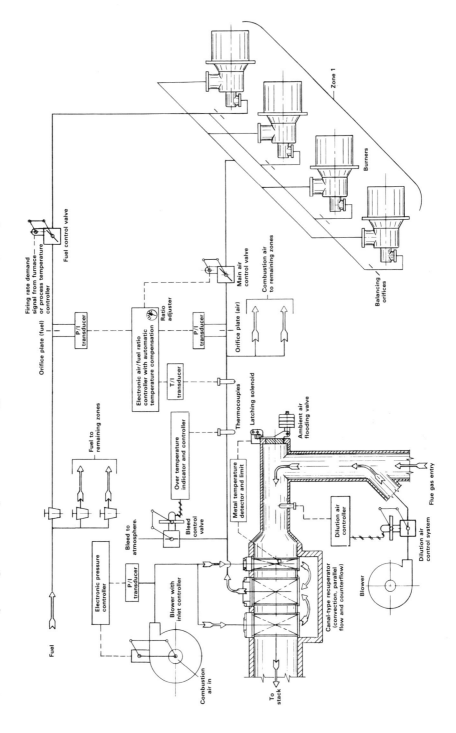

Figure 8.18. Air/fuel ratio control and recuperator protection systems for multiple zones of burners using preheated combustion air. This is not necessarily a design suitable for general use -- each application should be reviewed by an experienced combustion control engineer.

Figure 8.19. Air/fuel ratio control for integral regenerator burners (described in Parts 6 and 9) reclaiming their own exhaust heat in the form of preheated combustion air. These burner-regenerator units must be installed in pairs, but only the firing burner is shown in this simplified schematic diagram. Its "twin" (exhausting) burner would have all three cycling valves in opposite positions.

A fully metered pneumatic air/fuel ratio control system is shown. This is no necessarily a design suitable for general use -- each application should be reviewed by an experienced combustion control engineer.

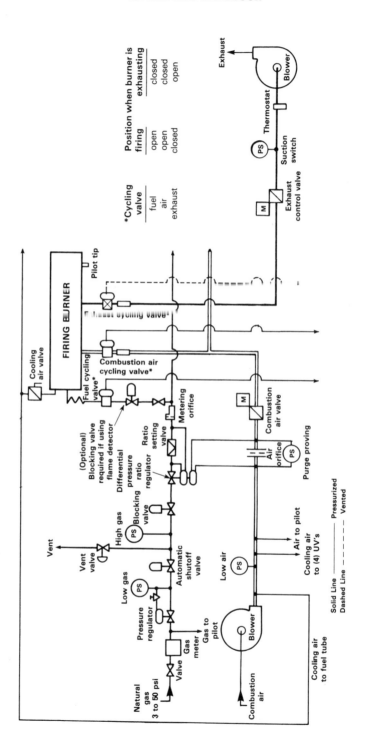

Air/fuel ratio control systems for multiple fuels. Figure 8.20 is a photograph of a tri-fuel boiler that is base-loaded with coal burned on an underfeed retort stoker while a packaged automatic dual-fuel burner takes the load swings with either gas or oil. Relatively simple controls can be used for this arrangement because the changes are not wide nor fast. For larger inputs, it is advisable to use an electronic air/fuel ratio control system with square root extractors so that the impulse signals from the various fuels can be added and a resultant air control signal generated. This was described in Part 7.

Figure 8.20. A tri-fuel fire tube boiler with underfeed retort stoker at left and combination gas-oil burner at right. Overfire air jets and the gas or oil flame assure low particulate emission. An induced draft fan and electric draft regulator are at the upper right.

ORIFICE BALANCING AND METERING SYSTEMS

It is usually desirable to fire all burners in one zone at the same firing rate so as to produce a uniform temperature throughout the zone. A reliable method for assuring equal flow to all burners in a zone utilizes an inexpensive but accurate orifice in each burner gas line with pressure taps and a portable manometer for measuring the pressure drop across that orifice. Figure 8.21 shows the piping for gas and air metering orifices for multiple burners. The orifice plates are calibrated; so measuring the pressure drop tells the operator the actual flow to each burner. A limiting orifice valve in each burner fuel line can be adjusted until the manometer indicates equal fuel flow to all burners.

Figure 8.21. Schematic piping for gas and air metering orifices for balancing or graduating burner firing rates. A single portable manometer can be used to check the flow at each orifice. If the gas flow or air/gas ratio at any burner needs to be corrected, the limiting orifice valve can be adjusted until the manometer shows the desired flow. See Reference 8.a (at the end of Part 8). Error may be as low as 2% on North American Series 8697 Orifices with only 10D straight clean pipe upstream and 4D downstream.

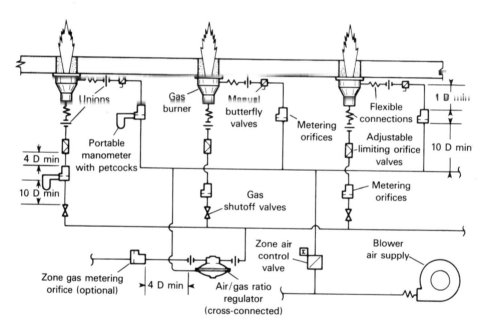

A similar but larger set of metering orifices can be installed in each burner air line for use with the same portable manometer.

For distillate oils of constant viscosity, variable-area, constant-head flow-meters (rotameters) can be used to accomplish the same purposes for oil. (See Figure 8.22.)

The orifice metering system has several advantages in addition to its accuracy. It provides an actual measure of fuel consumed. It shows whether the air/fuel ratio is rich, lean or stoichiometrically correct, and it makes it possible to set the desired ratio. It provides a means for setting a graduated firing rate down the length of a firing zone when that is desired.

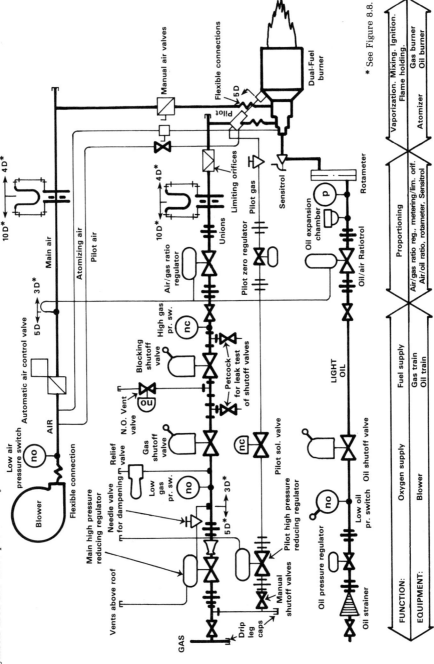

Figure 8.22. Generalized schematic piping for a dual-fuel combustion system using natural gas and distillate oil. Specific installations may require additional features and repositioning of components to comply with insurance regulations and applicable governmental codes. Symbols used are those preferred by Industrial Risk Insurers.

* See Figure 8.8.

FUNCTION:	Oxygen supply	Fuel supply	Proportioning	Vaporization. Mixing. Ignition. Flame holding.
		Gas train	Air/gas ratio reg., metering/lim. orif.	Gas burner
		Oil train	Air/oil ratio, rotameter, Sensitrol	Oil burner
EQUIPMENT:	Blower			Atomizer

TYPICAL PIPING ARRANGEMENTS

Figures 8.22 and 8.23 illustrate typical piping arrangements for a number of applications. Each of these diagrams represents only one of several possible ways in which the piping may be arranged.

Atomizing air lines and pilot air lines must be connected with the air supply line at a point upstream from all control valves. Shutoff valves should be placed wherever necessary to permit complete isolation of any particular burner from the rest of the system. Unions should be used liberally to permit easy removal of regulators or equipment requiring periodic maintenance.

Figure 8.22 shows a typical piping arrangement for a single dual-fuel burner with a premix pilot and with metering orifices. Insurance regulations and gas supply pressure determine the specific equipment to be included in the gas train. The air/gas ratio regulator for the pilots should be preceded by a pressure regulator if the supply pressure is greater than the allowable inlet pressure for the pilot ratio regulator. A manual air valve may be used in each pilot air line for shutoff purposes, or to aid in lighting the pilot. It should not be used as a permanent air pressure reducing valve. If the air pressure is too high for

Figure 8.23. Typical piping for a multi-burner dual-fuel system (gas and light oil). The two burners at the left constitute one control zone; the burner at the right, a second zone.

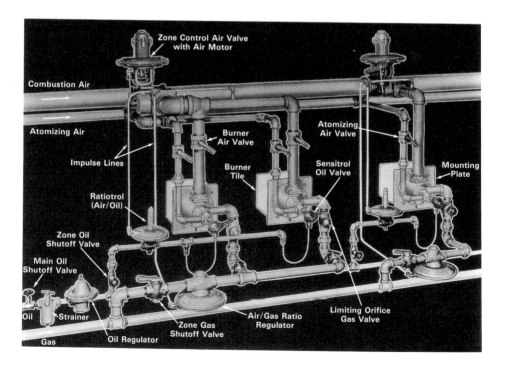

good pilot operation, use either an air pressure-reducing regulator, separate limiting air valve, or the air throttling adjustment built into many pilot air-gas mixers. Figure 8.23 illustrates a typical piping arrangement for *combination (dual-fuel)* burners having separate atomizing air and gas inlets and using *light oil*. The air/oil ratio regulators (Ratiotrols) are at the same elevation as the burners.

The oil line is of the "dead end" type which is permissible only when using light oil. An *expansion chamber* should be placed immediately downstream of the air/oil ratio regulator to protect it from thermal expansion of the oil after shutdown.

The number of burners to be supplied by one air/fuel ratio regulator depends on the length of piping involved and the relative capacity and cost of regulators and piping. Where long runs of pipe are involved, it may be cheaper to use several small regulators, but where there are many burners in one zone located close to one another, use of one or two large regulators is usually more economical. Light-up and servicing is easier with individual regulators.

Figures 8.24a and 8.24b are photographs of packaged fuel trains, gas and oil respectively. For a dual-fuel train, the oil train components would be mounted on the back side of the same frame that supports the gas train components. The many sizes and combinations of these trains can save the installer much engineering and assembly time.

Figure 8.24a. A pre-piped and pre-wired gas train. Gas enters at lower right through a pressure-reducing regulator; then flows through main and blocking fuel shutoff valves, rises and turns to the right through a pneumatic fully-metered air/fuel ratio control system (Figure 7.11). The items on the top rack (left to right: flame monitor cabinet, pressure switches for interlocks, and EPIC® furnace pressure control) were custom engineered into this unit, but are not normally considered part of a packaged fuel train.

Figure 8.24b.   Pre-piped, pre-wired oil train, NEMA 4. Flow is right to left. Top line is for oil; mid-. dle line is for atomizing medium (steam or compressed air); lowest line is for pilot gas. This special, non-catalog train is shown to illustrate the great variety of combinations that are possible. A dual-fuel train would consist of a gas train and an oil train back-to-back on a common mounting frame.

REFERENCES

8.a   **Miller, R. W.:** "Flow Measurement Engineering Handbook", McGraw-Hill Book Company, New York, NY, 1983.

8.b   **NFPA:** "Installation of Oil Burning Equipment" (NFPA 31), National Fire Protection Association, Batterymarch Park, Quincy, MA 02269.

8.c   **NFPA:** "National Fuel Gas Code" (NFPA 54, ANSI Z223.1), National Fire Protection Association, Batterymarch Park, Quincy, MA 02269.

8.d   **Walls, W. L.:** "Liquified Petroleum Gases Handbook" (includes NFPA 58 Standard for Storage and Handling Liquified Petroleum Gases), National Fire Protection Association, Batterymarch Park, Quincy, MA 02269.

ADDITIONAL SOURCES of information relative to combustion systems

Gill, J. H., and Quiel, J. M.: "Incineration of Hazardous, Toxic, and Mixed Wastes", North American Mfg. Co., Cleveland, OH, 1993

Monnot, G.: "Principles of Turbulent Fired Heat", Gulf Publishing Company, Houston, TX, 1985.

Remmy, G. Bickley Jr.: "Firing Ceramics", World Scientific Publ. Co., River Edge, NJ, 1994.

Thring, M. W.: "The Science of Flames and Furnaces", 2nd ed., John Wiley & Sons, Inc., New York, NY, 1962.

Trinks, W. and Mawhinney, M. H.: "Industrial Furnaces", vol. 1, 5th ed., John Wiley & Sons, Inc., New York, NY, 1961.

Trinks, W. and Mawhinney, M. H.: "Industrial Furnaces", vol. 2, 4th ed., John Wiley & Sons, Inc., New York, NY, 1967.

Part 9. HEAT RECOVERY

INTRODUCTION

The exhaust gases from most industrial heating processes contain considerable heat. To conserve energy, it is important to try to recover as much as possible of this "waste" heat, or, in other words, to usefully lower the final exhaust temperature at which the products of combustion (poc) are released to the atmosphere.

Heat Loss Reduction. An important aspect of *all heat recovery systems* is the minimization of: a) loss of heat from the products of combustion (poc) to the surroundings while the products of combustion are on their way to the heat recovery equipment, and b) loss of recovered heat from preheated combustion air, preheated load, or steam while on its way to its point of utilization.

Actions to be taken as a result of these admonitions are: *keep pipes and ducts short and straight*, and maintain very good insulation on them. The short, straight lines will also help keep pressure loss to a minimum, which will save both first costs and operating costs. If there is any choice in the matter, it is usually preferable to minimize pressure losses on the hot flue gas part of the system and make up for this by providing extra pumping power in the cold part of the system.

Insulation in hot poc ducts, hot air pipes, and hot air burners should be on the inside because the insulation must not only reduce heat loss, but also keep the containing wall cool enough that it will still have strength to support itself. (Adding insulation to the outside of a hot pipe or duct increases its metal temperature. Where it is not practical to insulate on the inside, it is usually necessary to invest in more expensive duct and pipe materials that can withstand higher temperatures.) Valves or other apparatus in the hot gas or air lines must likewise be made of higher quality materials.

Duct and Pipe Sizes for Hot Fluids. The furnace heat input is proportional to the weight of air, fuel, and poc -- not to their volume. Less weight of air and fuel should be required, because the whole object of the heat recovery exercise is to conserve fuel, but the ducts and pipes must be larger to convey the same weight of less dense hot fluids.

Pipes and ducts will have *much larger outside dimensions* because of the above-mentioned density effect combined with the necessity for insulation on the inside.

Pumping Power. Extraction of heat energy from a waste gas stream necessitates passing the gas over a heat transfer surface, and that requires pumping power to create the pressure head to force the hot gases to flow over the surfaces, such as the surface of the load being preheated, the surface of the waste heat boiler tubes, the surface of the recuperator heat exchanger, or the surface of the regenerator heat storage. Those few instances that recover heat without added pumping power utilize existing available pressure drop, but they may have minimal heat recovery effectiveness.

Condensation. When the higher or gross heating value (calorific value) of a fuel is measured, the products of combustion are cooled to the temperature of the air and fuel supplied to the burner. Heat recovery processes strive to reach this goal (lowest possible final poc temperature). As the final poc temperature is lowered, it reaches the dew point of some of its component gases (vapors at that condition).

As vapors condense to liquid, the latent heat of condensation is released to the heat recovery device, making an appreciable increase in the heat recovered (about 10% with natural gas). That added heat recovery (the difference between net and gross heating value) is very desirable, but it is always accompanied by liquid condensate, which may be undesirable because of corrosion problems or stream pollution. Even if there were no sulfur or other acid-forming ingredient in the fuel, the condensate could be troublesome if heat recovery were done by preheating a powdery load that might become pastelike, as in a lime kiln.

Dew points vary with the amount of O_2, CO_2, SO_2, SO_3, NOx, Fl, or Cl in the poc, as shown in Figure 9.1. Figure 9.2 shows the effect of sulfur from a fuel oil. The excess air curves are not equally spaced, because the extra oxygen tends to produce more SO_3, which has a catalyst-like effect in raising the dew point. For example, in a case where the H_2O dew point was only 110 F, 2 ppm SO_3 resulted in dew points of 126 to 250 F, depending on measurement method. See Reference 9.i.

In actual practice, most condensation appears to occur about 80 to 100 degrees (F) below the dew point. However, engineers usually apply some safety factor, often designing for final poc temperature somewhat above the dew point, especially where gases in slow-moving inside corners could encounter more cooling than the bulk of the stream.

With the development of better corrosion-resistant materials, some heat recovery devices are now intentionally designed to recover the heat of condensation. This may transfer some pollution problems from the stack gases to the drain liquids.

Figure 9.1.   Dew points of products of combustion of natural gas, adapted from data courtesy of B.C. Hydro. Acid dew points may be appreciably higher.

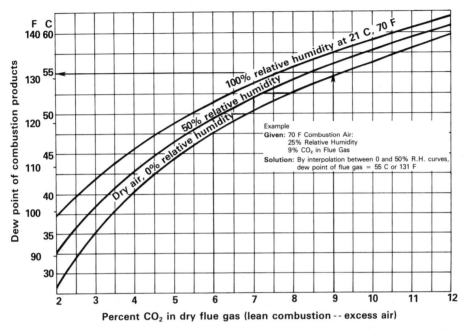

Percent CO_2 in dry flue gas (lean combustion -- excess air)

Figure 9.2. Effect of sulfur and excess air on acid dew points for 10-12 degree API crude oil. Adapted from C-E Natco. Even small concentrations of SO_3 in stack gas raise the dew point considerably.

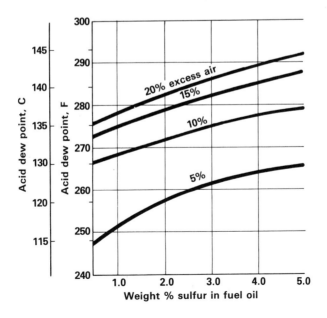

Weight % sulfur in fuel oil

Reduced Cleanup Costs. First costs and operating costs of cleanup equipment to remove pollutants from stack gases will be less if heat recovery has been used, because the poc temperature at the cleanup point will be lower. The lower temperature means that the gases will be *less voluminous* (more dense) and therefore can be handled with smaller equipment (baghouses or scrubbers, for example) and with smaller fans and piping. The smaller fans will also reduce operating costs.

A secondary effect of heat recovery on the cleanup operation is that the *lower temperature* will often allow purchase of less expensive components, such as fans, bags, and piping.

PREHEATING THE LOAD

On continuous furnaces, an added unfired preheat vestibule can preheat the load with free heat that would otherwise have been discharged up the stack. Examples: charging pigs, gates, and sprues down the stack of a die casting melter; a longer inlet tunnel on a conveyor furnace (Figure 9.3); an economizer on a boiler; a convection section on a petrochem heater.

Figure 9.3.   Adding an unfired preheat vestibule and lengthening the conveyor on any tunnel-like continuous furnace is an easy way to recover heat. No burners are added. No fuel is added. In fact, less fuel will be used. The poc are prevented from exiting through the old flue; so must pass over the cold load, thereby lowering their final exit temperature.

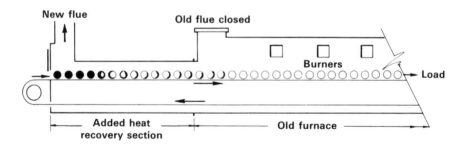

Less fuel is required to heat the same amount of load. The final exit poc temperature is lowered. This is a logical extension of the furnace, making it more efficient.

On batch furnaces, load preheating can be done, but it might be better to first consider making the process continuous. Figure 9.4 shows a preheating oven built next to an aluminum melter, from which the preheated loads must be transferred to the melter quickly to minimize cooling. Another way is to use a pair of batch furnaces, with A's flue gases preheating the load in B; then B's flue gases preheating the new cold load in A.

Figure 9.4.   Aluminum sow preheater beside the melter. To eliminate the heat losses between these two units, it would be ideal to feed the cold sows through a long narrow insulated chute directly into the melter -- a special form of dry hearth.

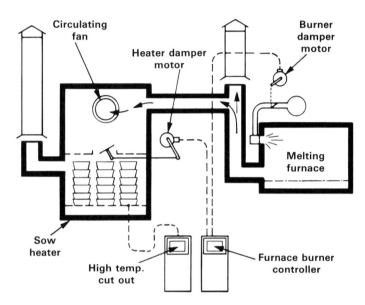

The load-related scheduling is perfect with heat recovery by load preheating. Care must be exercised to assure no condensation on the load if that would affect product quality.

GENERATING STEAM, HEATING WATER OR MAKEUP AIR

Recovered heat can be used to supply needed utilities within a short distance -- steam, hot water, makeup air. There must be good load-related scheduling, i.e. a double match between the time and extent of the need for the utilities and the time and extent of the availability of the waste heat from process furnaces.

Waste heat boilers, either steam or hot water, or makeup air heat exchangers can get their heat input from hot poc from the process furnaces; so they need no burners, no fuel. See Figure 9.5. The interdepartmental accounting may be more complex than if the heat recovery savings were fed directly back into the furnace that generated the hot poc (as with load or combustion air preheating).

Waste heat boilers are very safe because they are water-backed heat exchangers; so are less likely to suffer from thermal expansion or internal fires (and subsequent leaking) than are gas-to-gas heat exchangers. As with all forms of heat recovery, hot waste gas ducts and steam, hot water, or makeup air pipes must be short and well insulated.

Figure 9.5.   Waste heat boiler installed above a bank of heat treating furnaces. An induced draft fan provides pumping power, but uses far less energy than the boiler saves. This idea has been used for many years with both fire tube and water tube boilers in the chemical process industries.

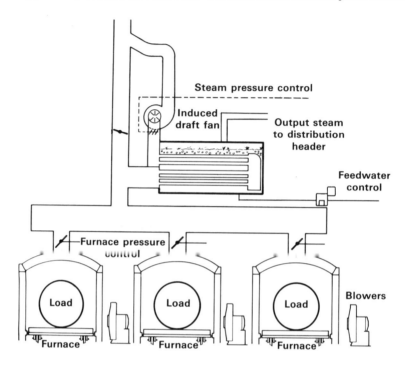

PREHEATING COMBUSTION AIR

When preheating combustion air in industrial heating processes, the waste flue gases may be as hot as 2800 F, depending on the process, and the preheated air may be at 1000 F to 2500 F; so minimizing heat losses from the conveying duct and pipe walls is especially critical. Waste gas ducts and combustion air pipes must be insulated and kept short, or the investment in heat recovery equipment may not be justified.

The other warnings and considerations discussed in the Introduction to this Part 9 must be heeded.

Benefits. *Available heat.* When the combustion air has been preheated, less of the chemical energy from the fuel is required to raise the temperature of the input ingredients (fuel and air) to the furnace temperature; so more chemical energy is *available* to do useful heating in the furnace. The percent available heat concept was discussed in Part 3 of Volume I. It is often looked upon as "best possible efficiency" (for a furnace with no significant losses other than flue loss). Refer to the Sankey diagrams, Figures 9.6a and 9.6b.

Figure 9.6a. Sankey diagram before addition of heat recovery equipment. Best possible efficiency = 100% (available heat)/(gross heat input). Heat recovery reduces the required gross heat input, as shown in Figure 9.6b.

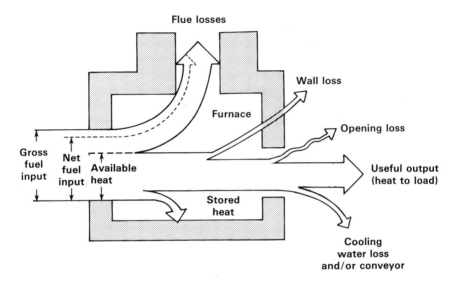

Figure 9.6b. Sankey diagram after heat recovery by preheating the combustion air or the load. Energy extracted from the flue gas is recycled into the combustion chamber, reducing the amount of gross heat input (fuel) required. Because this gross input is less, the *percent* available heat (formula 9/3) is greater.

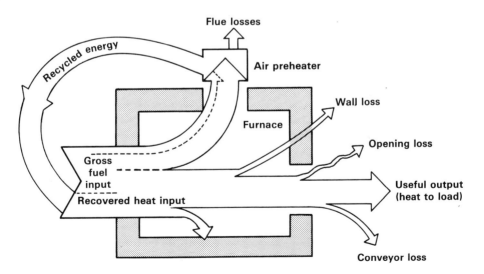

Formulas 9/1 through 9/3 review the significance of "available heat," which can be seen in the Sankey diagram. In these formulas, the units can be per unit of time, per unit of fuel, or per unit of load; but they must be consistent. Gross heat is derived from the gross or higher heating value of the fuel, which may be measured or calculated, but is usually available from the fuel supplier.

[9/1]   Net heat = gross heat − latent flue loss.

[9/2]   Available heat = net heat − sensible flue loss.

[9/3]   % Available heat = 100% × (available heat/gross heat).

Net heat can be derived from the net or lower heating value of the fuel; or from Formula 9/1 where the latent heat content of the flue gases, "latent flue loss," is calculated by multiplying the weight of water vapor from combustion of hydrogen in the fuel by the latent heat of vaporization (or condensation) of water at the partial pressure of water vapor in the flue.

The sensible heat content of the dry flue gases is the summation of the heat contents of all of the flue gas constituents except the water vapor. These heat contents are tabulated in Table 3.7 in Volume I, or they can be derived from polynomial expressions for specific heat of the flue gas components.

Formulas 9/1, 9/2, and 9/3 are generalizations applicable to a wide variety of conditions, including excess air, hot air, oxygen-enriched air, and oxy-fuel. Using subscripts "c" for cold air (assumed 60 F or 15.6 C), "p" for preheated air, and "o" for oxygen-enriched air or oxy-fuel,

[9/4]   % Available heat$_c$ = 100% × (available heat$_c$/gross heat$_c$),

[9/5]   % Available heat$_p$ = 100% × (available heat$_p$/gross heat$_c$),

[9/6]   Available heat$_p$ = available heat$_c$ + preheated air heat content,

where preheated air heat content can be determined from a polynomial formula for the specific heat of air, or from Table 3.7 in Volume I. See Figure 9.7, which is based on recent data and supersedes Figure 3.15 in Volume I.

The % fuel saved by preheating combustion air can be calculated using readings from Figure 9.7 via formula 9/7a.

[9/7a]   % fuel saved = 100% × [1 − (available ht$_c$/available ht$_p$)]

Tables 3.16a, b, and c in Volume I list percents fuel saved for natural gas, No. 2 fuel oil, and No. 6 fuel oil. These tables, Figure 9.7, and STOIC program printouts relate to Figure 9.8. It is necessary to know (a) the fuel, (b) the flue (not stack) gas exit temperature t$_3$ leaving the furnace, and (c) the air preheat

Figure 9.7. Available heat for 1025 Btu/ft³ natural gas with preheated combustion air at 5% excess air. Use this chart only if there is no unburned fuel in the products of combustion. Available heat percentages are computer-calculated with corrections for dissociation.

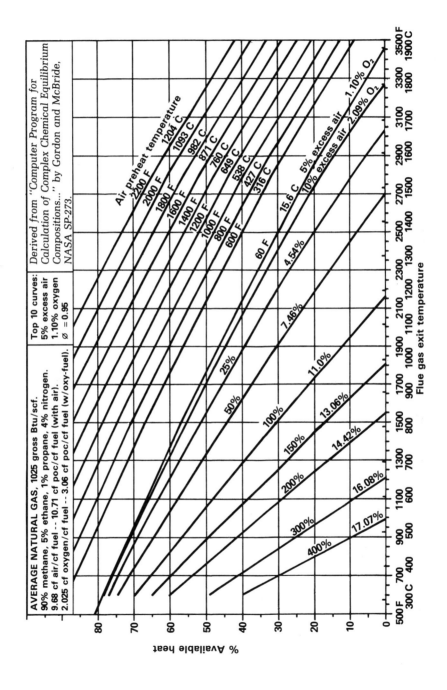

temperature t_2 entering the burner. If the fuel and/or combustion air temperatures, t_1, are appreciably different from 60 F (15.6 C), the results should be modified accordingly.

Similarly, the % fuel saved by oxygen-enriched combustion can be calculated using readings from Figure 13.4a or Table 13.4b via formula 9/7b, or consult North American Mfg. Co.'s Handbook Supplement 276.

[9/7b]   % fuel saved = 100% × [1 − (available ht_c/available ht_o)]

Figure 9.8.   Schematic diagram of a furnace with an air preheater (recuperator or regenerator). If t_5 is the maximum safe inlet gas temperature for the preheater, the minimum required weight flow rate of dilution air = (weight flow rate of combustion products) × (h_{poc_3} − h_{poc_5}) ÷ (h_{da_5} − h_{da_1}), where h_{da} = dilution air heat content per unit weight, and h_{poc} = poc heat content per unit weight.

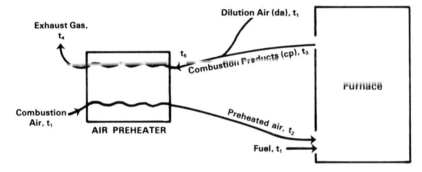

Example 9-1a.   US units. For a furnace with 2250 F flue gas exit temperature, burning natural gas with 5% excess air, find (a) the fuel saving with 1850 F preheated air, compared with 60 F air, and (b) the impact on NOx pollution if the burner emits 0.1 pounds NOx/million Btu with 60 F air and 0.57 pounds NOx per million Btu with 1850 F air.

(a) Enter the bottom scale of Figure 9.7 at 2250 F flue gas exit temperature. Move vertically to the 60 F combustion air curve for 5% excess air, then left to read 37% available heat with 60 F air.

Again, from the bottom 2250 F exit flue temperature, go up to 1850 F combustion air preheat temperature; then left to interpolate 73% available with 1850 F air.

% fuel saved = 100 [1 − (cold available heat/hot available heat)]
= 100 [1 − 37/73)] = 100 − 50.7 = 49.3%.

This is 50.7% as much fuel and scf air.

(b) Comparing cold vs. hot emission rates, with cold air, (0.1#NOx) (100% input rate) = 0.1 pounds NOx/unit time; with hot air, (0.57#NOx) (50.7% input rate) = 0.29 pounds NOx/unit time. (Lower firing rates because of higher flame temperature, and the heat exchange effect will lower the net NOx effect even further.)

Example 9-1b.   SI units. For a furnace with 1230 C flue gas exit temperature, burning natural gas with 5% excess air, find (a) the fuel saving with 1000 C preheated air, compared with 15.6 C air, and (b) the impact on NOx pollution if the burner emits 430 × 10^{-7} kg NOx/MJ with 15.6 C air, and 2450 × 10^{-7} kg NOx/MJ with 1000 C air.

(a) Enter the bottom scale of Figure 9.7 at 1230 C flue gas exit temperature. Move vertically to the 15.6 C combustion air curve for 5% excess air, then left to read 37% available heat with 15.6 C air.

Again, from the bottom 1230 C exit flue temperature, go up to the 1000 C combustion air preheat temperature; then left to read 73% available with 1000 C air.

% fuel saved = 100 [1 – (cold available heat/hot available heat)]

= 100 [1 – (37/73)] = 100 – 50.7 = 49.3%.

This is 50.7% as much fuel and scf air.

(b) Comparing cold vs. hot emission rates, with cold air, (430 × 10⁻⁷ kg NOx/MJ) (100% input rate) = 430 × 10⁻⁷ kg NOx/unit time; with hot air, (2450 × 10⁻⁷ kg NOx/MJ) (50.7 input rate) = 1242 × 10⁻⁷ kg NOx/unit time. While this is not good, it is not as bad as the initial 2450 × 10⁻⁷ figure might have appeared. (Lower firing rates because of higher flame temperature, and the heat exchange effect will lower the net NOx effect even further.)

Air/Fuel Ratio Control. The proportioning of air to fuel is often more critical in situations where heat recovery is used. Above all, excursions into soot-producing rich conditions must be avoided so as not to foul the heat transfer surfaces.

Before heat recovery is even considered, it is wise to upgrade air/fuel proportioning controls, as a less expensive, quicker fuel-saving measure. The range of acceptable lean-side air/fuel ratios should be narrower when using preheated air. If one chose to stay at 25% excess air instead of, say, 10%, the heat recovery equipment would have to be 1.25/1.10 = 1.136 times larger.

If the combustion chamber pressure remains constant, any of the 3 generations of air/fuel ratio control discussed in Part 7 (area control, pressure control, or flow control) may be used with heat recovery equipment. However, *if air preheating is used* as the heat recovery method, the rising air temperature in the recuperator or regenerator will have a "choking" effect, adding resistance to the air flow. This effect will change with throughput rate, upsetting air/fuel ratio. If the air flow sensor, usually a pair of pressure taps on an orifice plate, is located on the cold (upstream) side of the air preheater, the reduced air flow caused by choking will be detected and the ratio of air weight to fuel weight will not be changed by fluctuating downstream temperatures. See Figure 9.9.

With fully metered air/fuel ratio controls, all of the flow through the air flow meter must be delivered to the burner and combustion chamber. If there is any loss or diversion of a part of the measured air stream, the air/fuel ratio controller will be "deceived" and unable to correct for the change. Unfortunately, it is common for recuperators to leak because of thermal stresses; it is common to install an intentional bleed to try to protect recuperators from overheating during turndown conditions; and it is common to try to economize by using one recuperator for several zones of control.

Figure 9.9. Cold side air metering for air/fuel ratio control when an air preheater is used. This arrangement can be used only if there are no leaks and no bleeds between the air measuring orifice and the burner, and if there is only one control zone fed by the air heater.

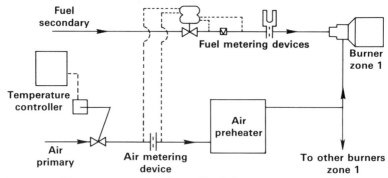

If it is impossible to guarantee that all of the metered air will be delivered to the burner, the air flow meter should be moved to the hot side of the recuperator, and temperature compensation must be added. Most air flow meters are volumetric; but the important flow factor for proper a/f ratio control is the weight rate of flow (actually mass rate of oxygen delivery to the burner). Temperature compensation on a volumetric flow meter converts a fully metered a/f ratio control into "mass flow control." This is easily done with digital electronic controllers. (See Figure 9.10.) If fuel-primary control were preferred over the air-primary control shown, positions (or functions) of the ratio-actuating valve and the temperature-actuating valve would be interchanged.

Figure 9.10. Hot side air metering for air/fuel ratio control when an air preheater is used. Air temperature compensation biases air flow volume to maintain the same ratio of air mass flow rate to gas flow rate as when the air was at standard temperature.

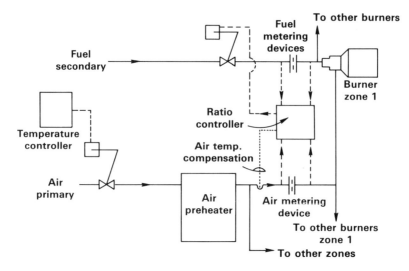

Recuperators are heat exchangers, generally gas-to-gas. In combustion systems, they extract heat from hot poc ("waste gas") and transfer it to cold combustion air. They are built in about as many forms as are heat exchangers...double pipe, shell and tube, plate type...with parallel flow, counterflow, cross flow, and many combinations.

By and large, recuperators are convection devices; so their effectiveness increases with higher Reynolds Number (more turbulence, more velocity, lower viscosity). Some large double pipe recuperators that simply replace a section of stack are named "stack" or "radiation" recuperators, the latter because their large cross section permits enough beam thickness that gas radiation may be a significant factor. Except for those recuperators specifically designed to utilize radiation, direct radiation from the furnace should be avoided because it may damage the heat exchanger.

For preheating air, the common policy is to take a minimum pressure drop on the poc side, and let a forced draft blower provide the greater pressure drop on the air side.

Ideally, recuperators are steady state devices, but must be able to handle changes in throughput to accommodate the turndown requirements of the process to which they are connected. However, they must never be turned to zero flow, i.e. on-off control should not be used. Strict compliance with the manufacturer's warnings regarding minimum flow is absolutely necessary to avoid overheating, warping, leaking, or burning out.

Parallel flow *vs.* counterflow. Figure 9.11 shows some typical temperatures for a double pipe (stack, radiation) type recuperator typical of those used on steel mill reheat furnaces, except that these normally vertical recuperators have been laid on their sides for easier graphics. Both of the selected cases start with the same 60 F combustion air and the same 2100 F poc entering the recuperator.

There would be little gain from lengthening the parallel flow unit, because the temperature difference at the exit end is approaching zero. The wider the temperature difference between the heat source and heat receiver (hot waste gas and cold combustion air) the greater the heat transfer rate. In contrast, lengthening the counterflow unit would achieve higher air preheat temperature and lower waste gas throwaway temperature. (This throwaway temperature is the T_2 for which the available heat charts show a lower number to be desirable for better fuel efficiency.)

The log mean temperature difference (LMTD) has a major effect upon the total heat transfer. This was discussed in Part 4 of Volume I. Values of LMTD are shown at the top of Figure 9.11, illustrating the efficiency advantage of counterflow over parallel flow. If the recuperator were lengthened, the advantage of counterflow would increase. A quick approximate comparison is possible

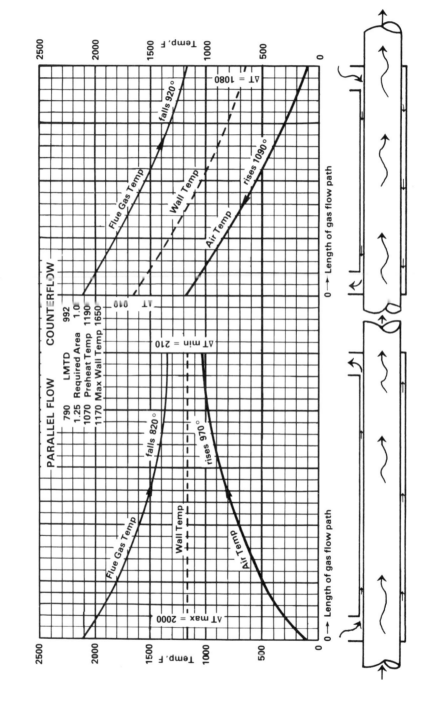

Figure 9.11. Comparison of **parallel flow** (left) and **counterflow** (right) recuperators. These principles could apply to shell and tube recuperators and plate type recuperators as well as the double pipe recuperators shown.

by visually comparing the integrated areas between the temperature curves. Counterflow units are more compact, thus reducing the capital investment; and they can develop higher air preheat temperatures.

For natural gas, the poc passages of a recuperator handle about 11 volumes per unit of time, while the air passages handle about 10 volumes. Because the specific heats of both the air and poc streams are almost the same (both largely nitrogen), the mean temperature of the dividing wall between air and poc passages will be approximately half way between the temperatures of the two streams (dashed lines in Figure 9.11).

The dividing wall temperature of the parallel flow recuperator is constant and well below the counterflow recuperator's maximum; so parallel flow units can safely be built with less expensive materials. The high temperature danger point is always near the hot poc inlet. The only coolant for a recuperator is the air that is to be heated; and in a counterflow unit, that air is already hot at the danger point (poc inlet). Figure 9.12 shows a practical compromise -- a counterflow unit stacked on top of a parallel flow unit, providing temperature protection at the danger spot, but taking advantage of the counterflow efficiency and compactness after the poc have been partially cooled.

Figure 9.12.   Two-stage recuperators combine the advantages of parallel flow (lower maximum wall temperature at poc entrance) with the advantage of counterflow (better heat transfer efficiency). The flow pattern at left provides more hot air volume for the same pressure drop with the same amount of heat transfer surface; the flow pattern at right produces higher air preheat temperature (but more pressure loss, or less hot air volume) for the same amount of heat transfer surface.

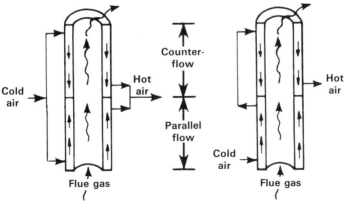

The principles of counterflow *vs.* parallel flow apply to all recuperator configurations. Figure 9.13 is a schematic flow diagram showing application to a shell-and-tube convection type recuperator (also termed canal type or channel type).

Figure 9.13.   Cross-flow convection recuperator with inlet air stream split to utilize advantages of both parallel and counterflow. Air stream a to c is cross-flow/parallel flow; air stream b to c is cross-flow/counterflow. Flue duct walls, W, and triatomic gases, G, radiate against recuperator shock tubes, S, creating thermal expansion and leak problems.

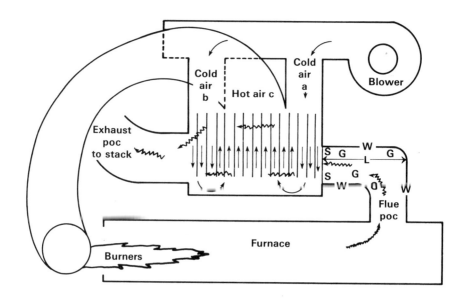

Protecting recuperators from overheating. See Reference 9.1. Figure 9.13 illustrates another problem with recuperators. The wall sections labeled "W" will probably be hot enough to radiate considerable extra heat against the front faces of the first row of convection surfaces ("shock tubes"), causing them to expand in inordinate proportion to downstream surfaces. If dimension "L" is more than about 3 feet or a meter, the gas beam will be sufficient to bombard the shock tubes with an additional dose of radiation -- gas radiation as well as wall radiation. It is therefore advisable to keep the poc approach connection as short as possible.

A recuperator's waste gas approach connection should also include at least one elbow so that furnace radiation cannot "see" directly into the recuperator and further overheat it. If straight-line radiation occurs between the furnace interior and the recuperator's surfaces, it may also chill the load in the furnace, resulting in product nonuniformity. Such a configuration also violates the objective of all heat recovery systems -- to utilize heat that would otherwise have been thrown away -- by stealing heat directly from the furnace and thereby necessitating more fuel input.

Uneven thermal expansion is the bane of recuperators, often stressing the materials to the point of failure. The resulting leaks either diminish valuable hot air supply or dilute the air stream with inert poc. Either way, the LMTD is reduced through the rest of the recuperator. Most systems are engineered so that the pressure on the air side of a recuperator is higher than on the poc side so that any leaks will be air into waste gas. Leakage of waste gas into the air stream could upset the air/fuel ratio and affect burner stability.

Reducing (fuel rich) furnace atmospheres are not generally friendly to heat recovery equipment, because of (a) fouling of heat transfer surfaces by pic (products of incomplete combustion, including soot), (b) possible fire in the recuperator, and (c) eventual loss of some recuperator materials from alternate exposures to reducing and oxidizing atmospheres. It is better to first incinerate the rich pic and then recover heat. (This, by the way, is also the philosophy of "starved air" multistage incinerators.) For all the above reasons, it is wise to incorporate a "lead-lag" (or crossover network) feature into the air/fuel ratio control system when adding a recuperator. Refer to Part 7.

Integral Air-preheater/Burners. Because of the problems cited above, and because hot air piping usually has to be larger than cold air piping for the same furnace production rate, it is highly desirable to use integral burner-recuperators or integral burner-regenerators, eliminating hot waste gas ducts and hot air pipes. These compact units combine the heat recovery device and the hot air burner in a 1-on-1 arrangement connected directly to the flue, with no piping between.

Integral burner-recuperators are limited in capacity because of the space required for the heat exchanger.

For integral burner-recuperators, the furnace exhaust port must be adjacent to the firing port (tile outlet). If the burner is turned down to a low firing rate without adequate control of the exhaust extraction rate (affecting furnace pressure), it is possible for hot products of combustion to short circuit to the exhaust and damage the recuperator section. This problem can be minimized by use of higher velocity burners, because the high momentum of the flame and poc stream prevents short circuiting.

Regenerators. Large checkerwork (refractory) regenerators have been in use since the middle of the 19th century -- originally to raise flame temperature with fuel gases of low calorific value, but coincidentally saving fuel. With the demise of open hearth furnaces, only large glass melting tanks still use checkerworks (which are usually larger than the furnace). Some incinerators now use several large silos filled with refractory "stones" or shapes to alternately collect waste heat from flue gases and return it to combustion air by use of sequentially programmed valve arrangements.

Integral burner-regenerators or *regenerative burners* eliminate hot poc ducting and hot air piping, while retaining the regenerator's advantages -- very high air preheat temperatures, with resultant high fuel efficiency. They are designed to be installed in pairs, each of the twin units being timed to alternately serve as burner, then flue. Because of space limitations, the small size of the heat recovering matrix necessitates a short time cycle, such as 20 seconds.

Whereas the traditional refractory checkerworks had a surface-to-volume ratio of about 4.5 ft²/ft³ (or 14.76 m²/m³), the small refractory balls used in the twin "beds" of the units described below have a ratio of about 300 ft²/ft³ (or 984 m²/m³); thus greatly multiplying their heat transfer capability.

Figures 9.14 and 9.15 explain how integral burner-regenerator systems work. An electronic sequencer opens and closes the 2 air inlets, 2 fuel valves, and two exhaust passages. The system includes provision for piloting, flame monitoring, and brief purging of the beds. Low or medium velocity flames are possible. To lower the NOx emissions, direct injection of fuel into the furnace chamber or flue gas recirculation is used. Special bed-dumping arrangements are built into the units when the process necessitates frequent bed cleaning or changing.

Figure 9.14.   Direct-firing integral burner-regenerators. These units are used in pairs, one firing while the other serves as flue and heat recovery device. They need not be opposite one another -- many burner arrangements are possible for either two- or multiple-burner furnaces.

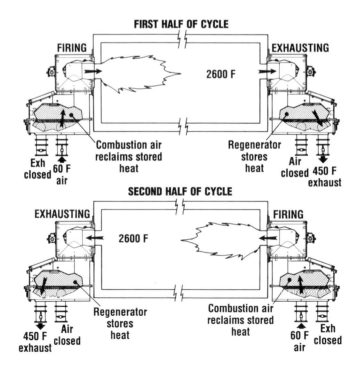

Figure 9.15. Indirect-firing integral burner-regenerators -- one pair of burners on opposite ends of one U-type radiant tube. These can also be applied to straight and "W" (4-pass) radiant tubes. Heat recovery makes up for the inherent inefficiency of indirect-fired operations. Only two pairs of cycling valves are used in this arrangement because the clapper valve that shuts off the eductor forces combustion air through the heat recovery bed and out the burner air nozzle.

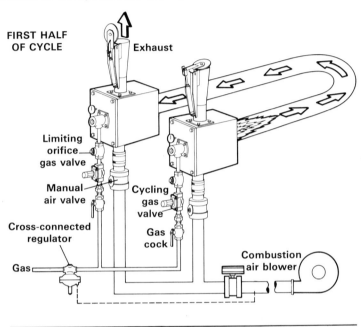

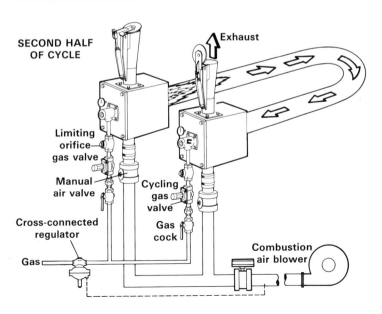

Regenerative burners have several advantages over integral burner-recuperator arrangements:

1) much higher air preheat temperatures are achieved;
2) final throw-away poc gas temperature is lower, usually about 300 F (150 C) at 80% extraction rate, permitting use of relatively low temperature exhausters (induced draft fans or eductors) and piping;
3) leaks and physical damage from thermal expansion and stress are minimized; and
4) the burners of a pair can be located to give good circulation and longer poc residence time in the furnace, and avoid poc short-circuiting.

Figure 9.16 shows some typical performance characteristics of integral burner-regenerators and recuperators. It also compares the "Heat Recovery Effectiveness" of the two devices -- an evaluator for comparing the performance of different types of heat exchangers with the same set of flow conditions. It is defined as "*the ratio of the actual heat transfer rate to the maximum possible heat transfer rate*".

The maximum possible heat transfer rate, q, in any heat exchanger will be defined by the product of the lower of the two heat carrying capacities, h, of the flowing fluids (the product of the mass, m, and specific heat, c, of the fluid) and the temperature difference between the maximum temperature to which the cold fluid might be heated and the cold fluid inlet temperature.

[9/8a,b,c] $q = h \times$ (temperature difference); $h_h = m_h \times c_h$; $h_c = m_c \times c_c$

[9/9]   Heat Recovery Effectiveness,

$$e = \frac{q}{q_{max}} = \frac{h_h \times (T_{hin} - T_{hout})}{h_{min} \times (T_{hin} - T_{cin})} = \frac{h_c \times (T_{cout} - T_{cin})}{h_{min} \times (T_{hin} - T_{cin})}$$

Figure 9.16. Regenerators have higher heat recovery effectiveness than recuperators.

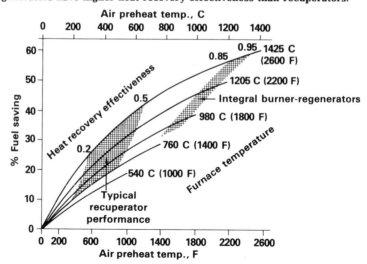

In a heat exchanger preheating combustion air with the combustion products of that air and a fuel, the heat carrying capacity of the air will always be lower than that of the combustion products (the mass is air, not air plus fuel) and the specific heat of air is lower than that of the combustion products, e.g. h_{min} = h_c = $m_c \times c_c$. Therefore,

[9/10]

$$e = \frac{q}{q_{max}} = \frac{h_h \times (T_{h_{in}} - T_{h_{out}})}{h_c \times (T_{h_{in}} - T_{c_{in}})} = \frac{h_c \times (T_{c_{out}} - T_{c_{in}})}{h_c \times (T_{h_{in}} - T_{c_{in}})} = \frac{T_{c_{out}} - T_{c_{in}}}{T_{h_{in}} - T_{c_{in}}}$$

In a regenerator, the maximum temperature to which the air can be heated will always be close to the temperature of the combustion products entering the regenerator because that is the temperature to which the leading edge of the packing is heated, whereas that for a recuperator will be the maximum temperature reached by the partition between the air and the combustion products (which will never be close to the incoming combustion product temperature as the partition is continuously cooled). For this reason, a regenerator will always have better heat recovery performance than a recuperator in typical industrial furnace applications.

Direct-fired compact burner-regenerators have greatly improved fuel efficiency and production when retrofitted to: aluminum melters (References 9.a and 9.b); barrel, box, and car-hearth forge and heat treat furnaces (References 9.e and 9.p); glass melters (Reference 9.d); ladle preheater-dryers (References 9.k, 9.m and 9.n); metal diffusion reactors; pit retort annealers; periodic kilns for ceramics (Reference 9.j); reheat furnaces (billet, slab) (References 9.i and 9.p); strip annealers (Reference 9.g).

Indirect-fired (radiant tube, immersion tube, or retort) integral burner-regenerators have been used to retrofit: chemical retorts, continuous steel strip annealing and galvanizing lines (Reference 9.f), and crude oil heater-treaters.

Compact burner-regenerators allow high efficiency fuel-fired furnaces to be constructed without heat recovery means such as counterflow preheat sections and external heat exchangers. Burner-regenerators effect considerable economies in size and cost. See Reference 9.o.

Many burner-regenerator-fired steel reheat furnaces have been built without conventional unfired preheat sections, i.e. fired "door-to-door". These provide more production per unit of floor space, with the regenerative burners recovering heat from the flue gases to give the same specific fuel consumption as traditional furnace designs. See Reference 9.q.

HEAT RECOVERY, OXYGEN ENRICHMENT, and OXY-FUEL are compared in Part 13.

REFERENCES

9.a **Bowers, J. D.:** "Heat Reclamation on Aluminum Melters", Die Casting Engineer, July/August 1988. Available as North American Handbook Supplement 240.

9.b **Bowers, J. D.:** "New Savings from Heat Recovery...in Aluminum Melting Furnaces", North American Handbook Supplement 246.

9.c **Chemical Engineering magazine:** "Process Heat Exchange", Section X. Waste Heat Recovery, McGraw-Hill Publications Co., New York, 1979.

9.d **Corneck, R. H.:** "Compact Regenerative Burners on a Unit Melter", Glass International, June 1987. Available as North American Application Report R-G-16A.

9.e **Ellwood City Forge:** "Car Bottom Forge Furnace...". North American Application Report R-Fge-96.

9.f **Kiraly, T. E. and Bugyis, E. J.:** "Application of Regenerative Burners to a Continuous Galvanizing Line", Iron and Steel Engineer, January 1989. Available as North American Application Report R-H-129.

9.g **Kondziola, James:** "Catenary Furnace Annealing Stainless Steel Strip". North American Application Report R-H-128.

9.h **Martin, R. R., Manning, F. S., and Reed, E. D.:** "Watch for Elevated Dew Points in SO_3-Bearing Stack Gases", Hydrocarbon Processing, June 1974, pp. 143-4.

9.i **Maynard, Mitchell:** "Twin Regenerative Burners Increase Energy Efficiency on New Reheat Furnace at Marion Steel", Industrial Heating, Dec., 1989. Available as North American Handbook Supplement 252.

9.j **McMann, F. C.:** "Regenerative Heat Recovery Applied to Periodic Kilns". Ceramic Engineering and Science Proceedings, Jan-Feb 1988.

9.k **North American Mfg. Co.:** "Application Report R-Steel-74".

9.l **North American Mfg. Co.:** "Sheet 8480-2 -- Recuperator Considerations".

9.m **Poe, L. G.:** "Ladle Preheaters -- TwinBed® *vs.* Oxy/Fuel and Oxy Enrichment". North American Handbook Supplement 241.

9.n **Poe, L. G.:** "Today's Ladle Preheaters and Dryers", North American Handbook Supplement 251.

9.o **Reed, R. J.:** "Future Consequences of Compact, Highly Effective Heat Recovery Devices" in "Heat Transfer in Furnaces", ASME Heat Transfer Div. Vol. 74, Book H00393. Available as North American Handbook Supplement 235.

9.p **Whipple, D. F.:** "Barrel Furnace" (for heating steel billets). North American Application Report R-Fge-99.

9.q **Newby, J. N.:** "TwinBed® Regenerative Burners", North American Handbook Supplement 254.

ADDITIONAL SOURCES of information relative to heat recovery

Hayes, A. J. et al (ed.): "Industrial Heat Exchangers", ASM International, Metals Park, OH 44073; 1985.

Karlekar, B. V. and Desmond, R. M.: "Heat Transfer", 2nd Edition, Published by West Publishing Co., Box 3526, St. Paul, MN 55165; 1982.

Kays, W. M. and London, A. L.: "Compact Heat Exchangers", McGraw-Hill Book Company, New York, New York; 1964.

Lukasiewicz, M. A. (ed.): "Industrial Combustion Technologies", ASM International, Metals Park, OH 44073; 1986.

Reay, D. A.: "Heat Recovery Systems", E. and F. N. Spon, Ltd., London, 1979.

Siegel, R. and Howell, J. R.: "Thermal Radiation Heat Transfer", Third Edition, Hemisphere Publishing Corp., (c/o Taylor and Francis), Washington, DC 20005-3521; 1992.

Thomas, Lindon C.: "Heat Transfer" (Professional Version); Published by Prentice-Hall, Inc., Englewood Cliffs, NJ 07632; 1993.

Part 10.   PROCESS CONTROL OPTIMIZATION

INTRODUCTION

Process control has moved from single loop analog controllers to digital, software-based, instrumentation. This transition has resulted in low cost, easy-to-use equipment that requires little maintenance. But the greatest benefit of microprocessor-based equipment is the power to allow system designers to focus on process interactions and optimization. The era of Personal Computers is here, and is causing another change in the control world.

BASIC PROCESS CONTROL

Much of our current control practice is based upon the process loop. The process is monitored by a sensor (thermocouple for example) which reports to a device (loop controller) that compares the actual value (process variable, PV) to the desired value (setpoint). Any resulting error is processed to determine the adjustment needed to bring the actual value to the desired value.

A control equation establishes the equipment response to an error signal. The proportional/integral/derivative (PID) algorithm is the most commonly used. Equipment suppliers identify this equipment as "Three Mode", "PID", and "Gain, Reset, and Rate". These names identify the three elements of the equation used to establish controller output based on the setpoint and the process variable:

$$\text{Output} = \text{Proportional} + \text{Integral} + \text{Derivative}$$
$$\text{where Error} = \text{Setpoint} - \text{Process Variable}.$$

Proportional, = Error × Gain, is:

an output that is the product of error and gain and occurs with no time delay. Large errors result in large outputs, and small errors result in small outputs. Thus, as the error nears zero, the output nears zero, and the controller cannot reach the setpoint.

Integral, = (Integration of Error) × Reset, is:

an output that increases in proportion to the time that the error is not zero; so, while proportional control cannot reach setpoint, the integral term adds small adjustments until the error is zero.

Derivative, = (Derivative of Error) × Rate, is:

an output proportional to the rate of change of the process variable. Therefore, for rapid changes, the rate term increases the output, causing faster response of the loop.

This is one of many loop implementations, but all share the common form of including P, I, and D terms with a user adjustable factor for tuning the loop response.

PID TUNING

Proper loop tuning is essential to system operation. While intuitive methods can work, a more formalized process gives better results.

Several methods can be used to calculate the P, I, and D parameters. Each method gives somewhat different results. Two of the methods are described here: the Ultimate Period and the Process Reaction Curve. The parameters calculated using these methods may be used as a starting point for further tuning, as briefly described later in this section.

Both methods calculate parameters that produce a Quarter Decay Ratio, which gives an initial overshoot to input changes, followed by decaying oscillation around the control point (Figure 10.1). The decay rate is such that each cycle of oscillation has an amplitude of one quarter that of the previous cycle.

Figure 10.1.   Quarter decay ratio response

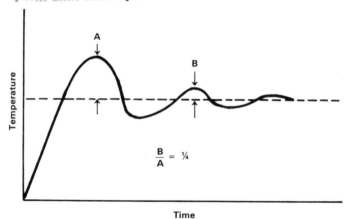

$$\frac{B}{A} = \frac{1}{4}$$

There is no single perfect set of tuning parameters, just as there is no single ideal response. Each system requires tuning for its specific application. A system set for fast response generally exhibits considerable overshoot, and a critically damped system may respond slowly. Changes in the process, loading, or setpoint cause some alteration in system response.

Ultimate Period Method. The Ultimate Period Method is also known as the Ziegler and Nicholls method. It is a closed loop procedure, where Rate and Reset are set to zero and the system gain increases until the system oscillates at constant amplitude (Figure 10.2). The period of oscillation is called the Ultimate Period (P_u), and the gain at which it occurs is the Ultimate Gain (K_u). Table 10.3 provides formulas for calculating the tuning parameters from this information.

Figure 10.2.   Ultimate period oscillation.

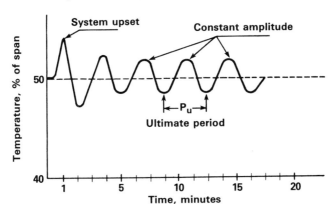

Procedure:

1) Stabilize the system at its normal operating temperature.
2) Set Rate and Reset to zero.
3) Set Gain to an arbitrary initial value.
4) Manually introduce an upset to the system by changing the setpoint.
5) Put the system into automatic.
6) Record the system response to the upset with a chart recorder. Adjust Gain until the system oscillates at constant amplitude. If, after an upset, the oscillation decays, increase Gain. If it grows, reduce Gain.
7) Use Table 10.3 to calculate the tuning parameters.

Table 10.3.   Ultimate period test (closed loop)

Algorithm type	Gain	Reset	Rate
Proportional only	$0.5K_u$	—	—
PI	$0.45K_u$	$1.2/P_u$	—
PID	$0.6K_u$	$2/P_u$	$0.2\ P_u$

(Where K_u is the ultimate gain, and
P_u is the ultimate period.)

Process Reaction Curve Method. The preceding Ultimate Period Method may require a number of test cycles to determine the values of K_u and P_u and it may be undesirable to allow the system to oscillate as required to establish the period. The Reaction Curve method may be used in these cases.

Process response to an input step change establishes the Process Reaction Curve. It measures the open loop (without feedback) response of the process.

Procedure:

1) Stabilize the system at a temperature about 5% below the average operating point.

2) Place the actuator in manual, holding last position established by the controller.

3) Manually increase the controlled variable by approximately 10% (ΔM). Record system reaction until a stable operating point is reached.

4) The recorded process reaction curve should resemble the curve in Figure 10.4.

5) Use the curve to calculate the reaction parameters R_r and L_r. The reaction rate of the system, R_r, is determined by drawing a line tangent to reaction curve at its point of maximum slope. The slope of this line is the reaction rate R_r in % per minute.

6) L_r is the process lag time, defined as the period (in minutes) from the introduction of the step change to the time at which the tangent line of the reaction rate crosses the temperature baseline.

7) ΔM is the amplitude of the controlled variable step change in units of percent (as established in step 3).

8) Calculate the tuning parameters using Table 10.5.

Figure 10.4.   Process reaction curve.

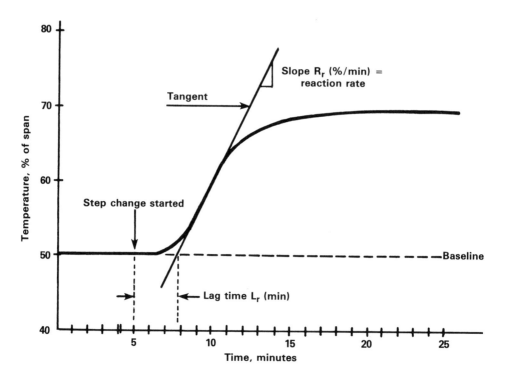

Table 10.5. **Process reaction rate test (open loop)**

Algorithm type	Gain	Reset	Rate
Proportional only	$\Delta M/(L_r \cdot R_r)$	—	—
PI	$0.9\Delta M/(L_r \cdot R_r)$	$0.3/L_r$	—
PID	$1.2\Delta M/(L_r \cdot R_r)$	$0.5/L_r$	$0.5\ L_r$

This method requires only one cycle to gather data and compute values. However, in systems with short lag times, the measurement of L_r may be difficult, requiring a high chart speed to determine lag time accurately.

Further Tuning. Although both methods described give Quarter Decay Ratio Response (Figure 10.1), each provides different tuning parameters. Since the response desired may be different from the theoretical, additional guidelines may be useful in fine tuning the system.

When less overshoot than the Quarter Decay Ratio is desired, either Gain or Reset may be lowered. However, reduced gain is the preferred method of overshoot reduction.

Increasing Gain or Reset increases system response and overshoot, while decreasing them slows response and increases stability. Adjust each for optimum control.

In fast responding systems, Rate can be used to limit overshoot and increase stability by allowing lower gain settings. However, care must be used, since Rate magnifies the effect of minor system fluctuations. Slowly responding systems can be operated with Rate set to zero.

Caution is advised in systems with large lag times. The effect of Rate may be delayed and become out of phase with the system response, causing oscillation.

Both self-tuning controllers and external loop tuning software packages exist and simplify the process.

ADVANCED PROCESS CONTROL

The single loop controllers of today are accurate, stable, and easy to use. They can also communicate with other equipment. Instead of independently controlling each process variable, these controllers interact to coordinate multivariable process applications.

Digital Era. Introduction of the microprocessor to loop control has changed the control world. We have shifted from "hardware" driven systems to software based systems. Whether in the form of a simple loop controller, a PLC, or a Distributed system, the process is operated by software. System architecture is driven by the flow of information, with networks connecting all parts of the process.

Figure 10.6. A dedicated ratio controller with extra circuits for process temperature control, O₂ trim, multiple air and fuel streams, furnace pressure control, and control of flue gas recirculation for NOx reduction.

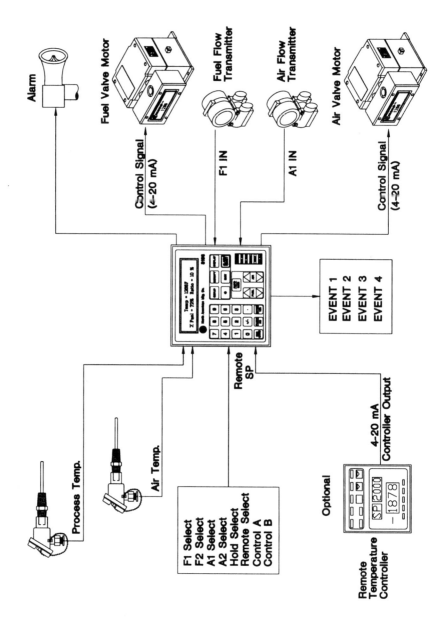

Because of differences between the continuous nature of analog equipment and the discrete function of digital designs, the process engineer must consider new factors when applying digital controls. For example, digital controllers calculate new output values at intervals that range from many times per second to once in several seconds. Slow variables such as temperature in a large furnace may be updated infrequently with no impact on control quality, while flow should be updated every second or faster. Ratio systems demand even faster response; dedicated ratio controllers (Figure 10.6) operate at 10 updates per second.

Cascade Control. Many applications have a long process lag time, where the measured variable responds slowly to a change in system input. Such systems often require heating a large mass to a closely controlled temperature. Conventional single loop control places a sensor on the work to control the furnace heat input. Since the load responds slowly to a change in heat input, the long lag time between an input change and detection at the work produces control overshoot. This lag makes tuning of the loop difficult, because it may be easy to overheat the work. Often maximum input must be limited during part of the heating cycle, to prevent damage to the furnace, resulting in longer cycle times. Better control can be achieved by using two loops in cascade (Figure 10.7).

The slower primary loop consists of a work temperature sensor that is compared to the target work setpoint, with the controller output connected as the remote setpoint of the faster second loop. The second loop monitors the furnace heat input and compares it to the output of the primary loop, establishing the furnace heat input. The result of this system is a self-limiting furnace input that maximizes the rate of heat transfer to the work.

Figure 10.7.   Cascade control loop.

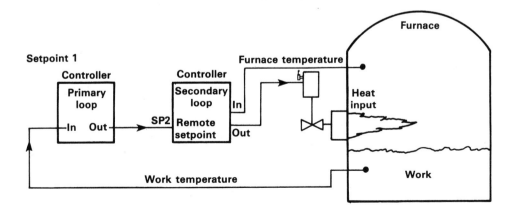

When the output of the primary loop is large, indicating a large deviation between setpoint and actual work temperature, the secondary loop setpoint is near its maximum. This provides the greatest heat input. As the primary deviation decreases, indicating that the work is approaching desired temperature, the lower secondary setpoint reduces the heat input. This declining temperature difference between the furnace and the work reduces the possibility of overshoot, allowing closer work temperature control.

Cascade control is effective only when the inner or secondary process response is faster than the outer or primary process response.

Programmable Logic Controllers. As computing power became less expensive, dedicated industrially-hardened computers were designed for shop-floor use. These systems replaced large hard-wired relay systems, and were built for easy configuration by using a relay-like ladder language. Initially designed to replace inflexible relay logic, PLCs now include loop capability and higher-level language support. The PLC is a powerful tool that integrates sequential logic and process control in a single package.

Sometimes, because of their universal nature, PLCs are used for tasks that are better accomplished by other equipment. Proper application of a PLC to industrial combustion systems requires understanding limitations of the control equipment, experience with the process, and a working knowledge of industry and national standards. It is not possible to convert a relay-based ladder diagram directly to a PLC system without understanding the inherent difference between the technologies. Easy configuration (the major PLC feature) is a double edged sword where security is important, such as in systems that control hazardous equipment.

Distributed Control Systems (DCSs). The trend to the lower cost computing power (that encouraged the growth of PLCs) allowed designers to distribute computing power throughout the system. The controller is placed close to the process and transmits information via a network to other controllers and operator interface equipment.

This allows a building block concept where each element is installed and debugged without affecting the remaining process elements. Tasks are concentrated at the lowest possible system level minimizing network communication, and reducing the impact of catastrophic failure. Local devices can operate in a "fall-back" mode that keeps the process running while repairs are made. The elements located at the process provide stand-alone basic control, while communication with a computer provides advanced strategies. This isolates the minimum day-to-day control operation from the complex. Computer intensive activities such as process graphics and data storage are focused at the top of this hierarchy.

Distributed control is moving even closer to the process with the advent of "Intelligent" transmitters. Intelligent field devices can improve sensor performance by correcting for temperature variations and linearizing transducer data. Corrected data can then be transmitted to the control equipment. Industry standard bus systems allow many field devices to transfer information over a single cable (wire or optical), eliminating separate wiring between the control equipment and each field device.

Personal Computers. Personal computers are common in the factory. High production volumes, standardized hardware and availability of robust operating systems have moved the PC from the office to the factory floor. With a graphical interface package, they are ideal as supervisors in distributed control systems. Standard configurable software allows manufacturing managers to gather and analyze process data using factory floor personal computers. Industrially-hardened personal computers are replacing PLCs in control applications.

FEATURES OF ADVANCED CONTROL

Systems generally are hierarchical in nature, including levels of equipment reaching from the process controller to a supervisory computer. Multiple levels are possible because these systems are software based. This software is often hidden from the user and designer by a configuration language, which frees the designer from complex programming tasks.

In addition to basic loop control, systems acquire data to display process trends. This information can be stored for analysis and used to establish performance benchmarks, or document batch history. Historical data can establish process capability, and highlight degradation from standards caused by equipment deterioration, process variation, or procedural change. Plant-wide information and product tracking is possible with systems equipped for network operation. The objective is *consistent* system performance.

Alarm logs, batch history, equipment performance reports, and maintenance records help diagnose equipment, process, or operator problems. Early warning can expedite maintenance and ensure uniform product quality. Computer supplied messages help service or operate the equipment. Step-by-step start-up and maintenance instructions are available, guiding operators through infrequently used procedures.

System-wide availability of information, coupled with distributed computing power allows alternatives to conventional PID control. For example, in a multi-variable process, the system could attempt to optimize desired conditions while operating within specified constraints. By monitoring the effect of small variations, the system could self-learn and continuously improve its operation.

Complex process systems are often described by a mathematical model to predict performance based on measured conditions. These predictions are compared to actual performance, or can run in "real-time," adjusting operation to keep performance at the calculated level.

Batch operations benefit from "on-line" storage of recipes, allowing quick change-over from one product to another. The system can calculate required adjustments to recipes based on measured conditions.

Often the method of communicating with a process operator is improved by computer technology. Visual displays using graphics and color compress large amounts of information into small, easy to understand formats. Computing power is used to simplify information and reduce language barriers.

Solid-state equipment is very reliable, and techniques of redundancy or fault-tolerant design result in systems with small probability of downtime. These specialized techniques increase system cost, and require detailed knowledge of both the process and the equipment selected. System performance is improved by designing in-fault tolerance and error trapping so that the equipment can recover from a fault or failure by accepting a degree of ambiguity and attempting to correct for it. "What-if" type logic tries to identify conditions that are inconsistent by comparing many variables and other system resources such as the action that the controller is requesting.

At its simplest level, error analysis recognizes deviations from the expected activity of a process. The deviation can contain clues to the problem and allow the system to identify the potential cause. The system then attempts to bring the process back into control or force it to a known safe state. For example, a system can verify that the magnitude and rate of change detected in a process variable is within reasonable limits, and within the physical ability of the process. Digital circuits can be equipped with error correction algorithms, and operator entry controls can check and limit entries to reasonable values.

Process safety considerations established by national standards limit and guide the application of control equipment to prevent hazardous conditions. Most vendors provide application notes to assist in proper use of their equipment in critical systems, but this is best handled by working with a supplier knowledge-able about the process, the control equipment, and applicable standards.

INTEGRATED PROCESS CONTROL

In all cases, proper system design begins with a detailed understanding of the process. While many procedures are available, most include the following basic steps:

1) Study the Process -- This includes monitoring and recording actual or simulated process data. Observations should include operating practices and manual actions that affect system performance. This study must determine safe process limits, and appropriate industry safety standards.

2) Evaluate results -- Use data and observations to graphically establish the correlation of process variables with desired process operation.

3) Consider available technologies -- Many system types are available to solve process control problems. Each system represents a mix of features that should be considered.

   • How is the system configured? Is custom software required, or can a process engineer configure the hardware? Is the language "self-documenting", or must complex manuals be maintained? Are key process functions built into the system, or must the designer start from the beginning? Can information be presented in engineering units and common language, or must the user interpret cryptic codes?

   • How reliable is the system? Is the equipment intended for shop-floor or control room installation? What are the failure modes, and are they predictable? Does the system accommodate mandated safety equipment?

   • Is the equipment easy to use and maintain?

   • Is the equipment flexible enough for future needs?

   • What are all costs associated with this equipment? Cost includes initial purchase, maintenance, and repair.

Application-specific control systems are often available. The manufacturer has completed the process study, evaluation and system engineering phase, and can offer a complete solution. Since the effort is spread over several systems, the cost is usually lower, and the technology is proven so that risk is reduced. Systems have been designed for many industrial applications, such as:

Kilns -- Innovative use of existing sensor and control technology allows improvement in performance, product uniformity, and efficiency in tunnel kilns, shuttle kilns, and rotary calciners.

Steel Reheat Furnaces -- Normal operation involves frequent load changes due to delays and product changes. By looking at the furnace as a system, a control package was developed to use the input of one zone to adjust the operation of an upstream zone. This cascaded control heats the steel as late as possible in the furnace, maximizing heat utilization without risking inadequately heated steel while preventing overheating and minimizing scaling.

Aluminum Melters -- Melter efficiency and performance is improved by controlling system input from different points in the melter at different times in the cycle. The control point and setpoints are changed, accommodating the differing heat transfer rates of aluminum as its phase changes, and taking advantage of energy stored in the furnace refractory to minimize fuel input and reduce dross formation and hydrogen absorption.

IMPLEMENTATION

When building on a base of loop control and using application-specific hardware, PLCs, and Personal Computers, successful system implementation requires experience with both the process and control equipment. The designer must integrate process understanding, control experience, and knowledge of industry standards to build a functional system.

ADDITIONAL SOURCES of information relative to process control optimization

Anderson, Norman A.: "Instrumentation for Process Measurement and Control", 3rd ed., Chilton Company, Radnor, PA, 1980.

Bibbero, Robert J.: "Microprocessors in Instruments and Control", Wiley-Interscience, New York, NY, 1977.

Cornforth, J. R. (ed.): "Combustion Engineering and Gas Utilization", 3rd ed., Chapter 11, British Gas and Chapman & Hall, 1992.

Murril, Paul W.: "Fundamentals of Process Control Theory", Instrument Society of America, Research Triangle Park, NC, 1981.

Ogata, Katsuhiko: "Discrete-time Control Systems", Prentice-Hall Inc., Englewood Cliffs, NJ, 1987.

Shinskey, F. G.: "Process-Control Systems", 2nd ed., McGraw-Hill Inc., New York, NY, 1979.

Part 11. POLLUTION CONTROL

REDUCTION OF NOx POLLUTION FROM INDUSTRIAL COMBUSTION

Although the media and the general public think of pollution control as a single problem, most engineers have learned that control of each of the several types of polluting emissions is really a chain of complex problems. The problems of the industrial heat processing industries are very different from the much-talked-about approaches for automobiles or for boilers and large power plants, each of which far exceeds the total emissions of industrial heating equipment.

TRACE SPECIES REPORTING

The primary products of combustion are carbon dioxide (CO_2), water vapor (H_2O), oxygen (O_2), and nitrogen (N_2). Percentages of these gases in the poc (products of combustion) are determined by simple chemical calculations, based on the composition of the fuel streams and stoichiometric ratio.

The concentrations of the trace compounds, or species, in the poc, measured in parts per million (ppm) instead of percentages, are path dependent. That is, the degree of their formation is determined by temperature and the concentrations of other species in the flame and in the burned mixture of fuel, oxidant, and poc. The small concentrations of the trace species do not significantly impact the larger concentrations of the primary products. Part 11 is primarily concerned with the trace species that are harmful pollutants, such as nitric oxide (NO), nitrogen dioxide (NO_2), carbon monoxide (CO), sulfur dioxide (SO_2), and particulates.

Trace Measurement Units. Table 11.1 lists the units of measurement of trace emission rates.

Correction for Dilution of Concentration Measurements. Concentration measurements are subject to dilution errors if excess air is increased in the combustion process. If a specified ppm emission level were not required to be corrected to a base oxygen level (such as 3% O_2 in the USA), any system could add dilution air to reduce the pollutant level and claim to be in compliance.

Old time boiler operators thought they could fool the "smoke inspector" by turning up the excess air, but *dilution is not the solution to pollution*. Figure 11.2 compares generation *vs.* concentration measurements.

Table 11.1.   Units of measurement of emission rates for trace species fall into 3 classes.
(See Appendix Table C.6 or the Glossary for equivalent units.)

US units	Metric Units	Explanations and comments
1) tons/year	tonnes/yr	POLLUTANT GENERATION rate. Primary permit value. Total annual allowed or measured mass flow rate.
pounds/hr	mg/h	POLLUTANT GENERATION rate. Total hourly allowable, or measured, mass flow rate.
2) pounds/million Btu	grams/gigajoule	POLLUTING "EFFICIENCY". Factors out the size of the combustion system. Generally mass/hr divided by gross input/hr.
3) ppm$_V$	ppm$_V$	POLLUTANT CONCENTRATION. Volume fraction of a given pollutant contained in an exhaust stream. The ppm$_{vd}$ (dry) is reported by most trace gas analyzers
ppm$_m$	mg/nm³*	POLLUTANT CONCENTRATION. Weight fraction of a given pollutant contained in an exhaust stream.

Figure 11.2.   A concentration measurement loses its meaning unless corrected to a standardized % O$_2$. A generation measurement is not affected by excess O$_2$.

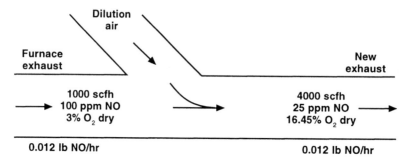

0.012 lb NO/hr   0.012 lb NO/hr

Within any given enforcement jurisdiction, all specified trace emission allowances and all measured actual trace emissions must be referenced to the same base oxygen level. (3% O$_2$, which is about 15% excess air, within the U.S. EPA's jurisdiction.)

An actual trace measurement in ppm$_{vd}$ (parts per million, by volume, dry basis) can be converted to the reference level as follows:

$$[11/1a] \quad ppm_{vd \, ref \, O_2} - \frac{20.9\% - O_{2ref}}{20.9\% - O_{2act}} \times ppm_{vd \, act \, O_2}$$

_____

* normal cubic metre. See Glossary.

For areas where the enforcement base is 3% O_2, the above becomes:

[11/1b] $ppm_{vd\ ref\ O_2} = \dfrac{17.9\%}{20.9\% - O_{2act}} \times ppm_{vd\ act\ O_2}$

Table 11.3 provides factors based on Formulas 11/1a and 11/1b. The use of those factors is illustrated in Example 11.1. It is important to remember that the base oxygen correction does not imply a ratio condition within a furnace. Data recorded at any excess air level must be corrected to the same oxygen basis for comparison of the total emission level.

Example 11-1. If a reading of 80 ppm_v NOx is measured while operating with 4% oxygen by volume, dry, in the flue gas, find the standardized base equivalent ppm_v at 3% O_2 dry.

Solution: The corrected ppm NOx (dry basis by volume) = 80 $ppm_{vd} \times \dfrac{20.9 - 3}{20.9 - 4}$

= 80 × 1.059 = 84.7 ppm_v.

Implementation Standards, Units. In the United States to date, enforcement has been left to the states or local agencies, but they must set up standards that are within U.S. federal guidelines.

Space considerations prevent publication here of legal limits on pollution emissions of various enforcing jurisdictions. Such rulings and allowable exceptions thereto may change at any time. It is the equipment owner's responsibility to keep abreast of applicable regulations. See Reference 11.a at the end of this Part 11. Information on US EPA standards for measurement can be found in Appendix A of 40 CFR 60.

The basic specification is for ambient conditions: e.g. on the sidewalk, where people are exposed--at 0.053 ppm (by volume) NOx, or 100 micrograms per cubic metre of air. This is an allowed average over a 24-hour period, not to be exceeded more than once per year. (U.S. Natl. Ambient Air Quality Standard, 1992.)

Relating ambient standards to "source" (in stack) values is complex. Source review programs include:

(a) new source performance standards

(b) prevention of significant deterioration (PSD) in the environment in attainment areas--no more than 40 tons NOx per year; and

(c) non-attainment area bubbles, offsets, tradings, or banking.

Table 11.3. Converting wet %O₂, dry %O₂, and % excess air; and multipliers* for correcting NOx flue gas analysis readings to a 3% oxygen basis (per example 11-1) for a typical natural gas.

% Excess oxygen		% Excess air	
dry	wet	% XSAir	Multiplier
0	0	0	0.86
1		4.53	0.90
1.10	0.82	5	0.90
1.22	0.90	5.57	0.91
2	1	9.54	0.95
2.09	1.66	10	0.95
2.41	1.73	11.7	0.97
2.98	2	15	1.00
3	2.49	15.1	1.00
3.57	2.51	18.6	1.03
3.80	3	20	1.05
4	3 20	21.4	1.06
4 54	3.38	25	1.09
4.71	4.03	26.2	1.10
5	4	28.4	1.13
5.22	4.26	30	1.14
5.83	5.02	34.8	1.19
5.85	5	35	1.19
6	5.02	36.3	1.20
6.43	5.16	40	1.24
6.92	5.55	44.6	1.28
7.46	6	50	1.33
8	6.50	55.9	1.38
8.38	7.01	60	1.43
9.04	7.35	68.7	1.51
9.83	8	80	1.61
10	8.77	82.6	1.64
11	8.93	100	1.81
11.1	9.92	102	1.83
12	10	121	2.01
13	10.9	148	2.26
13.1	12.0	150	2.29
13.8	12.0	175	2.52
14	12.8	183	2.59
14.4	13.0	200	2.75
15	13.5	230	3.03
	14.1		

---

* These multipliers may be used to correct any volumetric gas concentration reading to a 3% oxygen basis. They correct for dilution only, not for the effects of excess air on the chemical kinetics phenomena.

Example 11-2. If the specified PSD level is 40 tons NOx per year, that is equivalent to

$$\frac{(40 \text{ tons/yr}) (2000 \text{ lb/ton})}{(24 \text{ hr/day}) (365 \text{ days/yr})} = 9.13 \text{ lb NOx/hr.}$$

If you were considering adding heating equipment that emit 1 lb NOx per million Btu, the maximum input that could be added would be:

$$\frac{9.13 \text{ lb NOx/hr}}{1 \text{ lb NOx/million Btu}} = 9.13 \text{ million Btu/hr.}$$

But if a unit that emits only 0.25 lb NOx/million Btu becomes available, then units totaling 4 (9.13) or 36.5 million Btu/hr could be added.

The last two columns of Appendix Table C.6 (Pollutant Concentration) give conversion factors for six typical fuels, for changing ppm to lb NOx per million Btu and for changing lb NOx per million Btu to ppm. All these conversion factors are for ppm$_{vd}$ (of NO_2, CH_4, CO, or SO_2, at 3% O_2 by volume dry); and per million gross Btu.

If it should be necessary to determine such conversion factors for specific fuels, North American Mfg. Company's Handbook Supplement 248 and STOIC Computer Program provide help in making the calculations. It is first necessary to ascertain the following from the fuel supplier: the analysis of the fuel (volumetric for gases, by weight for liquids and solids), and the gross heating value of the fuel. In some parts of the world, the lb NOx per million Btu is based on *net* or *lower* heating value. When making comparisons, both should be on the gross basis or both on the net basis.

NOx EMISSIONS

Why NOx Emissions are a Problem. Nitrogen oxides, or NOx, emissions are generated by combustion systems where nitrogen and oxygen are present within a locally high temperature region of the flame. The abbreviation NOx is chemical shorthand for the combined species of NO and NO_2. These species of emissions pose a significant health hazard in ambient air. Other detrimental environmental effects of NOx emissions are photochemical smog and acid rain, both found in industrial areas around the world.

In the lower atmosphere, NO reacts with oxygen in the air to form both NO_2 and ozone, O_3. Ground level ozone is a health hazard, blocking air passages and impairing respiratory performance. As a pollutant, ozone is far more widespread than NOx, with over one hundred U.S. counties measured as out of compliance by the EPA ambient air guidelines. In 1991, only Los Angeles county was out of compliance for NOx emissions.

Mechanisms of NOx Formation. NOx can be formed in three distinct ways in combustion systems, the most common of which is the thermal NOx pathway. The elementary reactions responsible for this type of NOx generation are frequently called the "Zeldovich Mechanism" after Y. B. Zeldovich who detailed their contribution in a 1946 paper (Reference 11.b). The reactions are

$$O_2 + M \leftrightarrows O + O + M$$

$$N_2 + O \leftrightarrows NO + N$$

$$N + O_2 \leftrightarrows NO + O$$

$$N + OH \leftrightarrows NO + H.$$

A fifth reaction for nitrogen dissociation is often included, but is not a significant contributor to the N atom pool at normal air-based combustion temperatures.

The equilibrium concentration of NO is strongly temperature-dependent. Figure 11.4 shows the exponential relationship between temperature and equilibrium levels of NO for three mixtures: air with 20.9% O_2, and products of combustion with both 2% and 5% oxygen by dry volume. These three are typical gas analyses that occur in industrial combustion applications.

An equilibrium analysis of the global reaction,

$$N_2 + O_2 \rightarrow 2NO,$$

shows that the concentration of NO is proportional to $K_p(T)$, the equilibrium constant for this reaction at constant pressure, and the square root of both the oxygen and nitrogen concentrations at equilibrium,

$$[NO] = K_p(T) \sqrt{[O_2] [N_2]}.$$

Glassman (Reference 11.c) has also shown by an analysis of the thermal NOx mechanism that the rate of NO growth with time is proportional to the square root of the oxygen concentration,

$$d[NO] / dt \approx \sqrt{[O_2]_{eq}} \, [N_2]_{eq} .$$

Therefore to minimize equilibrium concentration and the rate of growth of NO, both the temperature and oxygen concentration need to be controlled.

An examination of the reaction kinetics of the thermal NOx mechanism shows that the growth of NO is much slower than the completion of the other hydrocarbon reactions. In a shock tube study of the CH_4-O_2-N_2 reaction system, Bowman and Seery [Ref. 11.d] found that hydrocarbon reactions were complete in less than 0.0001 seconds. NO growth was found to be much slower. In

Figure 11.4.   Equilibrium NOx emissions *vs.* gas temperature (poc = products of combustion).

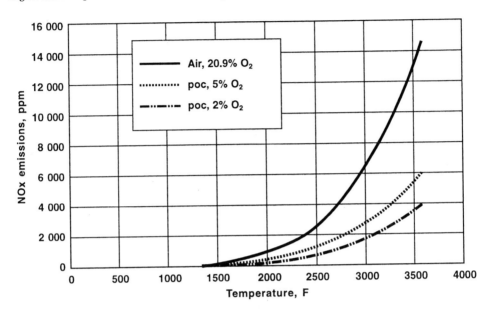

fact, at a time of 0.01 seconds, concentrations were still far short of equilibrium. The extended times required for equilibrium indicate that NOx can be controlled by reducing residence time.

The slow reaction rates, low concentrations, and small effect on overall product temperatures and concentrations allow the problem of NOx generation to be decoupled from the other chemical reactions. Researchers such as Bracco (Ref. 11.e) and Williams (Ref. 11.f) have used this fact to numerically model the production of NO in flames. These same factors are extremely important in the creation of a low NOx emission burner.

The two other pathways that can produce NOx emissions are the chemical (or fuel) mechanisms, and the "Prompt" or "Fenimore" NOx mechanism. Chemical NOx is created when nitrogen atoms are located in bonds within the fuel molecule, commonly referred to as "fuel-bound nitrogen." C-N and N-H bonds are common in liquid and solid fuels and can produce significant contributions to the total NOx emission level. During the reaction of these compounds additional N atoms are released into the radical pool, increasing NO formation. Prompt NOx is in some ways similar to chemical NOx, except that the C-N bonds are created through the reaction

$$CH + N_2 \rightarrow HCN + N$$

The HCN is then oxidized, potentially leading to increased NOx levels in the product gases. HCN generation in flames is strongly ratio-dependent, with peak concentrations formed at sub-stoichiometric or fuel-rich conditions where there is reduced oxygen availability.

Figure 11.5. Conversion of chemical NOx (from N).

$$\text{NOx ppm}_{vd} \text{ at 0\% O}_2 = \frac{\left(\dfrac{ft^3\ N}{lb\ fuel}\right) \times 10^6}{\left(\dfrac{ft^3\ dpoc}{lb\ fuel}\right)} \times \text{conversion factor}$$

$$= \frac{\left(\dfrac{370}{14}\right) \times \%N \times 10^6}{\left(\dfrac{ft^3\ dpoc}{lb\ fuel}\right)} \times \text{conversion factor}$$

$$= \frac{(0.2707) \times \%N \times 10^6}{\left(\dfrac{ft^3\ dpoc}{lb\ fuel}\right)} \times \text{conversion factor}$$

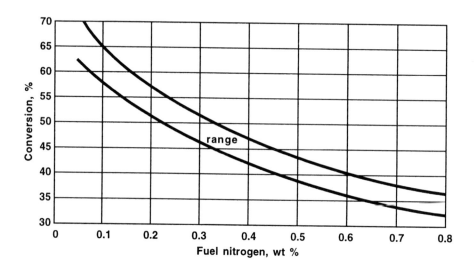

Figure 11.6. Fuel bound nitrogen (wt %) for #2 fuel oil.

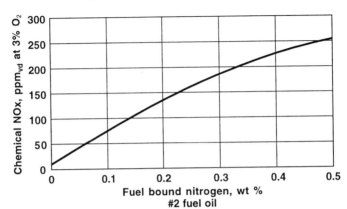

Low NOx Combustion. The preceding section described how NOx is formed in combustion systems. However, the real issue is: How can the overall NOx emission from a combustion process be controlled? When a fuel and air mixture at a given ratio is burned, it produces predictable levels of water vapor and carbon dioxide. This mixture must always have the same adiabatic flame temperature, primary product concentrations, and equilibrium NOx concentrations.

Why should the NOx emission level be different for flames with the same fuel and air/fuel ratio? All product levels would be the same if combustion were an equilibrium process. But fortunately for equipment designers, industrial combustion is governed by non-equilibrium and kinetic effects. In industrial applications the primary goal of combustion is to heat some product

Figure 11.7. Equilibrium NOx production *vs.* equivalence ratio.

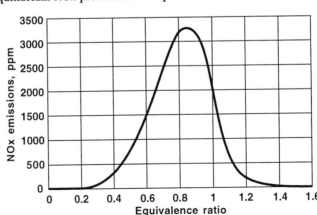

or load. The furnace where this heating takes place is maintained at a temperature significantly below the adiabatic flame temperature of the fired air/fuel mixture. When chemical reaction takes place, products are both formed and rapidly released into this "reduced" temperature environment. This allows the NOx forming reactions to be quenched, essentially "freezing" their concentrations at levels far below equilibrium at flame temperature.

Reduced temperature greatly inhibits the rate of NOx formation. While the time to equilibrium is short at flame temperature, the time to the lower equilibrium concentrations at typical furnace temperatures can be hours, days, or even years. The total gas residence time in an industrial furnace is only a few seconds, after which gases are released to the ambient atmosphere. Therefore the NOx producing mechanisms never reach equilibrium, and concentrations are frozen at a much lower level.

Figure 11.0.   Actual NO concentration profile in a furnace environment.

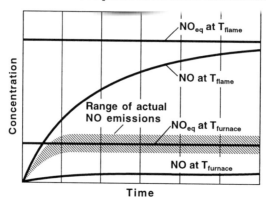

Combustion equipment manufacturers take advantage of these non-equilibrium effects in designing low NOx burners. These effects can be maximized by using combustion strategies that limit peak flame temperatures, limit localized in-flame oxygen concentrations, and reduce residence time at peak temperature.

In combustion systems, water vapor and carbon dioxide can be considered as "state" variables. If the fuel composition and air/fuel ratio are known, then the concentrations of these products after combustion is also known. NOx production in a flame behaves as a "path" variable. To determine the final NOx concentration, the time, temperature, and local concentration profiles must be known. The optimization of one or more of these three factors gives low NOx equipment manufacturers flexibility in designing low emission systems.

Low NOx Strategies and Equipment. NOx production from combustion sources is strongly dependent on four primary variables: temperature, concentrations of oxygen and nitrogen, C-N bonds within the chosen fuel, and residence time. If any one or more of these variables are properly controlled, NOx production will be reduced. When evaluating the following NOx reduction techniques on a given combustion system, the most important constraint must be the process itself. Low emission burners can have significantly different flame patterns and internal chemistry; so a plant engineer must examine these features and determine their effects on the fired process.

NOx reduction techniques developed over the last two decades fall into four basic categories:

- Modification of Operating Conditions
- Modification of Combustion System
- Modification of Burner Internals
- Post Combustion Cleanup

Modification of Operating Conditions. The first step in achieving reduced NOx emissions in a given combustion installation is to examine the current operating conditions. If a small reduction is required, on the order of 5-15%, proper burner tuning and furnace operation may be sufficient.

The NOx emission rates of many burners are sensitive to air/fuel ratio. As excess air is decreased toward stoichiometric firing, less oxygen is available for conversion to NO, despite increased flame temperatures. Equilibrium NOx calculations (Figure 11.7) show that the maximum emission rate occurs at 20-30% excess air. While the rate of air-fuel mixing may move this peak in either direction, the majority of burners have reduced emissions at 10% excess air.

Infiltration of air into a furnace can also have a noticeable impact on system NOx emissions. Recently, an industrial furnace was found to emit 50% more NOx than a laboratory furnace equipped with the same identical burner. The burner was re-tested in the lab furnace where the original measurements were found to be correct. A thorough investigation of this mysterious chain of events showed that the lab furnace was in good condition, sealed tightly, whereas the industrial furnace had leaks allowing infiltration of considerable "tramp air" (excess air), which made much more oxygen available for production of NOx.

When areas of negative furnace pressure occur, cold air is drawn in through furnace ports and doors, leaky seals, and other openings. If this infiltrated air mixes with the flame envelope it can create locally high oxygen concentrations in the highest temperature flame zones. This can result in higher levels of NOx emissions at any burner air/fuel ratio. Proper furnace pressure control can be a very effective technique for not only limiting furnace NOx production but also increasing furnace efficiency (see Part 7 of this book).

When chemical NOx is a problem, fuel switching can be a viable option. Different fuels contain different levels of atomic nitrogen (N). A switch to a similar fuel with lower %N can greatly reduce NOx emissions. When operating condition adjustments are not sufficient to achieve the desired emission rate, further steps must be taken.

Modification of Combustion System. Further reduction of NOx emissions can be achieved with changes to the combustion system. A common combustion modification is the addition of flue gas recirculation (FGR). FGR systems bring cooled products of combustion from the exhaust stack back into the combustion zone. This flue gas can be added to either the air or fuel streams or supplied through its own burner connection. The method of introduction can change the NOx reducing effectiveness of a given quantity of flue gas.

The addition of flue gas to a burner will reduce both peak flame temperature and local oxygen concentration. Typical NOx reductions achieved with FGR are shown in Figure 11.9. FGR is most effective in combustion systems that have both a low flue gas exit temperature and a low oxygen level, maximizing both NOx reduction mechanisms.

FGR rate can be specified in many different ways, such as % of stoichiometric air volume, % of total flue gas volume, and % of total oxidant. When comparing two FGR systems it is important to put them on the same dilution basis. When FGR is added through the combustion air blower, the best measurement of FGR rate is the burner inlet oxygen concentration. This measurement is indicative of the NOx, reducing mechanism and provides an easy checkpoint during combustion system tune-ups.

Figure 11.9. **NOx reduction in natural gas flames for varying oxidant oxygen levels.**

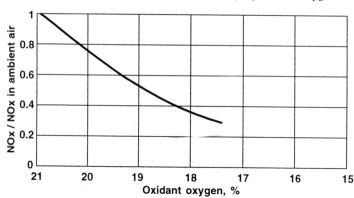

Unfortunately, the NOx reduction achieved with FGR is not without cost. Quite often the addition of an FGR system will require larger burners, inlet air piping, and combustion air blowers to accommodate the increase in oxidant volume. Operating and maintenance costs typically increase when FGR is added to a combustion system. Changes to the combustion air blower, or the use of a separate FGR fan, increase power consumption and flue gas exit temperatures. This results in a reduction of the system's thermal efficiency and increases operating costs. Water condensation in the recirculation system can be corrosive, reducing the operating life of piping, valves, and burner internals. Also, operation of the combustion system with high levels of FGR tends to reduce the stability of the burner, potentially resulting in control difficulty and increasing frequency of burner outages.

Many hidden costs of FGR can be avoided by using the largest "free" source of FGR available, the furnace itself. Many types of burners develop in-furnace recirculation zones that bring low oxygen, low* temperature gases directly into the flame envelope. These gases again reduce peak flame temperature and reduce local oxygen levels. High swirl burners, flat flame burners, and high velocity burners typically have lower NOx emissions than other flame types because of their ability to pull furnace gases into the flame.

Recently developed techniques extended this type of NOx reduction. Separated high velocity air and gas jets can reduce the combustion system NOx emissions by as much as 90%. Operation of such a system on a furnace involves heating with standard burner air and fuel flows to a safe autoignition temperature (typically above 1450 F) to satisfy flame supervision requirements. At that time, fuel flow is diverted from the burner internals to raw gas injectors located outside the burner tile, as shown in Figure 11.10. Fuel and air streams mix thoroughly with furnace gases, becoming extremely dilute before combining in front of the burner tile. The streams autoignite achieving complete combustion within the furnace environment. This technique is particularly effective in reducing emissions from preheated air burners. Careful consideration of furnace geometry, desired flame pattern, and emissions requirements is required before conversion to a dilute gas injection system, because injector placement is critical for optimum performance.

---

* Although furnace gases are not low temperature, they are significantly lower than typical flame temperatures.

Figure 11.10.   Schematic of dilute jet burner system.

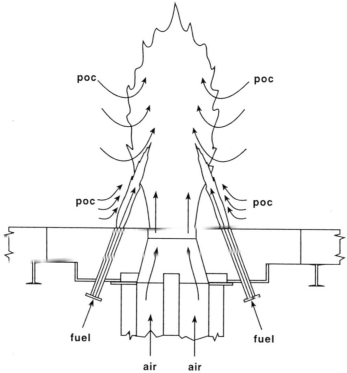

Figure 11.11.   Equilibrium NOx production *vs.* oxygen concentration and temperature.

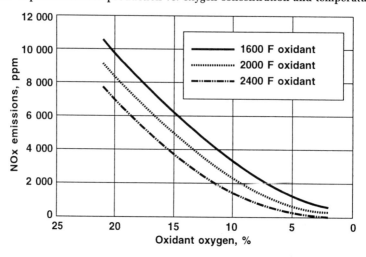

Modification of Burner Internals. Another very effective combustion modification is the addition of staging to the burner internals. The term "staged" applies to burners or systems that split one of the reactants, either fuel or air, into two streams. The first of the two streams is combined with the entirety of the second reactant and burned under design conditions in the "primary zone". The second stream is mixed with both the products of the primary zone and the furnace gases to complete combustion. See Figure 11.12.

When the air stream is split, the resulting system is referred to as *air-staged*. Air-staged systems fire fuel-rich in the primary zone. Since little oxygen is available there, very little NOx production occurs. The secondary air stream completes combustion, burning out all the rich products from the first stage. Peak flame temperatures are avoided in both stages, again limiting NOx emissions. Air-staging is one of the best techniques to reduce chemical NOx since little oxygen is present when atomic nitrogen is liberated from the fuel.

The main drawback to air-staged combustion is the complexity of the burner installation. Most systems require an extended tile or burner quarl to allow for complete reaction in the primary zone. If the secondary air is mixed with the first stage products before reaction is complete, many of the emission benefits can be lost. In many systems, the tile must contain complex refractory baffles to divert the secondary air stream; these baffles can be difficult to install or replace and can be prone to failure.

Figure 11.12.   Schematic of staged combustion systems.

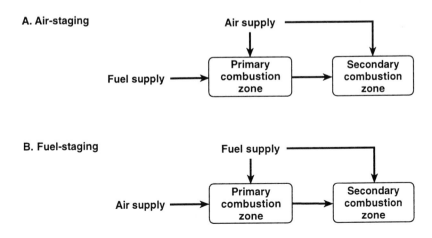

In *fuel-staging*, the fuel is split into two streams, resulting in a fuel-lean primary zone. Even though the oxygen concentration is high, fuel-lean combustion zones produce relatively low flame temperatures, greatly limiting NOx production. The combustion products from the primary zone behave as an air-FGR mix, helping to minimize NOx emissions in the second combustion zone.

Fuel-staged systems do not generally have the drawbacks associated with air-staging. The first stage uses lean combustion, where a high level of excess air is present, that occurs much faster than the rich combustion found in air-staging. This permits a tile geometry similar to conventional combustion systems. Bypass fuel passages tend to be small, as only one-twentieth of the total reactant volume must bypass the primary zone when using natural gas. The only disadvantages are secondary fuel injector integrity and the potential requirement of two distinct gas connections on each burner.

Further improvement can be made to fuel-staged systems by changing the mixing in the combustion zones. Fully premixing the reactants in the primary stage produces an extremely uniform flame with no high temperature zones (Figure 11.13). An emission analysis of a lean premixed flame may show NOx levels below 5 ppm (on a 3% oxygen basis). Mixing improvements in the secondary zone may reap benefits as well. [Reference 11.g.]

The addition of dilute gas injection techniques to a lean premix primary core can produce NOx emission levels below 15 ppm (on a 3% oxygen basis) for burners firing at 10% excess air. Such a system provides NOx emissions below nearly all regulated levels without any additional operating expense or combustion system compromises.

Figure 11.13. Schematic of a partially premixed fuel-staged burner.

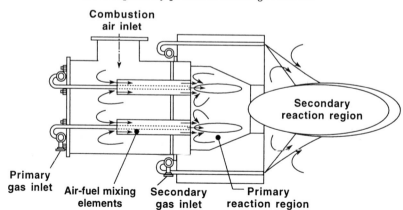

Post Combustion Cleanup. When use of the previously described techniques fail to produce an acceptable NOx emission rate, post combustion cleanup is the final alternative. The two most common techniques are selective non-catalytic reduction (SNCR) and selective catalytic reduction (SCR).

SNCR relies on the injection of ammonia, urea, or other nitrogen-containing compounds into the flue gas stream. The flue gas must be within a specified temperature window at the point of injection and injection systems must be carefully designed to fully mix the NOx reducing agent and the flue gas stream. When proper conditions are achieved the agent will break down the NOx present in the flue gas stream, converting the N back to molecular nitrogen, N_2, with water vapor as the other product. Many case studies show NOx emissions can be reduced by 50% to 80% using SNCR. In some installations, it is possible to move the temperature window with small quantities of other gases, such as hydrogen.

SCR is similar to SNCR, but uses a catalytic material such as platinum or palladium to make the reaction proceed more completely at a lower flue gas temperature. A predetermined amount of ammonia is mixed thoroughly with the flue gas upstream of the catalyst monolith. This results in as much as 95% NOx reduction in the exit gas. Catalyst life is typically 3-5 years but may be shortened by a variety of factors.

While both are extremely effective in reducing the total NOx emissions, several factors must be taken into account when weighing a decision to use either SNCR or SCR. These systems tend to have extremely high capital and maintenance costs. Good control is required to match reagent injection rate to combustion system firing rate through all turndown conditions. Reagent "slip streams" are regulated down to extremely low levels and improper injector placement or deviation in the temperature window can cause NOx emission increases. In addition to the high capital cost of catalysts, disposal questions and potential catalyst poisoning by the fired process must also be addressed. A complete economic analysis focusing on the cost of NOx reduction per pound of NOx reduced should always be undertaken before implementing any NOx reducing system on a particular furnace or application.

On-going NOx Reduction. This is not the end of NOx reduction technique development. New methods, as well as combinations of the above techniques and improved burner tuning, will provide continued reduction of NOx emissions. Much attention is being paid to individual applications for simultaneous improvement of process efficiencies and NOx emission levels. It is hoped that these will meld into a general solution to the problems raised by combustion in every industry.

The incident described in the preceding section on "Modification of Operating Conditions" shows that a burner manufacturer may be able to guarantee *burner* performance, but cannot often guarantee *furnace* performance, which must be conditioned upon very specific circumstances. Not just burners, but furnace temperature, pressure, and configuration--design, construction, maintenance, the whole combustion environment and process--affect NO formation.

CARBON MONOXIDE (CO),
UNBURNED HYDROCARBONS, VOCs

Why CO and Other Unburned Hydrocarbons (UHC) are Problems. Carbon monoxide or aldehydes may result from a rich air/fuel ratio, poor mixing, or flame quenching. Quenching or cooling can cause aldehyde formation as a result of the temperature reducing effects of too much cold excess air.

Carbon monoxide, CO, is odorless, invisible, and it can cause death by asphyxiation. Aldehydes cause eye, nose, and throat irritation, headaches, and/or nausea. They serve as a precursor of CO troubles.

Off-gases from incineration of gaseous, liquid, or solid wastes may include volatile organic compounds (voc), poly-nuclear aromatics (pna), and inorganic compounds. Any of these may be pollutants. Most result from a rich ratio of the waste being incinerated to air, although an interruption of the combustion process may also produce these pollutants. An interruption might also be caused by an excess of cold air that (a) cools the mixture* below its minimum ignition temperature, (b) dilutes the mixture below the lower limit of flammability, or (c) increases the mixture velocity above its flame speed. Obviously, too rich or too lean an air/fuel ratio may cause these forms of pollution.

The occurrence of any of these events presents a no-win situation for the operator of an incinerator that is not equipped with modern burners and controls. To prevent smoke and gaseous pollutant formation, he avoids operating on the rich side. But if he swings very far to the lean side, he may again produce gaseous pollutants because of the cooling effect of the excess air. Unless the charged waste flows very steadily and is of constant calorific value and density, there will be sudden changes in the stoichiometric air requirement, which may be difficult to track.

---

*of vaporized waste material with air. (Most liquid and solid wastes must be vaporized in an incinerator before they can be burned.)

Limiting CO and Other Unburned Hydrocarbon (UHC) Emissions. CO and UHC emissions can be limited by controlling the same three basic factors that influence NOx emissions: temperature, oxygen concentration, and residence time at elevated temperatures. Unfortunately, each of these must be controlled in the opposite direction from that of NOx reduction. If all three factors are increased (higher temperature, oxygen concentration, longer residence time), CO production can essentially be eliminated. CO oxidation can take place in the furnace atmosphere where NO concentrations are already frozen. This distinction can be very important since CO emission levels can rarely be sacrificed for reduced NOx. A low emission system must keep both pollutants to a minimum.

SULFUR DIOXIDE EMISSIONS

Why Sulfur Oxides Emissions are a Problem. Sulfur dioxide, SO_2, and sulfur trioxide, SO_3 (collectively "SOx"), will be present in the products of combustion of any fuel containing sulfur or sulfur compounds. When such fuels are burned, they are oxidized to sulfur dioxide, and to a much lesser extent, to sulfur trioxide. Both contribute to smog and to acid rain.

The final production of SOx by the combustion process cannot be altered. All sulfur bound within the fuel will be converted to SOx, unless the combustion process is somehow bypassed. Condensation of sulfur-bearing product streams can cause serious corrosion problems within the flue system of any combustion process. Applications that use heat recovery systems should avoid fuels with any appreciable sulfur content.

Table 11.14. Calculation of final sulfur content

$$SO_2 \text{ ppm}_{vd} \text{ at } 0\% \ O_2 = \frac{\left(\dfrac{ft^3 \ SO_2}{lb \ fuel}\right) \times 10^6}{\left(\dfrac{ft^3 \ dpoc}{lb \ fuel}\right)}$$

$$= \frac{\left(\dfrac{379}{32}\right) \times \dfrac{\%S}{100} \times 10^6}{\left(\dfrac{ft^3 \ dpoc}{lb \ fuel}\right)}$$

$$= \frac{(0.1184) \times \%S \times 10^6}{\left(\dfrac{ft^3 \ dpoc}{lb \ fuel}\right)}$$

SOx levels can be reduced by post combustion treatment, such as wet or dry scrubbers, limestone injection, and new membrane technologies. Large power plants that combust coal as their primary fuel frequently use two or more of these techniques to limit their SOx emission levels.

PARTICULATE EMISSIONS

Why Particulate Emissions are a Problem. Soot from flames is made up of particles of carbon and heavy hydrocarbon particulate (sometimes called "coke" and "char"). Smoke and some plumes consist of concentrations of these soot particles. They most often result from some form of incomplete combustion. They may combine with other pollutants to make smog. They are not only dirty but harmful to plant and animal life, including humans.

Causes of Soot Formation. Rich air/fuel ratio (excess fuel, reducing atmosphere) combined with high temperature is the most common cause of soot formation. Some processes, particularly metallurgical, require a reducing atmosphere to protect the required metallurgical reaction from the effects of oxidation. In such cases, an afterburner, scrubber, or other after-treatment may be necessary.

Delayed mixing* is a technique used in some burners to form soot intentionally for the purpose of increasing flame radiation--from the luminosity of the soot particles. The rich mixture (from delayed mixing, a form of staged air addition), plus the high temperature to raise the soot particles to glowing condition (above about 1400 F, 760 C), provide just the combination that creates a luminous flame. It also happens to create a long flame. The same luminosity effect is achieved with the very high air preheat that is possible with modern regenerative burners, but without the long flame--that is, with less mixing delay.

---

* Air and gas parallel, laminar, and at equal velocities (type F flame, see Figure 6.2), or air and gas parallel with one fluid at much higher velocity than the other (type G flame). A more common problem occurs when burners are turned down to such low firing rates that their fuel and air streams lack sufficient flow energy to accomplish reasonable mixing.

In all the delayed mixing situations, regardless of degree of mixing, it is important to assure incineration of the luminous soot particles so that insignificant particulate pollution results. Incineration is accomplished by providing:

- co-annular fuel and air supplies through the burner;
- temperature above the ignition temperature of the soot particles; and
- sufficient combustion chamber space (residence time) for the slow soot combustion to go to completion.

Poor mixing (incomplete mixing) is not excusable. It fails to follow through with the requirements for complete combustion. A revamp of the system is required to assure complete combustion. Overfiring of a furnace may result in pushing air and fuel through a furnace in such large quantities that they do not have time (or space) in which to mix thoroughly.

Quenched flames are an example of a situation in which adequate temperature is not maintained for enough time or until enough air is mixed with the rich mix. The resultant cooling of the flame below its minimum ignition energy results in "freezing" the soot particles; so the soot formation becomes irreversible without further heat addition, such as with an afterburner.

Two common examples of quenched flames are: those that impinge on boiler tubes, coating them with soot, and those in direct-fired air heaters where relatively cool duct air chills the flame, producing pic (products of incomplete combustion), usually aldehydes, and (rare) soot particles.

The solution for quenched flames is faster mixing. Prompt combustion provides more residence (burnout) time; and also raises the flame temperature, thus minimizing the likelihood of locking in the soot particles.

Unsaturated fuels are often prone to soot formation because of their molecular structure, which may have a higher C/H (carbon to hydrogen) ratio than that of the saturated fuel. LP (liquefied petroleum) gas from a refinery source, rather than from a natural gas source, may contain higher percentages of propylene and butylene mixed with the propane and butane, and therefore produce a sooty, smoky flame. This seemingly innocent substitution may present burner/ flame performance problems in some critical situations such as radiant tube burners, some direct-fired air heater burners, raw gas fume incineration burners, and atmosphere generator burners for special atmosphere heat treat furnaces.

Table 11.15.   General conversion factors (relating to air pollution control)

a) 100 tons per year of flue gas results from 28 900 Btu/hr input (natural gas)

b) 1 megawatt of **power out**put corresponds roughly to 8-11 million Btu/hr **fuel in**put;
   1MW = 3.414 million Btu/hr

c) 1% sulfur in heavy fuel oil yields 700 ppm SO_2 in the dry stoichiometric products
   of combustion

d) 20%   opacity =   Density #1 ~ Ringleman #1
   40%   "   =   Density #2 ~   "   #2
   60%   "   =   Density #3 ~   "   #3
   80%   "   =   Density #4 ~   "   #4
   100%   "   =   Density #5 ~   "   #5

e) 200 ppm = 0.02%
   10 000 ppm = 1%

f) 10 000 ppm$_V$ = 1.0% by volume
   1 ppm$_V$ NO_2 = 1948 µg/sm³
   1 ppm$_V$ NO   = 1272 µg/sm³
   1 ppm$_V$ CH_4 = 676 µg/sm³
   1 ppm$_V$ CO   = 1183 µg/sm³
   1 ppm$_V$ SO_2 = 2707 µg/sm³

g) 0.2 to 2.0% N in fuel may yield 60 to 2100 ppm NO

h) 1 g/ft³ (gram per cubic foot) = 35.3 × 10⁶ µg/m³
   1 µg/ft³ (microgram per cubic foot) = 35.3 µg/m³
   1 lb/ft³ (pound per cubic foot) = 16 × 10⁹ µg/m³
   1 µg/m³ = 4.37 × 10⁻⁷ grains/ft³
   1 microgram per liter = 1 milligram/m³

i) For US EPA, NOx is always calculated as NO_2 even though it may be largely NO.
   Similarly, combustibles are figured as though CH_4.

REFERENCES

11.a **US National Archives and Records Administration:** "Code of Federal Regulations 40, Protection of Environment, Parts 1-799.5055 and 1500-1517.7", Washington, DC, 1989.

11.b **Zeldovich, Ya. B.:** Acta Physecochem USSR 21,557,1946.

11.c **Glassman, I.:** "Combustion", pp. 318-382, Academic Press, 1987.

11.d **Bowman, C. T. and Seery, D. V.:** "Emissions from Continuous Combustion Systems", pg. 123, Plenum Press, New York, 1972.

11.e **Bracco, F.:** "Nitric Oxide Formation in Droplet Diffusion Flames", Fourteenth Combustion Symposium, pp. 831-842, The Combustion Institute, 1972.

11.f **Williams, A. et al:** "Prediction of NOx Emissions from Oxygen-Enriched Low NOx Burners", International Conference on Environmental Control of Combustion Processes, AFRC, 1991.

11.g **Johnson, Gregory L.:** "Premixed High Velocity Fuel Jet Low NOx Burner". Patent #5201650, April 13, 1993.

ADDITIONAL SOURCES of information relative to pollution control

American Conference of Governmental Industrial Hygienists: "Industrial Ventilation", Committee on Industrial Ventilation, Cincinnati, OH, 1990.

American Gas Association: "Natural Gas Application for Air Pollution Control", Prentice-Hall, Inc., Englewood Cliffs, NJ, 1987.

Brunner, C. R.: "Incineration Systems--Selection and Design", Van Nostrand Reinhold Company, New York, 1984.

Cheremisinoff, P.: "Waste Incineration Pocket Handbook", Pudvan Publishing Company, Northbrook, IL, 1987.

Gas Research Institute: "Analytical Models for Industrial Gas Burner Design", US National Technical Information Service, Springfield, VA, 1983.

Gill, J. H. and Quiel, J. M.: "Incineration of Hazardous, Toxic, and Mixed Wastes", North American Mfg. Co., Cleveland, OH, 44105, 1993.

Hawksley, Badzioch, and Blackett: "Measurement of Solids in Flue Gases", The Institute of Fuel, London, 1977.

Los Angeles Air Pollution Control District: "Air Pollution Engineering Manual, AP-40", 2nd ed., US Environmental Protection Agency, Washington, DC, 1973.

Niessen, W. R.: "Combustion and Incineration Processes, Applications in Environmental Engineering", Marcel Dekker Inc., New York, 1978.

Rich and Cherry: "Hazardous Waste Treatment Technologies", Pudvan Publishing Company, Northbrook, IL, 1987.

Robertson, Thomas F: "Development of a Partially Premixed Low NOx Burner" © Master's Thesis, Department of Mechanical Engineering, Case Western Reserve University, Cleveland, OH 44106, 1996.

Shell Development Company: "Afterburner Systems Study, PB-212 560", US National Technical Information Service, Springfield, VA, 1972.

Singer, Cook, Harris, Rowe, and Grumer: "Flame Characteristics causing Air Pollution Production of Oxides of Nitrogen and Carbon Monoxide", US Dept. of Interior, Bureau of Mines, Washington, DC, 1967.

US Environmental Protection Agency: "Compilation of Air Pollutant Emission Factors AP-42", 4th ed., Environmental Protection Agency, Research Triangle Park, NC, 1985.

US Environmental Protection Agency: "Glossary of Environmental Terms and Acronym List, OPA-87-017", Office of Public Affairs (A-107), Washington, DC, 1988.

Part 12.   NOISE MINIMIZATION

by Robert E. Schreter, Schreter Associates, Roswell, GA 30075

FUNDAMENTALS OF SOUND

To control noise or sound, it is important to comprehend its basic characteristics. Sound is any cyclical pressure variation in an elastic medium (gas, liquid, or solid) that is perceived and interpreted by the ear. Unlike electromagnetic waves, sound cannot travel through a vacuum. Sound may be useful or useless; wanted or unwanted.

Noise is generally considered to be unwanted sound. Another distinction is that sound carries desirable information, such as speech, music, or even the warning of an automobile horn. It is often difficult to differentiate between the two -- to decide whether a sound carries information or is just noise. As an example of both, a machine with a screeching dry bearing carries a message to the operator of an impending bearing failure; but to a neighbor, it is disturbing and he would rather not be exposed to the noise. What is "noise" to one person may be a desirable "sound," or at least an understandable message, to another.

Sound is generated whenever the molecules of an elastic medium are disturbed and are caused to vibrate in a cyclical fashion. Such a disturbance may be created by the motion of a solid object, such as the beating of a fly's wings, an oscillating piston, the vibration of a loud speaker diaphragm (Figure 12.1a); by the motion of a liquid, such as a crashing liquid wave; or by the motion of a gas, such as the sudden expansion and contraction of hot gas in an oscillating flame envelope (Figure 12.1b) or a clap of thunder. If the pressure does not fluctuate, sound will not be produced. (But the pressure fluctuations that make sounds are **very** small, relative to the ambient pressure.)

Sounds have characteristics that permit us to recognize their sources. The general characteristics of individual sounds must be completely understood in order to be able to control them. The sound of a pipe organ is quite different from a wailing siren, yet both are generated by a jet of air, and can even have identical frequency and intensity. In the organ pipe (Figure 12.1c), the air causes a column of air in the pipe to vibrate at a characteristic tone or overtone. In the siren (Figure 12.1d), the air jet is separated into individual pulses by a rotating perforated wheel.

Figure 12.1. Sounds generated by (a) a loud speaker, (b) an oscillating flame front, (c) an organ pipe, and (d) a siren.

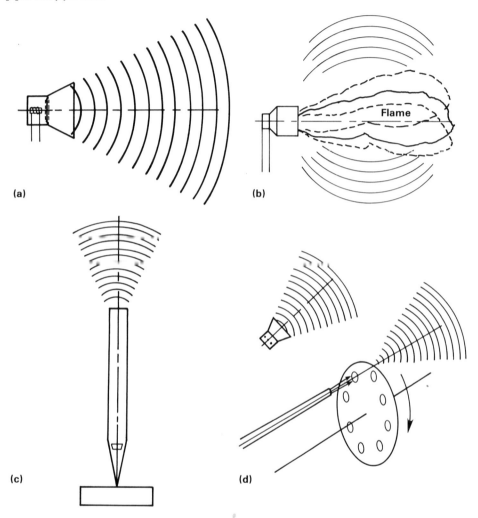

(a)

(b)

Flame

(c)

(d)

To understand and develop a mental picture of a sound wave, consider (Figure 12.2) a piston being driven by a crank running at constant speed. As the piston moves to the right, it compresses air. This is represented by the positive loop of a sine wave, or by the closely spaced lines illustrating air compression. As the piston moves to the left, it creates a low pressure, often called a rarefaction, illustrated by the negative loop of the sine wave, or by sparsely spaced lines. This sound wave moves to the right at a uniform speed, the velocity of sound, and will have a constant rms (root mean square) pressure. Once out of the tube, the wave will expand, and the pressure dissipate.

Figure 12.2. **Compression and rarefaction waves in an elastic medium (usually air) generated by a moving piston or a vibrating surface, such as a speaker diaphragm or a flame.**

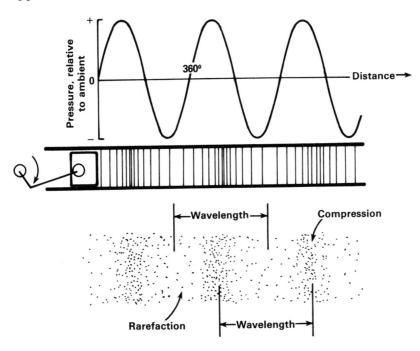

Wavelength is the distance traveled by a sound as the pressure varies through one complete cycle. See Figure 12.2.

[12/1] $\lambda = V/f$

where λ = wavelength, ft
$\quad\quad$ V = velocity of sound, ft/sec
$\quad\quad$ f = frequency, Hz

Low frequency sounds have long wavelengths; high frequency sounds have short wavelengths. Understanding wavelength is important when making sound measurements and in the design of noise suppression systems and attenuators.

Frequency of a sound is the speed with which its cyclical pressure variation occurs, dp/dt (the differential of pressure relative to time). It is simpler to recognize the characteristic frequency of a pure tone such as the pure musical "A" tone shown in Figure 12.3a, which has a frequency of 440 cycles per second, or 440 Hz. (Most standards organizations have agreed on using "hertz" or "Hz" in place of "cycles per second.") Many terms are used for frequency, but the most common in acoustics is kHz or kilohertz, meaning thousand hertz.

Figure 12.3a.   A pure sound wave, such as might be generated by a carefully-made tuning fork struck lightly on a rubber block. The A tone, in the middle of a piano keyboard, has a frequency of 440 Hz and a period of 0.00227 second. At 60 F, the speed of sound in air is about 1117 ft/sec; so the wavelength, λ, is (1117 ft/sec)/(440 cycles/sec) = 2.538 ft/cycle, or 0.7742 metres.

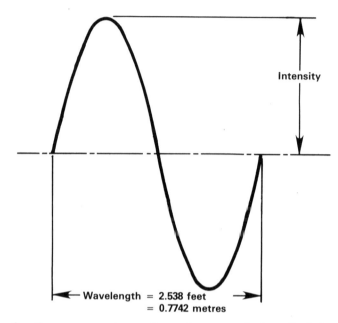

Intensity

Wavelength = 2.538 feet
= 0.7742 metres

The time for the pressure to go through one cycle is called the **period**, or **periodic time**, which is the reciprocal of the frequency.

[21/2] $T = 1/f$

where T = periodic time, seconds
     f = frequency, Hz (cycles per second)

The frequency is often the result of a complex wave or waves made up of a composite of different waves. An oscilloscope depiction of such a sound might look like Figure 12.3b.

Low-frequency sound is classified as having a wavelength longer than 5 feet, or a frequency below 240 Hz. It radiates in all directions; travels around barriers, partitions, and corners; and easily penetrates holes, continuing to travel freely in all directions. Low-frequency sound is difficult to attenuate. Human hearing is less sensitive to low-frequency noise and can tolerate much greater intensity without ear damage. At midlife, humans can hear sounds down to about 15 Hz, although they can perceive, through feeling, music and low frequency vibrations with frequencies as low as 1 or 2 Hz.

Figure 12.3b.   Real sound waves are complex composites of a variety of frequencies and intensities.

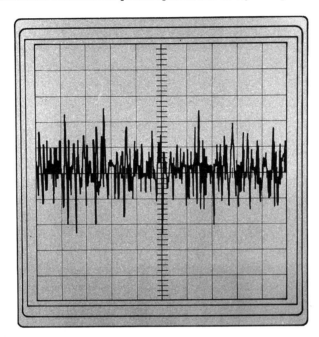

Medium-frequency sounds, "mid-frequency," are those having a wavelength from 5 feet to 1 foot, or frequencies from 240 Hz to 1200 Hz. They are more directional than low-frequencies, yet they disburse more than high frequencies. Mid-frequency sounds are relatively easy to attenuate, using absorptive techniques. Because they are within the normal range of speech sounds, medium-frequency noises are extremely important considerations in noise control. The ear is most susceptible to permanent damage at the upper part of this range.

High-frequency sounds are those with wavelengths of less than one foot and frequencies of 1200 Hz to 20 000 Hz. These sounds travel in straight lines and can be reflected much like light. They do not easily go around corners nor change direction. High frequency sound is relatively easy to attenuate, especially through air.

High frequency sounds are included in the upper part of speech frequencies, so loss in these areas can be significant to good hearing. Humans in midlife rarely hear sounds above about 15 000 Hz.

Sound pressure (intensity) is measured in microbars. One microbar (or μ-bar) = one millionth of the normal barometric pressure. One microbar equals 1.45 $\times$ 10^{-5} psi, or 1.02 $\times$ 10^{-2} micropascals.

Sound pressure level, L_p, is measured by most sound and noise instruments in **decibels**, which are ratios, and are always relative to some base pressure or power, using a logarithmic function to reduce an otherwise burdensome long scale. The current accepted base pressure for L_p is the threshold of human hearing at 1000 Hz, which is 20 micronewtons per square metre (20 μ-N/m², or 20 micropascals (20 μ-Pa), or 0.0002 microbars (20 μ-bars). This is discussed in more detail two sections later, under "The Physics and Math of Sound."

Velocity of sound (or speed of sound) varies with the medium through which the sound travels, and depends on the temperature and density of the medium (unlike particle velocity, which varies with frequency). The speed of sound in standard air (60 F, 14.696 psia) is 1130 ft/sec, or 344.4 m/s. However, sound velocity may be considerably different from this when dealing with combustion, where the composition, temperature, and density of the gases are significantly different from those of standard air, and change rapidly.

EFFECTS OF NOISE ON HUMANS

Physiological and psychological aspects. Sound has an enormous effect on our lives -- our moods, our attitudes, our health, our job performance. When we lose our hearing, it has a profound effect on every aspect of our lives.

The ear is the sensory organ that permits us to detect the air pressure variations called sound. The ear is the most sensitive organ in the human body. It has the ability to detect microscopic pressure changes. On the low end of the hearing range, the threshold of sound for humans is 20 micropascals, which is equal to a sound pressure of 2.9×10^{-9} psi, which is far less than that created by a fly landing on a sheet of paper.

The highest sound that the human ear can hear is unknown, primarily because it would result in permanent ear damage, but the maximum level is assusmed to be about 150 dB. The human ear's resolution of range of sensitivity is in excess of 31 600 000 to 1 on a pressure basis. Any organ that sensitive can be damaged easily without proper care and protection. The human ear has the ability to detect an enormous range of sound intensities, and at the same time, it can also discern a wide range of frequencies...from as low as 10 Hz to greater than 16 000 Hz.

The human ear, coupled with the brain, has the amazing ability to discern tiny changes in sound patterns that permit picking out one voice or sound from thousands of similar sounds.

Burner and furnace operating personnel may incur selective hearing damage from high sound intensity in the range from 100 to about 400 Hz, which is quite broad and normally affects speech recognition. The hearing loss is permanent. Even hearing aids do not alleviate the condition.

Most hearing damage is the result of repeated exposure to:
    a) loud noise,
    b) intense pure tone sound,
    c) long exposure duration at intensities above 85 dBA,
    d) loud impulse noise, such as from gunfire, and/or
    e) loud repetitious noises in narrow frequency bands.

The ear can tolerate more intense low frequency noise. It is most susceptible to damage in the range above 3500 Hz.

Noise afflictions. Hearing loss is common in most industrialized countries. In the United States of America, one out of five people has sustained a significant hearing loss. One out of ten people are deaf or have a severe hearing loss. The incidence of occupational hearing loss started to decline in the 1960s and 1970s, when OSHA, MSHA, and EPA were having an impact on industrial noise exposure. The relaxation of occupational noise standards in the 1980s has resulted in increased hearing loss.

The physical effects of intense sound or noise are not to be ignored. Exposure to intense sound fields tends to cause increased blood pressure, headaches due to dilation of blood vessels in the brain, dilation of the pupils of the eyes, increased secretion of stomach acid, confusion, and the inability to think. It has been known for years that watch makers or persons doing precise, delicate work, cannot function in intense noise. People report problems relating to noise ranging from inability to focus their eyes to difficulty concentrating on complex tasks.

The problems associated with not being able to communicate or hear also lead to physical problems and danger. Failure to hear warning signals, machines running, alarm horns, or vocal warnings by co-workers, all too frequently result in accidents, and even loss of life.

Ironically, the psychological effects of deafness are far worse than the physiological ones. As far back as recorded history, there has been a stigma attached to being deaf.

The newly deaf person, or one who becomes hard of hearing, experiences profound changes in his life, which frequently result in loss of friends and social contact. Because he has difficulty hearing, he frequently misunderstands or misinterprets what people are saying; so he tends to withdraw, become suspicious, and associate only with those having the same disability (unless he is able to correct his hearing disability). It is said that the deaf suffer even greater impairment in life than do the blind.

Community response to noise. How do our neighbors and the community in general respond to noise? What are the factors that cause the community to react negatively toward noise and to those responsible for its production? The answers to these important questions will show us exactly what must be done to avoid a negative community or employee response to our projects. It is well known that paying attention to a noise will often satisfy the person complaining, but when it is allowed to go unattended, a complaint to a public agency is far more difficult to correct.

Some of the characteristic noise factors that play an important role in the way the community responds to noise are:

1. Discrete frequencies and pure tones stand out from background noise. These tones may even be of a lower intensity than the accepted background, yet they may be the primary cause of complaints.

2. High intensity noises.

3. Warbling noises -- fluctuating intensity. Beating noise with cyclical intensity variations.

4. Sudden noises, such as blowoff of a high pressure relief valve.

5. Noise that would normally be acceptable in a metropolitan area will be offensive in a secluded rural area where the ambient noise is low.

6. Low frequency noises or vibrations that cause resonant effects such as rattling windows in local structures, or that communicate feeling to an observer.

7. Noise of a frequency distribution that interferes with verbal communication.

8. Visual effects that convey to the viewer the idea that a plant or piece of equipment is noisy, even though the noise level is at an acceptable level.

9. Other negative characteristics of the plant or equipment, such as appearance, odor, or dust, which call attention to the equipment and to the noise, will frequently result in a noise complaint even though they may not be truly responsible for a noise problem.

Liability for noise reduction. The U.S. Occupational Safety and Health Administration (**OSHA**) is not concerned with ambient noise, but with the exposure of workers to that noise. Their present standard limits an employee's noise exposure to 90 dB as an 8-hour time-weighted average (TWA). Employee exposure to noise above the permissible exposure limit (PEL) must be reduced by feasible engineering controls or administrative controls. Where such controls cannot reduce employee exposure to within permissible limits, employees are to be furnished with personal protection equipment.

The OSHA noise standard also requires that the employer administer a continuing effective hearing conservation program if exposure exceeds the PEL (85 dB). OSHA has interpreted this to require employers to provide audiometric testing for those employees exposed above the PEL without regard to the use of personal protective equipment.

The U.S. **Bureau of Mines** Mine Safety and Health Administration has powers similar to OSHA. Although their activities are generally limited to mining operations, they may, in some cases involve themselves in operations not directly connected with mining. Details of current regulations should be checked with local OSHA or Bureau of Mines offices to ensure compliance with the latest rulings.

The U.S. **EPA** is concerned with noise transcending property lines; not with worker exposure to noise. The EPA has been active in writing model legislation, and encourages the individual states to adopt, adapt, or write their own regulations. Many states, cities, and towns have taken up the challenge, writing their own versions of noise legislation. Some states have relegated the responsibility for writing noise legislation to municipalities.

Municipal ordinances cover a wide variety of environmental noise topics. They vary from very comprehensive to copy-cat or one-up legislation. Some are so confining that they are not enforceable. Such legislation is so widely varied that it is impossible to summarize. Local noise ordinances should be consulted on a job-by-job basis to avoid conflict with ever-changing legislation.

THE PHYSICS AND MATH OF SOUND

Sound pressure. Units commonly used in making sound measurements are not generally encountered in engineering; so require some explanation. To establish a perspective from which to study sound measurements, a range of representative values is shown in Table 12.4.

Table 12.4. Typical sound pressure levels, L_p, or SPL
(2.9 – E9 = 2.9 × 10^{-9} = 0.000 000 0029 psi)

Source	microbars	psi	micropascals	L_p, dBA
Reference level	0.0002	2.9 – E9	20	0
Sound studio	0.002	2.9 – E8	200	20
Quiet office	0.02	2.9 – E7	2 000	40
Conversation, 3 feet	0.2	2.9 – E6	20 000	60
Noisy restaurant	2	2.9 – E5	200 000	80
Printing press	20	2.9 – E4	2 000 000	100
50 hp siren, 100 feet	200	2.9 – E3	20 000 000	120
Jet plane	2000	2.9 – E2	200 000 000	140

Table 12.4 shows a huge variation in sound pressure -- ten million to one. Few, if any, scientific instruments are capable of linearly measuring seven orders of magnitude, and it is difficult for the human mind to perceive such a broad span. To obtain a manageable range of figures, acoustic engineers take the base 10 logarithm of the ratio of the actual sound pressure to some arbitrary standard pressure. The formula for performing this mathematical manipulation is

[12/3a] $dB = \log_{10}(A/B)$

where dB   = decibels
    A, B = sound pressures (actual, base) with consistent units

A modified form of the dB relationship is used to represent sound power as well as sound pressure.

Sound Pressure Level, L_p, or SPL, which is what will actually be measured by a microphone at a specified distance, r, from the sound source, may be expressed in decibels, dB. The equation relating sound pressure in micropascals to dB is

[12/3b] $L_p = 20[\log_{10}(p/p_o)] = 20[\log_{10}(p/20)]$

where L_p = SPL = sound pressure level, dB
    p   = actual sound pressure in micropascals.
    p_o = reference sound pressure = 20 micropascals

The notations for sound readings have changed over the years. Because both old and new forms appear in current literature, it is well to be aware of both, as shown in Table 12.5.

Table 12.5. Notations for sound terms

Term	Old form	New form
Sound Pressure, dB	SPL	L_p
Sound Power, dB	PWL	L_w
Equivalent Sound Pressure, dB		L_{eq}
Day-Night Sound Pressure, dB		L_{dn}

Sound power, L_w or PWL, is the total acoustic energy rate radiating from a point source of origin of a sound, in watts. It may be thought of as the source energy required to produce a sound.

It is helpful to think of sound pressure level as the pressure, L_p, measured at a distance "D" from a source of L_w watts power. Sound pressure is what is actually measured by a microphone and sound level meter. See Tables 12.6a, 12.6b, and 12.9.

Table 12.6a. Combining decibels from multiple sound sources

dB difference between two source levels being added	dB increment to be added to the higher level	dB difference between two source levels being added	dB increment to be added to the higher level
0	3	7	0.79
1	2.57	8	0.62
2	2.15	9	0.51
3	1.76	10	0.42
4	1.44	11	0.35
5	1.17	13	0.23
6	0.96	15	0.13

Ambient noise is treated as another source, and should be added to the equipment sound level. If the difference between the ambient level and the equipment level is zero, then the total noise level is 3 dB higher than either.

Table 12.6b. Ambient noise correction for sound level measurements. For a given dB difference, subtract the dB error from the total noise.

dB difference	dB error	dB difference	dB error
0.6	10	6	1.25
1	6.75	7	0.93
2	4.25	8	0.72
3	2.95	9	0.55
4	2.17	10	0.46
5	1.65		

When the total sound at a point has been measured, and it is desirable to isolate the sound level for specific equipment, this "dB error" is to be subtracted from the measured total noise.

When the sound source is nondirectional and there are no reflections or disturbances, the sound will propagate in a spherical pattern. All the energy is radiated outward, so the intensity at any distance can be calculated by dividing the watts of energy by the total surface of the sphere enclosing the sound source.

[12/4] $I = W/S$

where I = intensity, in watts/ft²
       W = sound power, in watts
       S = surface area of the sphere, in ft²

The intensity varies inversely with the square of the distance from the source

[12/5] $I_2 = I_1 (r_1/r_2)$

where r_1 = original distance from sound source
       r_2 = new distance from sound source

To bring this into more manageable numbers, it is expressed as a logarithmic ratio. The **sound power level**, or L_W, or PWL (also expressed in decibels, as referenced to 10^{-12} watts) is

[12/6] $L_W = PWL = 10 \{\log_{10} [W/(1 \times 10^{-12})]\}$

dBA = "A"-weighted network factor. The human ear can detect frequencies from about 10 to 16 000 Hz. Unfortunately, it is not equally sensitive to all frequencies; so it gives a distorted impression of actual sound intensity. To correct this misimpression, acoustics engineers developed electronic networks that enable representation of sounds as humans hear them.

Figure 12.8 depicts the weighting characteristic, which represents human perception of sounds at low loudness by the "A" network (or scale); at medium loudness by the "B" network; and at high loudness by the "C" network. Distortion is greatest in the lower frequencies, especially at low and medium loudness.

Sound level meters have been designed to take into account these idiosyncrasies, so as to give an indication of how the human ear perceived sound.

Figure 12.8.   Response characteristics of standard frequency weighting scales for sound load meters and of the human ear at threshold.

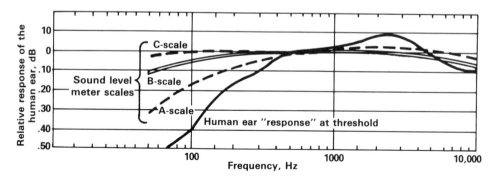

Tests have indicated that the human ear is less likely to be permanently damaged by low frequency high intensities than by high frequency high intensities. This damage risk is quite similar to the "A" network. Figure 12.8 shows that the ear is less likely to sustain damage from a sound level 30 dB higher at 50 Hz than at 1000 Hz. Because of the similarity of the "A" network and the damage risk criterion, agencies such as OSHA and NIOSH use the "A" network for most sound testing.

SOUND PROPAGATION, AND SOUND MEASUREMENT

These topics are too complex for thorough coverage in this handbook. Readers are referred to vendors' literature.

High frequency sound travels in straight lines. It can be reflected much like light and does not easily change direction, nor go around corners. It is relatively easy to attenuate, and is attenuated by air. High frequency sound is the principal offender in hearing loss.

Low frequency sound travels in all directions -- around barriers, partitions, and corners. It easily penetrates holes, continuing to travel in all directions. Low frequency sound is difficult to attenuate. Human hearing is less sensitive to low frequency noise and can tolerate much more intensity without ear damage.

NOISE CONTROL

After the noise source is identified and after its intensity at various wavelengths is determined, the remaining task is to reduce the noise intensity to acceptable limits. Numerous methods and combinations of methods can be used. There are five basic methods for controlling noise. It can be: MOVED, ABSORBED, BLOCKED, REDUCED, or SHORTENED IN EXPOSURE TIME.

Determination of the best method(s) required weighing one method against the others by comparing their effectiveness and the ratio of attenuation to cost (in dBA/$).

Moving the noise source. The most obvious but most frequently overlooked solution is to eliminate the noise source, or to move it. Table 12.8 shows how the sound pressure level can be diminished by increasing the distance between the source and the receiver.

Table 12.9. Reduction in dB with increasing distance (based on hemispherical radiation)

Distance, feet	10	20	40	80	160	320	640	1280	2560
Distance, metres	3	6.1	12.2	24.4	48.8	97.5	195	390	780
Reduction, dB	17.5	23.5	29.5	35.6	41.6	47.6	53.6	59.6	65.7

Sound pressure level drops by 6 dB each time the distance is doubled, as shown in Table 12.9. This provides a convenient way to estimate the effect of moving the noise source and receiver farther apart.

The same goal can be accomplished by one or more of the following:

a) eliminating the most intense noise source,
b) moving the noise source to another area,
c) increasing the distance from receiving location to source,
d) relocating the receiver, or
e) installing walls or partitions that increase the effective distance between source and receiver.

These methods often prove more economical than use of attenuators.

Absorbing noise. Materials such as fiber glass, mineral wool, or cork felt are good sound absorbers -- effectively reduce sound passing through them or passing over their surfaces. Much of the incident sound energy is converted to heat energy. Only a small percentage is reflected.

Sound barriers, reflectors, and enclosures must be designed for the frequencies to be controlled. A barrier that resonates in sympathy with the sound source will act as a loudspeaker, retransmitting the sound energy that was to be blocked. Similarly, reflector walls must effectively reduce the transmission of sound through the reflector.

When sound passes through walls or barriers, its sound pressure level is reduced. This type of sound reduction, "transmission loss," specified in dB, is generally proportional to the barrier weight and density.

Porous silencers utilize the absorption of mid-frequency and high frequency sound waves as they pass through permeable materials.

Resonators use a time delay to cause phase reversal and cancellation of all or part of the original wave. A resonator acts like a capacitor, storing pressure, then releasing it 180 degrees out of phase with the original wave, thus causing its cancellation. An application of this idea is the 1/4 wave tube, shown in Figures 12.10a and 12.10b.

As the peak pressure passes the mouth of a **quarter wave tube**, the pressure pulse travels 1/4 wave down the tube and reflects up the tube (another 1/4 wave length), emerging from the mouth of the tube 180 degrees out of phase with the original wave; so the resultant sound pressure is largely negated. Resonators and reflectors use this phase displacement principle.

Figure 12.10a.   Quarter wave tube ("Spook tube") applied to a radiant tube fired with a compressed air and propane burner, wherein an oscillating flame front may drive the tube into violent oscillation at a discrete frequency unique for an organ pipe of that configuration and length.

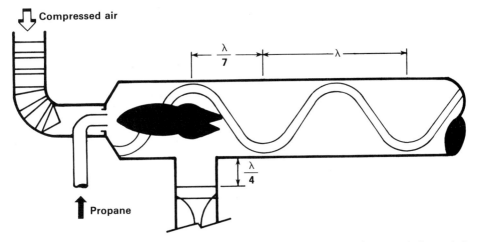

Figure 12.10b.   Wave cancellation in a quarter wave tube, showing (from top to bottom) the original wave, the reflected wave, and the resultant wave.

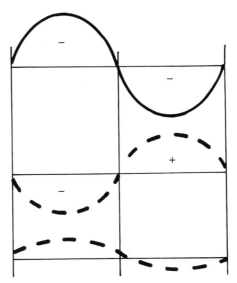

A volumetric device such as a **Helmholtz resonator**, or an "apple jug," can be used to replace the 1/4 wave tube with the same effect. See Figure 12.11. Phase reversal may be accomplished by varying the impedance of the restricting neck, or by changing the volume of the jug. The resonator can be tuned on the job by simply varying the chamber volume.

Figure 12.11. **Helmholtz resonator** can be tuned by filling the jug with water until anti-resonance occurs, or by replacing the bottom of the jug with a movable piston.

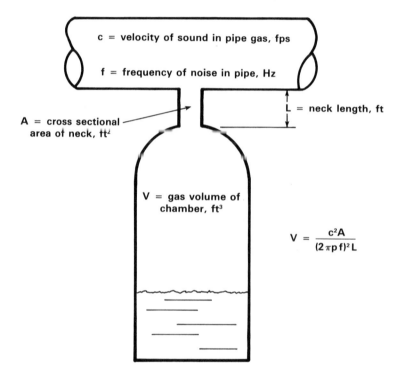

c = velocity of sound in pipe gas, fps

f = frequency of noise in pipe, Hz

L = neck length, ft

A = cross sectional area of neck, ft²

V = gas volume of chamber, ft³

$$V = \frac{c^2 A}{(2\pi p f)^2 L}$$

Helmholtz resonators can result in maximum reduction over a narrow range of frequencies, but they can be arranged to overlap so as to attenuate over a wider range.

Reflectors, somewhat similar to Quincke tube reactive mufflers, Reference 12.a, achieve attenuation by phase reversal. Figure 12.12 illustrates an arrangement for silencing a duct burner, where the duct oscillates as an open ended pipe. The reflected wave travels 1.5 wavelengths; so it is 180 degrees out of phase with the original wave, thereby canceling, or at least attenuating, the original wave.

Figure 12.12.   Reflecting tube silencing a duct burner

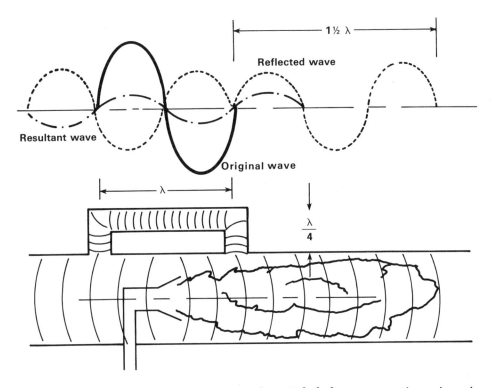

Detuning is the best way to silence or reduce Helmholtz or organ pipe noise...in much the same way that ac electrical resonant circuits are detuned. Resonance occurs when the inductive and capacitive reactances are equal and resistance is negligible. The impedance of the circuit approaches zero; so the current approaches infinity.

[12/7]   current = electromotive force/impedance

[12/8]   impedance = [(resistance)² + (inductive reactance − capacitive reactance)²]⁰·⁵

To reduce the current, make a change that will give the impedance a finite value, which will reduce the current. The easiest way to do this is to add resistance to the circuit. Similarly, to reduce the sound intensity of a tuned resonator, add impedance to the acoustic circuit. *For example, a gas-fired radiant tube oscillating at its natural frequency* can be detuned by simply adding a restricting orifice (impedance) at either end of the tube or pipe. These same theories can be applied to long air piping that tends to oscillate.

Reactive silencers, Figure 12.13, are among the most commonly used in air flow systems. They are highly effective, relatively inexpensive, and readily available in many sizes and shapes. They can be designed to filter out discrete frequencies as well as broad band noise. They are most effective in the medium to high frequency ranges. Although some are designed for low frequencies, they are quite large and heavy.

The inner part of a reactive silencer consists of walls of perforated metal backed by mineral wool or fiber glass insulation. Actual silencing occurs by absorption in the mineral wool and by reactive silencing. The volume behind the perforated plates acts like the volume in a Helmholtz resonator, and the perforations act like the throat of the bottle. Silencers of this type can contain multiple tuned sections. Tuning is generally done by varying the volume of the cavity and the diameter of the holes in the perforated plate.

Figure 12.13. **Reactive silencers** reduce noise by absorbent filler material and by many miniature Helmholtz-like resonators.

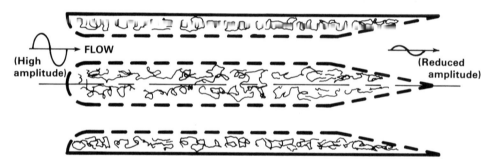

Expansion chambers, Figure 12.14, make the sound pressure wave expand, thereby reducing its intensity. The muffler shown also acts as a reflective silencer.

Figure 12.14. **Expansion chambers** (mufflers) allow sound pressure waves to expand and dissipate their energy.

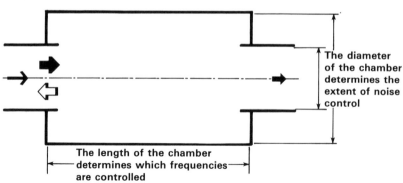

Blocking the noise source. This method, Figure 12.15 encompasses anything that would impede or eliminate the passage of noise from the source to the receiver...a source enclosure, a wall, or a receiver enclosure.

Figure 12.15. **Blocking sound** can be accomplished by a source enclosure, a barrier wall, or a receiver enclosure.

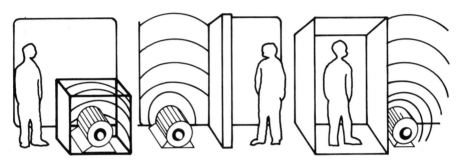

If an enclosure cannot be sealed, as with a motor or compressor that requires cooling air or process air, sound absorbing attenuators should be applied to all intake and discharge pipes and ducts. See Figure 12.16. Flexible connections should be used in pipes, ducts, and conduits passing through source and receiver enclosure walls to prevent transmission of vibration to the enclosure. Otherwise, the enclosure may act as a sounding board, actually amplifying the noise. Any noise source must be isolated from the floor and structural supports to prevent transmission through the floor and supports, thus bypassing the sound-containing effectiveness of the enclosure.

Figure 12.16. **Flexible connections** are necessary to isolate any sound transmitted by pipes, ducts, or conduits passing through sound enclosures. Ventilation openings without connecting ducts should be fitted with stub ducts to reduce direct sound radiation through the openings.

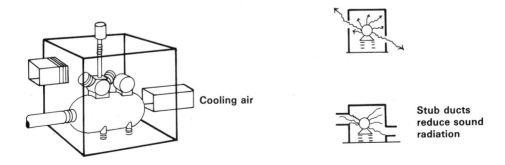

Cooling air

Stub ducts reduce sound radiation

Enclosures should be lined with absorptive material. Transmission loss through the wall should be maximized. Openings or leaks in the enclosure defeat the purpose of an otherwise well-designed enclosure.

The use of earplugs, or better yet, earphones, to block sound from workers' ears is a variation on the receiver enclosure principle. This is a less desirable solution, but it is acceptable when other solutions have been exhausted.

Reducing the noise source is the preferred method for abating noise. Some methods of reducing noise at its source include:

1. Substituting low-noise-level motors or machines for standard motors or machines.

2. Balancing a machine to eliminate vibration, or installing vibration mounts to prevent transmitting vibration into large surfaces that act as sounding boards.

3. Redesigning air nozzles and jets to include a boundary layer (Figure 12.17), which will greatly reduce shear effects and turbulence with ambient air, thereby reducing turbulence-generated noise.

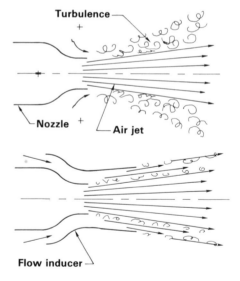

Figure 12.17a. **A high velocity jet shooting into slower moving surrounding gases can create viscous shear and turbulence.**

Figure 12.17b. **A flow inducer reduces shear velocity and turbulence.**

4. Introducing various mutes and silencers at the noise source can reduce the transmitted noise.

5. Changing the size or length of resonating cavities and pipes to eliminate or reduce resonant noises.

6. Separating large, high speed jets into a number of smaller jets can significantly reduce gas jet noise.

7. Breaking large, high speed jets into a number of smaller jets to shift the generated frequencies to higher frequencies, which are easier to attenuate.

8. Adding damping material and web plates to large surfaces to increase resonant frequency, making it much easier to attenuate the remaining sound.

9. Reducing the size or reshaping rotating cavities can reduce siren-induced types of noise.

10. Equipping large cylindrical surfaces, like chimneys, classically associated with air-flow-generated moaning sounds, with an uneven spiral or other eddy generators.

Whatever the method, every effort should be expended to stop the noise at the source; or at least to shift its frequency to a range where it can be attenuated.

Reducing exposure time. In severe cases where it is impractical or ineffective to use the methods previously described, reduction of exposure time can prove helpful. This is a viable option as long as managers enforce the regulations and limit exposure to the prescribed levels and times.

In the USA, OSHA permits exposure to sound levels in excess of 90 dBA, but the time must be reduced in accordance with Table 12.18.

Table 12.18. USA OSHA-allowed exposure of persons

L_p = SPL, dBA at the person's ear	Time duration, in hours
80	32
85	16
90	8
95	4
100	2
105	1
110	0.5
115	0.25

Equation 12/9 is the source for Table 12.18, and can be used to calculate exposure times at other sound pressure levels.

[12/9] Time, hours $= 8/2^{(SPL - 90)/5}$

Alternating personnel is another approach to reducing exposure time. Each operator works in the higher intensity area for a portion of his work shift.

NOISE SOURCES IN COMBUSTION SYSTEMS

Fan or blower noise can be quite significant in itself, or it can drive other acoustic systems such as chambers, cavities, ducts, or pipes to high noise levels. The varieties of fans and blowers on the market make it difficult to accurately estimate noise levels. Each has its own characteristic noise, and they range from crude home-made fans to sophisticated high efficiency machines.

Sound power levels for fans are normally stated at full load. This level varies considerably depending on the machine's efficiency. For most fans, the efficiency curve shows wide variation from the rating point. Because sound power level depends on efficiency, one would expect the level to depend on the operating point as well as the rating point. One can assume that the sound power level will remain reasonably constant within 10 to 15% of the fan's full load rating because efficiency varies little in that range. Below that, sound power levels can be expected to increase.

When the manufacturer's data is not available, sound power level can be estimated by:

a) using sound data for a machine of similar design produced by the same manufacturer,

b) a mathematical method developed by J. Barrie Graham, consulting engineer (Reference 12.a), or

c) using homologous fan information, compensating for different operating parameters.

Method b is detailed, with examples, in References 12.c and 12.f. Method c is detailed, with examples in Reference 12.f.

Fan noise is directional. The following individual components contribute to total fan noise:

a) intake noise   d) shaft seal leakage noise

b) outlet noise   e) motor or engine noise

c) casing noise   f) vibration noise (balance, mounting).

If noise from other sources is transmitted through inlet and outlet ducts, it may contribute to the noise radiated by the fan or blower housing.

Open fan intake and exhaust noise levels are equal for all practical purposes. The individual sound power level of the intake or discharge will be about 3 dB less than the sound power level of the machine.

[12/10] L_w, open intake or outlet = $L_{w, total}$ − 3 dB

Radiated fan casing noise is generally low compared to levels emitted by the inlet or outlet. The casing noise may become a significant consideration, however, if the inlet and outlet are ducted or provided with silencers. If the noise transmitted through the fan casing is less than 10 dB below that of the open intake and discharge, it will have no effect on the overall sound level. In the absence of actual casing radiation readings or data, the casing radiation may be estimated as the overall sound power level at the inlet or outlet minus the correction dB from Table 12.19.

[12/11] L_w, radiated = $L_{w, fan}$ − Correction$_{from Table 12.19}$

Table 12.19. Correction for casing radiation L_w or PWL (from Reference 12.c)

Casing thickness, gauge	14	12	10	8				
Casing thickness, inches					1/4	3/8	1/2	3/4
Correction, − dB	16	18	19	20	22	25	27	30

Example 12-1. The total sound power of a blower fabricated with 10 gauge steel is 124 dB. The inlet and outlet are connected to sound-insulated ducts. Using Table 12.19, the sound power radiated by the casing will be 124 − 19 = 105 dB.

Piping system noise. *Resonance in fans, pipes, ducts.* Acoustic surge, pulsing, puffing, and pumping are terms frequently used to describe a phenomenon that occurs in air systems. The rhythmic puffing generally happens in large air systems with long air conduits.

The magnitude of the surge may consist of barely perceptible pulsing that creates little more than an annoyance. It may attain levels of such intensity that it is difficult for workers to safely or legally remain in the area. Surging can reach an intensity severe enough to flex the fan casing violently, causing fatigue failure. The pressure pulsations can cause flexing and additional stress on the wheel and blades, resulting in damage and failure.

In combustion applications, the pressure excursions caused by the surging, can create burner instability that results in flame-holding problems. The phenomenon is dangerous and should be avoided.

The basic causes of surge in air systems are: a) blower or fan instability, and b) duct, pipe, or cavity resonance.

Fan instability is a function of the pressure/volume characteristic of the fan, and it is usually not a problem if operation is confined to the portion of the curve where pressure rises with decreased flow, points A to B in Figure 12.20a. However, instability becomes inherent when flow is further reduced to a point of operation on the back side of the curve, points C to D in Figure 12.20b, where pressure decreases with decreased flow. The problem is generally magnified when the ducts are long and large in diameter.

Figures 12.20a and 12.20b show how pulsations happen in blowers as they are turned down to lower flows. As the operating point moves from A to B to C to D, the downstream pressure in the duct is suddenly greater than the pressure at the blower outlet, causing a backflow. Then the operating condition jumps back to B or A. Pulsation is oscillation between the high flow and low flow sides of the curve's peak.

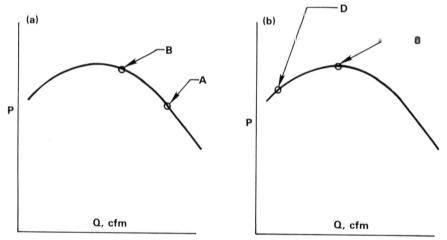

The basic difficulty is not acoustic in nature, but one related to proper fan selection or design. For details on fan design, see texts on that subject.

Pipe resonance, sometimes termed an "organ pipe effect," was discussed briefly in the previous section under "Reflectors" in connection with Figure 12.12. More detail on this subject will be covered in Reference 12.f.

Simple textbook-type pipes are rarely encountered in actual practice. The more complicated systems consist of many parallel or series branches with takeoffs at many places. These complex systems can be handled much as one does the computations in complex ac electrical circuits, by merely breaking the system into a number of simple series and parallel circuits.

Oscillation occurs when a susceptible pipe or cavity is excited by a periodic disturbance or driving force such as the pulses of fan blades passing the fan housing cutoff, the motion of an unstable air or gas jet, the fluttering of a valve or even a loose inlet guard on a blower. The magnitude of the sound or oscillation can be quite small, or it can reach levels capable of creating intense noise problems, severe pressure pulsations, or even structural damage. In severe cases with long ducts, standing waves can cause varying pressure along the length of the duct. This, in turn, can result in non-uniform burner capacities, depending on the positions of the burners relative to the standing waves.

Pipe systems obey the same physical laws as musical wind instruments, and resonance can be predicted based on those relationships. Several factors can affect the frequency and intensity of oscillations.

The simplest way to detune a resonating burner piping system is to increase its impedance, by adding some resistance (an orifice or sliding plate) across the exit end. Computations of the resistance size are difficult, but simple trial and error provides a quick, practical solution. The pressure loss across the outlet of a combustion system can be estimated by formulas [12/12a] and [12/12b].

[12/12a] Pressure loss, $"wc = \left(\dfrac{acfh}{1658.5 \ K \ a} \right)^2 \times G$

where   a = open area of the orifice, in square inches
acfh = actual flow of all products of combustion, including excess air, at the exit gas temperature
G = the gas gravity of the exit flue gases at temperature t, relative to air = 520/(460 + t)
K = flow coefficient of the added resistance (orifice)

A convenient way to figure the acfh of the poc (products of combustion) when using natural gas is

[12/12b] $acfh = \dfrac{Btu/hr}{1000 \ Btu/cf \ gas} \times [10 \ (1 + \%XSAir/100)]$

$acfh = 0.010 \times (Btu/hr) \times [1 + (\%XSAir/100)]$

where   Btu/hr = actual gross heat release of the burner(s) with exit resistance in place

%XSAir = percent excess air

Example 12-2.   A 3″ ID tube is fired with 1000 Btu/cf natural gas at 100% excess air. The exit gas temperature is 410 F. By trial and error, the resonance was found to be minimized when the exit was restricted to 3.60 square inches of open area (2.14″ diameter orifice, assume K = 0.7). If the burner input rate at this new condition is 164 000 gross Btu/hr, what will be the pressure drop across the exit restriction?

By formula [12/12b], acfh = 0.010 × (164 000) × (1 + 100/100) = 3280 acfh.

By formula [12/12a], ″wc = [3280/1658.5 (0.7) (3.60)]² × [520/(410 + 460)] = 0.37″wc.

This exit restriction may have actually reduced the flow capacity of the system, so care must be taken to use an acfh in formula [12/12a] that actually represents the new restricted flow.

Immersion and radiant tubes present unique problems when a tube becomes resonant. Because of its length, the tube can easily fall into low-frequency oscillation. The heat at each oscillation drives the excursion, increasing its intensity. The resulting pressure antinode, which occurs at the burner, can become so high that it exceeds the gas pressure at the burner, momentarily shutting off the supply of gas, and reducing or extinguishing the flame.

As the reflected wave subsides, gas flow is re-established. Because of the valve inertia in the gas regulator, that flow may exceed the normal rate. The air-gas mixture then re-ignites (from a continuous ignition system or the hot tube), creating an explosion. Then the cycle repeats. The tube is usually strong enough to contain the explosion, but exhaust systems, usually of lighter construction, may not.

Air and gas jet noise is significant in most applications where medium or high pressure gases are discharged into stagnant air. Jet noise is often a problem in combustion applications using atomizers, inspirators, inductors, or high velocity burners.

The higher frequencies of jet noise are generally caused by the shearing action of a high velocity air or gas jet as it exits from the solid boundary of the nozzle, creating turbulence in the adjacent low velocity gases. (Figure 12.17a shows turbulence by an unconfined jet.) Buffeting large scale turbulence, generally further downstream of the primary jet, creates lower frequency sound.

Formula [12/13] is useful in estimating a jet's sound power level dB. For other than circular nozzles, the noise will increase in proportion to ratio of actual periphery to the round periphery because the noise increases with the surface area of the jet.

[12/13] $L_w = 10 \cdot (P_a/P_r) \cdot \log_{10}[e \cdot M^5 \cdot \rho \cdot V^3 \cdot (A/2) \cdot 10^{12}]$

where L_w = PWL, sound power level, dB
   e = a constant, about 1×10^{-4}
   M = Mach number = V/speed of sound, in ft/sec, at jet temperature and pressure
   ρ = jet gas density, pounds/ft^3
   V = mean jet velocity, ft/sec
   A = nozzle area, ft^2
   P_a = perimeter of actual nozzle, same units as P_r
   P_r = perimeter of round nozzle of same A, same units as P_a

A free air or gas jet impinging on a solid surface increases the above normal jet noise by an additional 7 dB. (See Reference 12.h.)

COMBUSTION NOISE

Combustion noise and the previously-discussed combustion system noises are often difficult to distinguish from one another, as each can seem to come from the other's source. They may be excited by, or resonating in sympathy with, combustion reactions. Noises around combustion processes may come from flows or equipment such as blowers, motors, valves, dampers, regulators, furnaces, and general mechanical equipment.

Figure 12.21 depicts the typical spectrum of combustion noise. The authors of Reference 12.a classify four overlapping combustion noise sources in combustion systems: 1) direct flame noise, 2) furnace response, 3) burner tile response, and 4) flow noise amplification.

Figure 12.21. A combustion system noise spectrum, illustrating the complexity of identifying combustion system noise sources.

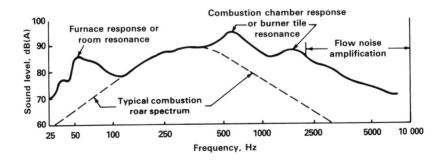

Combustion noise is an inclusive term, including sounds that can be attributed directly or indirectly to the fuel-burning process. Its intensity ranges from barely perceptible to power levels in excess of 150 dB. Freqencies vary from 1 or 2 Hz to greater than 15 kHz. "Combustion roar" seems to imply noises in the low frequency range, but generally means the sounds attributable directly to the flame, such as the parabola in Figure 12.21 extending from about 25 to 2500 Hz and reaching a peak near 300 Hz. This spectrum includes burners of all sizes. However, intensities vary widely and the peak intensity shifts to lower frequencies as burner size increases.

Furnace response (left end of the combustion noise spectrum shown in Figure 12.21) can be predicted in the same way as room response or room resonance, which will be covered in Reference 12.f.

Burner tile response, or combustion chamber resonance (Figure 12.21), is usually a high dB mid-frequency noise. For example, a 12" diameter × 18" long ignition tile for a 1 000 000 Btu/hr burner has a natural (fundamental) frequency of 620 Hz, with its first two harmonic frequencies at 1240 and 1860 Hz. For larger burner sizes, the size of the tile also increases, in turn lowering the natural frequency and the harmonics, but never to the low frequencies associated with room response.

Flow noise (right, high frequency, end of the combustion noise spectrum shown in Figure 12.21) can originate in air, gas, and flue gas lines, and is transmitted through the burner and flame. It may be either amplified or suppressed by the burner and flame, depending on when heat is added. (See Reference 12.i.) If heat is added at the point of greatest compression of the flow sound waves, or abstracted at the point of greatest rarefaction, the vibration will be amplified. The converse is also true.

Flame flow noises (due to turbulence in the flame itself and the shear turbulence that occurs at the interface between the high velocity gases leaving the burner tile and the stagnant gases in the furnace) are generally at higher frequencies and of moderate intensity.

The principal factors that affect combustion noise and combustion-driven noise are discussed below.

a) The **type of fuel** greatly affects the sound power generated by a flame. Faster burning fuels generate higher noise levels, so flame velocity or rate of flame propagation is a good indicator of the possible noise level. Natural gas (1.0 fps = 0.3 m/s) and fossil fuels are slower burning. Man-made fuels usually contain fast burning carbon monoxide and hydrogen. (See Table 1.10 of Volume I of the North American Combustion Handbook.) However, fuels with large percentages of inerts (such as nitrogen, carbon dioxide, water vapor) have their burning velocities slowed by the diluent.

b) **Reduced particle size** increases combustion intensity, therefore noise intensity. Liquid fuel droplet size affects the rate of evaporation. Vaporization speed is inversely proportional to droplet size, and burning speed depends on this evaporation rate. For solid fuels, noise level is similarly related to particle size.

c) **Firing rate** is probably the most important factor in noise generation. It is logical that a higher energy liberation rate will have a greater tendency to produce noise.

d) **Burner size.** Because of the assumed proportionality between burner size and firing rate, combustion intensity has been thought to have little effect on noise level (other than the firing rate previously discussed). The author has found that the proportionality between burner size and noise generated holds reasonably true for small burners in the 100 000 Btu/hr to 1 000 000 Btu/hr range (105.5 to 1055 MJ/hr or 0.0293 to 0.2929 MW range). For burners in the 10 000 000 to 200 000 000 Btu/hr (10 550 to 211 000 MJ/hr or 2.929 to 58.58 MW), the proportionality did not apply -- considerably reduced from that predicted by equations for smaller burners. The reduced noise generation appears to be tied to reduced mixedness, lower turbulence, and lower combustion intensity associated with the larger geometry. [See "Predicting sound power levels of burners" and Figure 12.22, later in this Part 12.]

e) **Pressure drop across a burner.** The air and fuel pressure drops across a burner are known to affect the sound power emitted from the burner. They control the velocities of the air and fuel streams that affect turbulence, and ultimately, mixing of air and fuel as well as the distribution of hot gases within the flame, all of which control the rate of reaction. The effect of burner pressure drop will be quantified under "Predicting sound power levels of burners".

f) **Flame shape and size.** The relationship of a flame envelope shape, size, and motion to the production of noise is similar to the functioning of a loudspeaker. The flame envelope (or "surface") acts as the diaphragm of a loudspeaker, and the flame motion as the excursion of the diaphragm. See Figures 12.1a and 12.1b. For a speaker, the sound power generated is a function of the area of the diaphragm and the amplitude of its excursion. For the flame, the sound power is associated with the surface area of the flame and the motion of the flame envelope.

Flame size also is related to heat release and combustion intensity (Btu/ft^{3}hr or MJ/m^{3}h). Because noise is related to surface area, it follows that the relationship between flame surface and combustion intensity is significant in understanding noise generation. If one thinks of the flame volume and surface area as being spherical, it is obvious that the ratio of surface area to volume decreases as the diameter increases, so the sound intensity of larger flames will increase at less than a 1:1 proportionality to their volumes.

g) **Combustion intensity**, or rate at which heat is liberated within a given volume, along with firing rate and burner size, influences the sound power generated. The more energy liberated in a unit volume, the greater will be the pressure, and in an oscillating system, the greater the oscillating pressure or the acoustic power.

High-intensity combustion generates intense sound fields. About the only thing that can be done to reduce this noise is to change the spectrum, making it easier to attenuate the noise. For example, use of multiple, smaller high-intensity nozzles will shift the spectrum to a higher frequency and make it easier to attenuate.

h) **Air/fuel ratio** also affects sound power level, the highest level being generated when a/f ratio is slightly on the rich side of stoichiometric. The noise level drops dramatically as the flame is turned richer, gets longer and more stable. On the lean side, as excess air increases above stoichiometric, noise drops off slowly until the flame becomes unstable, and then the noise increases.

I) **Fuel-air mixedness**. Mixing of fuel and air plays an important role in flame-generated noise, as well as fuel-burning rates. Poorly mixed fuel and air tend to burn more slowly, producing less noise.

Burners that introduce the fuel concentrically and in the center of an annular ring of air inevitably have lower noise levels than those that introduce the gas peripherally, either concentrically or in multi-jet fashion. Factors that tend to reduce mixing will definitely reduce combustion noise. An illustration of this principle is the radiant tube burner or immersion tube burner that swirls secondary air as it enters the tube, with gas injected axially down the center of the air stream. The centrifugal force of the air causes it to hug the tube's interior wall, and the gas lazily stays within the core of the tube. There is little interface turbulence, and consequently mixing is reduced to a minimum. The combustion resembles a closely controlled, long, lazy, slow-burning diffusion flame that produces little noise.

j) **Type of fuel and air mixing system**. Fuel and air mixedness also relates to this. The emitted sound power will be affected directly or indirectly by:

The mixing or combining zone, and the combustion zone, which determine the mixedness of the fuel and air.

The design of the air and fuel ports (their shape, size, velocity, velocity relative to one another, proximity to one another, and physical boundaries relative to the individual streams.

The aerodynamic considerations are beyond this handbook, but the following guidelines illustrate the factors involved.

Designers of low noise burners should strive for slow mixing, approaching that of a diffusion flame. Air and fuel passages must be designed with little difference between their fluid velocities. Gas should be introduced in the center of the burner to reduce its surface or mixing area. The combining zone must have solid boundaries. To avoid creating vortices at the interface of the streams and the ambient gas, the combining cone must not permit jetting into a stagnant area. The burner designer's goal should be orderly and complete mixing of air and fuel.

k) **Aerodynamics of the ignition port and combustion chamber.** The ignition port (flame holder, burner nozzle), and combustion chamber have enormous effects on the noise produced by a burner. Also to be considered is noise generated by resonation in response to burner excitation.

The ignition port and combustion chamber, if used, are the primary means for stabilizing the flame and providing the necessary reradiation to bring the air-fuel mixture to the reaction temperature critical for complete combustion. The shape of the refractory tile (quarl), or port provides the aerodynamics that stabilize a portion of the mixture in a relatively quiescent zone, where feed speed is slowed down to equal flame speed. This environment allows the mixture in that zone to come up to temperature and provide piloting to the main stream. Other tile shapes cause toroidal recirculation zones within the tile, thereby creating the necessary temperature and piloting function. The combination of burner port and tile shapes provides the basic piloting and stabilizing functions.

For low noise level considerations, the tile design must have stabilizing regions that aerodynamically avoid zones of high or large scale turbulence. Furthermore, the stabilizing region must itself be stationary, with little or no movement. The tile and the combustion chamber must provide an environment where the flame, beyond its stabilizers, does not tear or separate from the main flame or main stream. The interface between the stabilizing zones and the main stream must be such that the resultant flame does not shed periodically. If shedding does occur, a periodic noise will be generated having the same, or a multiple of, the shedding frequency.

PREDICTING SOUND POWER LEVELS OF BURNERS

Most of the many theories proposed for predicting sound power levels were developed for specific types of burners and for a limited capacity range. When one looks at the number of factors that affect sound power, it is not surprising that no single formula or procedure permits computation of sound power levels over a broad range of burner types, fuels, and capacities.

In the hope of finding a better way to predict sound power level, an analysis was done of a large data base of information on low pressure air burners of different sizes and types. The data was obtained from six different types of burners, and four to five sizes of burners of each type, using oil, gas, or pulverized coal as fuels. Burner capacities ranged from 100 000 Btu/hr to 110 000 000 Btu/hr (105.5 to 116 050 MJ/h, or 32.2 to 32 230 kW). Noise levels were measured for 100% and 30% firing rates.

Figure 12.22. Average sound power level data for a number of types and sizes of gaseous, liquid, and solid fuel burners over two broad ranges of q' (corrected) firing rates.

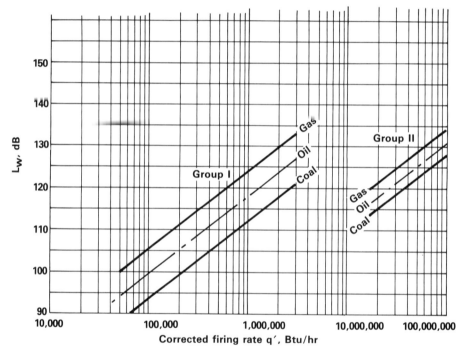

Figure 12.22 is a semi-log plot of the results of the above-mentioned analysis. The horizontal scale is q' = (gross heat release rate, in Btu/hr) × (pressure drop across burner)/(abs. air pressure).

Individual data points have been replaced with bands for clarity. It is not possible to find one formula to cover all cases, but an approximation is possible for nozzle-mix, low-pressure-air burners.

Two distinct groupings of points were apparent. A curve-fitting procedure gives the approximate formulae of equations 12/14a and 12/14b, which might be usable for modeling purposes.

For Group I, heat release range up to $q' = 4\ 000\ 000$ Btu/hr:

[12/14a] $dB = 27.4 + 7.66 \ln q'$

For Group II, heat release from $q' = 18\ 000\ 000$ to $110\ 000\ 000$:

[12/14b] $dB = 45.43 + 5.2395 \ln q'$

where $\ln q'$ means the natural logarithm of the value of q'. (The natural logarithm is the log to the base e as opposed to common log, base 10, used in all prior sound formulae in this Part 12.)

The geometry and types of flame holding, among other factors, are distinctly different for the small and large groups of burners. For any given burner, noise levels during turndown followed a slope parallel to the lines shown in Figure 12.22.

Some other conclusions can be drawn from the data:

a) The slope of the turndown lines was similar for all burners within a homologous series.

b) For the four series of burners tested, the slopes of the turndown lines were reasonably consistent.

c) The individual burners were all operated through the same range of pressure drops for the Group I burners, and at 1.5 times that drop for the Group II burners. The flame patterns and flame characteristics and shapes were different for each series of burner.

d) Burners operating on natural gas were at 5% excess air.

e) Burners operating on distillate (#2) fuel oil were at 20% excess air.

f) Burners operating on powered coal were at 50% excess air.

g) Within a group, burners firing on natural gas generally had sound intensities 3 to 5 dB higher than when firing oil. Those operating on pulverized coal had noise levels 2 to 3 dB lower than those on oil. In general, a higher combustion intensity (shorter flame) resulted in a higher noise level.

h) The frequency of the combustion roar and the burner tile response shifted downward as the burner capacity was increased.

i) Flow noise seemed relatively constant for different capacities, burner sizes, and burner types. There were only minor variations from type to type. The empirical data of Figure 12.22 is far from definitive, but it can provide an approximation of sound power levels for use when precise data is not available, provided it is used with full understanding of its limitations.

Example 12-3.   Estimate noise level for a moderate-to-high-intensity distillate fuel oil flame from a nozzle-mix burner operating at 5 000 000 Btu/hr capacity with 16 osig (1 psig) air pressure drop and 14.7 psia + 1 psig = 15.7 psia air line pressure.

Calculate $q' = 5\ 000\ 000 \times (16/16)/15.7 = 318\ 500$. This value of q' classifies the burner in Group I. Enter Figure 12.21 at 318 500 on the scale across the bottom. Moving vertically, one would normally turn left at the centerline of Group I because it is being fired with fuel oil, but its moderate-to-high flame intensity suggests interpolating about 2 dB higher, at a sound power level of 118 dB.

REFERENCES

12.a   **Diehl, G. M.:** "Machinery Acoustics", John Wiley & Sons, Inc. New York, NY, 1973.

12.b   **Giammar, R. D. and Putnam, A. A.:** "Guide for the Design of Low Noise-Level Combustion Systems", American Gas Association Basic Research Project BR-3-5, pg. 9, January, 1971.

12.c   **Harris, C. M.:** "Handbook of Acoustic Measurement and Noise Control", Chapter 41, McGraw-Hill Book Co., 3rd ed., 1991.

12.d   **Jorgensen, R. (ed.):** "Fan Engineering", Chapters 4 and 16; Buffalo Forge Co., Buffalo, NY, 1983.

12.e   **Putnam, A. A.:** "Combustion-Driven Oscillations in Industry", American Elsevier Publishing Co., Inc., New York, NY, 1971.

12.f   **Thurmann, A. and Miller, R. K.:** "Fundamentals of Noise Control Engineering", pp. 73-75; Fairmont Press/Prentice-Hall, Lilburn, GA, 1990.

12.g   **Vincent, S., Mills, J. S., and Petersen, A. C.:** "Industrial Noise Control Manual"; US Department of Health, Education, and Welfare; Printing Office, Washington DC, Stock –017-033-00073, Catalog –1975.HE20.7108:N69.

12.h   **Rayleigh, B. and Strutt, J. W.:** "The Theory of Sound"; vol. 2, pg. 322; Dover Publications, New York, NY, 1982.

ADDITIONAL SOURCES of information relative to noise

Beranek, L. L. and Ver, I. L.: "Noise and Vibration Control Engineering", John Wiley & Sons, Inc., New York, NY, 1992.

Berger, E. H. et al: "Noise & Hearing Conservation Manual", 4th ed., American Industrial Hygiene Assn., Fairfax, VA, 1986.

Blevins, R. D.: "Flow-Induced Vibration", 2nd ed., Van Nostrand Reinhold, New York, NY, 1990.

Peterson, A. P. G. (GenRad, Inc.): "Handbook of Noise Measurement", 9th ed., 1980; QuadTech, Inc., Bolton, MA 01740-1107.

Part 13.   OXYGEN ENRICHMENT AND OXY-FUEL FIRING
largely contributed by Hisashi Kobayashi, Praxair, Inc.

Air contains only about 20.9% oxygen, and the balance is primarily nitrogen. See Table 13.1, repeated from Part 1 of Volume I of this handbook.

Table 13.1.   Composition of air[1]

Component	% by Volume (mols) — % by Weight —	Dry Bulb Temperature (db) and Relative Humidity (rh)					
		60 F db 0% rh[2]	60 F db 80% rh	60 F db 100% hr	90 F db 20% rh	90 F db 80% rh	90 F db 100% rh
Oxygen, O₂		20.99 23.20	20.70 23.00	20.62 22.94	20.79 23.06	20.19 22.63	19.99 22.50
Nitrogen, N₂		78.03 75.46	76.94 74.86	76.67 74.63	77.29 75.01	75.06 73.61	74.32 73.18
Argon, Ar		0.94 1.30	0.93 1.29	0.92 1.29	0.93 1.29	0.90 1.27	0.90 1.26
Other[3]		0.04 0.04	0.04 0.04	0.04 0.04	0.04 0.04	0.04 0.04	0.04 0.04
Water, H₂O		0.00 0.00	1.40 0.87	1.75 1.10	0.95 0.59	3.81 2.45	4.76 3.02

In combustion systems that use blower air as their oxygen source, the large nitrogen content of the air absorbs heat and increases the volume of the furnace and flue gases. Oxygen enrichment, oxygen lancing, or oxy-fuel firing is often used to improve combustion characteristics.

In *oxygen enrichment*, the oxygen concentration is increased above that of normal air by blending commercial oxygen with the blower air to raise the oxygen content of the mixed air-oxygen stream to 22% to 35%.

In *oxygen lancing*, a special form of oxygen enrichment, commercial oxygen (typically 90% to 100% purity) is injected near an air-fuel flame to improve the combustion characteristics -- higher flame temperature, and higher % available heat. The most common reason for using oxygen enrichment is to increase the available heat input to the load in the furnace. This can improve productivity or fuel economy, or both.

---

[1] For easy computation, it is convenient to remember these ratios:
   Air/O₂ = 100/20.99 = 4.76 by volume (mols); Air/O₂ = 100/23.20 = 4.31 by weight
   N₂/O₂ = 3.76 by volume (mols); N₂/O₂ = 3.31 by weight.
[2] From International Critical Tables; all other columns calculated from I.C.T. data and from Reference 13.r at the end of this Part 13.
[3] CO₂ (about 0.03%), H₂ (about 0.01%), Neon, Helium, Krypton, Xenon.

In *oxy-fuel firing*, no blower air is used. Fuel is mixed with commercial oxygen and burned. The use of oxy-fuel firing is increasing in glass melting and in metallurgical furnaces to save fuel for higher productivity, to reduce NOx emissions, and to lower the volumes of furnace and flue gases.

This Part 13 discusses the commercial sources of oxygen, calculation of the oxygen requirement, changes in combustion characteristics, and methods of combustion with oxygen.

OXYGEN SOURCES

The oxygen required for combustion can be delivered as cryogenic liquid, or generated on site from an air separation system. Oxygen supply systems range in size from less than one ton per day (1000 scfh) to over one thousand tons per day. The purity of the oxygen ranges from better than 00.9% with cryogenic processes, to 90-95% with adsorption processes, to 28-35% with membrane units.

The most economic supply system depends on the volume, purity, and pressure required for the oxygen combustion process. Table 13.2 compares various oxygen sources.

Cryogenic System. Cryogenic air separation plants produce most of the commercially used gaseous and liquid oxygen. Air is liquefied at a very low temperature, and then distilled to separate the oxygen from the nitrogen. Purity may range from 70% to 99.5 + %. Regional facilities produce liquid oxygen for local distribution by truck or rail. Large volume users may have plants built on their sites, typically for oxygen requirements greater than 100 000 scfh.

Bulk Liquid Supply System. For relatively small oxygen requirements, such as 1000 to 10 000 scfh, delivered liquid oxygen is often the most economic supply option. High purity (99.5 + %) liquid oxygen produced in a regional air separation plant can be delivered and stored at users' sites in special tanks designed for cryogenic liquids. A vaporizer can then change the liquid oxygen to gaseous oxygen at about 100 to 200 psig pressure, and it can then be piped to points of use.

Adsorption Systems. For oxygen requirements from 10 000 to 200 000 scfh, an adsorption system is often the most economic supply. Pressure Swing Adsorption (PSA) or Vacuum Swing Adsorption (VSA) processes are used to separate oxygen from air on a user's site. In a typical PSA or VSA process, nitrogen from air is preferentially adsorbed in a bed of synthetic zeolites and removed by cyclic fluctuation, or "swing," of pressure. The oxygen product stream typically has an oxygen purity ranging from 90 to 95%, about 5% argon and some nitrogen.

Membrane System. Air is passed through a thin film or membrane material through which oxygen permeates faster than nitrogen. Oxygen enriched air with 28 to 35% oxygen content is typically produced from a single stage separation. Commercial membrane systems have been available since the 1980s. Further improvements in their technology are expected. A blower or compressor is required for most applications.

Because of their small size, simplicity, and modular design, membrane systems are attractive for producing low purity oxygen products in the 1000 to 15 000 scfh equivalent pure oxygen flow rate range.

Table 13.2.   Comparisons of Oxygen Sources *(Adapted from Reference 13.i at the end of this Part 13.)* (1 ton per day ~ 1000 scfh for pure oxygen.)

Source	Tons/day* typical capacity	% oxygen purity	psig dlvry press	kWh/ton* power reqd	
Cryogenic plant	50-2000 +	70-99.5 +	3 250	230-250 350-400	Low cost for large volume users. Maybe use of co-products.
Bulk liquid	0-50	99.5 +	250†	700-800	Tank and vaporizer required. Best use flexibility.
Pressure-swing adsorption	10-100	80-95	3-20	230-600	Generally more economical than low capacity cyrogenic.
Membrane	1-15	28-35	1	350-600	Limited purity. Very simple process.

Calculating Equivalent Pure Oxygen. In comparing oxygen products of different purities, the content of the excess pure oxygen above that contained in air becomes an important measure, and is termed "equivalent pure oxygen." An oxygen-enriched stream is assumed to be a mixture of normal air and pure oxygen.

When a fuel is burned with an oxygen-enriched stream, the overall combustion characteristics can be viewed as combined effects of normal air combustion and pure oxygen combustion. Because the enhanced characteristics of oxygen-enriched combustion are due to the excess pure oxygen, the costs of oxygen products need to be compared by their contents of equivalent pure oxygen.

---

* Ton of equivalent pure oxygen.
† Typical gas pressure obtained by vaporizing liquid oxygen.

Formula [13/1] gives the portion of the equivalent pure oxygen in the total oxygen contained in an oxygen-enriched stream.

[13/1]   % Equivalent pure oxygen $= \dfrac{\%O_2 - 20.9}{79.1 \times \%O_2} \times 100\%$

Example 13-1. A 35% purity oxygen product from a membrane system would be valued at

$\dfrac{(35 - 20.9)}{(79.1 \times 35)} \times 100 = 50.9\%$ equivalent pure oxygen.

If pure oxygen cost \$3/scf, the membrane oxygen would be worth $(50.9/100) \times \$3 = \$1.527/scf$.

EFFECTS OF OXYGEN ON COMBUSTION

Oxygen and air requirements, per volumes. Formula [13/2] calculates the required fuel flow rate, Ff, in scfh or m³/h.

[13/2]   Ff = reqd gross heat input rate ÷ gross heat value per unit fuel
         = reqd gross Btu/hr ÷ gross Btu/scf fuel, or
         = reqd gross GJ/h ÷ gross GJ/m³ fuel.

Formula [13/3] gives the % fuel saved by situation 2 over situation 1, using % available heat figures, %AH, from Figure 13.4a.

[13/3]   % fuel saved $= 100 \times \left(1 - \dfrac{\%AH_2}{\%AH_1}\right)$.

Formula [13/4a] determines the volume of pure oxygen required per unit volume of fuel for stoichiometric (perfect) combustion, from the volumetric (molal) analysis of a gaseous fuel:

[13/4a] V_{rpo}/V_{fuel} = (%CH₄ × 0.02) + (C₂H₆ × 0.035)
          + (%C₃H₈ × 0.05) + (C₄H₁₀ × 0.065)
          + (%H₂ × 0.005) + (%CO × 0.005) − (%O₂ × 0.01).

Formula [13/4b] tells how to evaluate F_{rpo}, the required volume flow rate of pure oxygen, in scfh or m³/h.

[13/4b] F_{rpo} = Ff × volume pure oxygen reqd/volume fuel
          = Ff × V_{rpo}/V_{fuel}.

Formula [13/5] shows the volume flow rate of air-oxygen mixture, F_{aom}.

[13/5] $F_{aom} = F_{rpo}/(\%e/100)$, where e = desired volume concentration of pure oxygen in the total enriched mixture.

By definition, %e, the volume concentration of pure oxygen in the total volume of enriched mixture is

[13/6a] $\%e/100\% = \dfrac{0.209 \times F_a + (\% \text{ purity}/100\%) \times F_{oc}}{F_a + F_{oc}}$.

The 0.209 in this formula is the decimal equivalent of 20.9% oxygen in normal air. F_a is the volume flow rate of blower air; F_{oc} is the volume flow rate of commercial oxygen. A prior engineering decision has usually been made as to the desired value of e; so it is possible to solve formula [13/6a] for the ratios, F_a/F_{oc} and F_{aom}/F_{oc}, as follows:

[13/6b] $F_a/F_{oc} = \dfrac{(\%p/100) - (\%e/100)}{(\%e/100) - 0.209}$,

and the flow of enriched air-oxygen mix is

[13/6c; 13/6d] $F_{aom} = F_a + F_{oc}$;   so $F_{aom}/F_{oc} = (F_a/F_{oc}) + 1$; and

[13/6e] $F_{oc} = \dfrac{F_{aom}}{(F_a/F_{oc}) + 1} = \dfrac{[13/5]}{[13/6b] + 1}$; and

[13/6f] $F_a = (F_a/F_{oc}) \times (F_{oc}) = [13/6b] \times [13/6e]$.

The volume of the poc (products of combustion) is often important for sizing stack gas cleanup equipment, or for determining adequate in-furnace circulation for fast, uniform heat transfer. Formulas [13/7] and [13/8] calculate this. The volume of stoichiometric poc, per unit volume of fuel,

[13/7] $V_{poc}/V_{fuel} = V_{CO_2}/V_{fuel}$...from [3/13 of Volume I]
            $+ V_{H_2O}/V_{fuel}$...from [3/15 of Volume I]
            $+ V_{N_2}/V_{fuel}$.... from [13/8]...where this last term may be read from the inset graph of Figure 13.3.

[3/13 of Volume I] $V_{CO_2}/V_{fuel} = (\%CO \times 0.01) + (\%CH_4 \times 0.01) + (\%C_2H_6 \times 0.02) + (\%C_3H_8 \times 0.03) + (\%C_4H_{10} \times 0.04) + (\%CO_2 \times 0.01)$.

[3/15 of Volume I] $V_{H_2O}/V_{fuel} = (\%H_2 \times 0.01) + (\%CH_4 \times 0.02) + (\%C_2H_6 \times 0.03) + (\%C_3H_8 \times 0.04) + (\%C_4H_{10} \times 0.05) + (\%H_2O \times 0.01)$.

[13/8] $V_{N_2 \text{ in poc}}/V_{fuel}$ = (%N$_2$ in fuel/100) + [1 + (equivalent % excess air/100)] × [(1 − e/100)/(e/100)] × (V_{rpo}/V_{fuel})], where the last term may be from Formula 13/4a.

Figure 13.3 is a plot of some of the above flue gas components for a range of e-values from 20.9% to 100% for "average natural gas." When adding these gas volumes, it is important to remember that if a total volume is being determined for a flow downstream of a scrubber, the water vapor should be omitted.

Figure 13.3.   Products of combustion for various percents of oxygen concentration in air-oxygen mixes for Average Natural Gas -- plotted on both Cartesian and log-log coordinates. A curve fit for the top left portion of the log-log plots (e = 20.9 to 50%) gives the approximate empirical formulas:

cf N$_2$/cf fuel, with equiv. 10% excess air = $650/e^{1.433}$.
cf N$_2$/cf fuel, stoichiometric (0% excess air) = $560/e^{1.433}$.

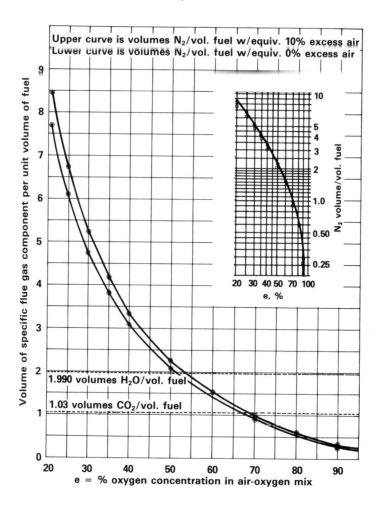

Upper curve is volumes N$_2$/vol. fuel w/equiv. 10% excess air
Lower curve is volumes N$_2$/vol. fuel w/equiv. 0% excess air

1.990 volumes H$_2$O/vol. fuel
1.03 volumes CO$_2$/vol. fuel

Volume of specific flue gas component per unit volume of fuel

N$_2$ volume/vol. fuel

e, %

e = % oxygen concentration in air-oxygen mix

Figure 13.4a. Available heat, expressed as a percentage of the gross heating value, for an "average natural gas," 1025 Btu/ft³, 5% excess oxidant, with standard air, and with various degrees of oxygen enrichment. This data is applicable only if there is no unburned fuel in the products of combustion.

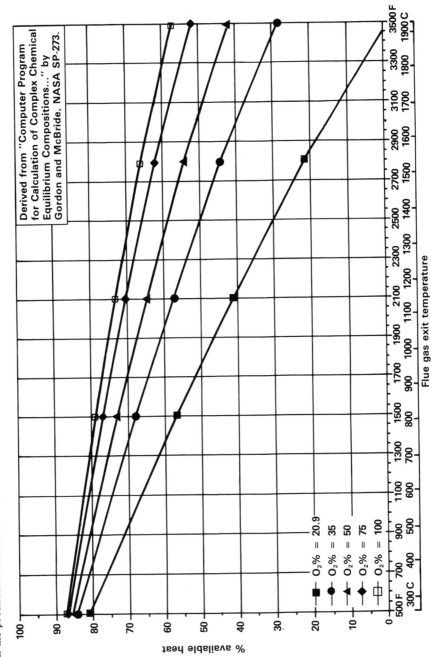

Example 13-2. A furnace with a flue gas exit temperature of 1800 F requires 1 000 000 available Btu/hr with an average natural gas (V_{rpo}/V_{fuel} = 2.025) and air (case a). Consideration is being given to enriching the air to e = 35% (by volume) concentration of pure oxygen in the total volume of enriched mix (case b), or to all commercial oxygen, no blower air (case c). The commercial oxygen available is 95% pure oxygen.

FIND, for each of the three modes of operation:

i   % available heat
ii   required gross heat input rate, gross Btu/hr (kk = million)
iii   required fuel volume flow rate, scfh, Ff = [13/2]
iv   % fuel savings (b over a, c over b, c over a)
v   reqd pure oxygen volume flow, F_{rpo} = [13/4b]
vi   resultant volume flow rate of enriched mixture, F_{aom} = [13/5]
vii   ratio of F_a to F_{oc} = [13/6b]
viii   ratio of F_{aom} to F_{oc} = [13/6d]
ix   required commercial oxygen volume flow rate, F_{oc} = [13/6e]
x   required air volume flow rate, F_a = [13/6f]
xi   resultant poc volume flow rate, F_{poc}, = N_2 + H_2O + CO_2 [from Figure 13.3].

RESULTS of these example calculations are listed in the following table in US units:

Step	Case a [e = 20.9]	Case b [e = 35]	Case c [e = 95]
i	(Fig.13.4a)%AH = 49.0	62.8	75.3
ii	1kk/.490 = 2.04kk*	1.59kk	1.33kk
iii	2.04kk/1025 = 1990	1550	1296
iv	NA†	100% × [1 − (49.0/62.8)] = 22.0 for b over a	c over b, 16.6% ; c over a, 34.9%
v	NA	[Fig. 13.4] 1550 × 2.025 = 3140	1296 × 2.025 = 2624
vi	NA	[13/5] 3140/.35 = 8970	2694/.95 = 2762
vii	NA	(.95−.35)/(.35−.209) [13.6b] = 4.255	0
viii	NA	[13/6d] 4.255 + 1 = 5.255	1.0
ix	NA	[13/6e] 8790/5.255 = 1710	F_{aom} = F_{oc} = $F_{rpo}/(p/100)$ = 2624/0.95 = 2762
x	1991 × 9.68 = 19273	[13/6f] 4.255 × 1710 = 7280	0
xi	1991 × 10.71 = 21324	(3.800 + 1.990 + 1.030) × Ff = 6.82 × 1522 = 10380	(0.05 + 1.99 + 1.03) × Ff = 3.07 × 1296 = 3978

* kk = million   † NA = not applicable

Available Heat and Fuel Savings. The significance of "% Available Heat" is explained in Chapter 3 of Volume I. Figure 13.4a is a % available heat chart for an average natural gas, with various degrees of oxygen enrichment -- all the way to 100% oxygen, which is termed oxy-fuel firing. Table 13.4b provides the same data in a form that can be read more accurately for a limited number of points. These recent data, computer-calculated with corrections for variable specific heats and for dissociation, supersede that of Figure 3.17 in Volume I.

Table 13.4b. Available heat, expressed as a percentage of the gross heating value, for an "average natural gas," 1025 Btu/ft³, 5% excess oxidant, with standard air, and with various degrees of oxygen enrichment. This data is applicable only if there is no unburned fuel in the products of combustion. *Derived from "Computer Program for Calculation of Complex Chemical Equilibrium Compositions..." by Gordon and McBride, NASA SP-273.*

% available heat relative to gross heating value

Equivalence ratio, Ø	% excess oxidant	% oxygen in O₂-air mix	Flue gas exit temperature, F					Adiabatic flame temperature		
			500	1500	2100	2800	3500	Kelvin	Celsius	Fahrenheit
0.95	5	20.9	81.03	56.95	41.16	21.59	—	2178	1905	3460
0.95	5	35	84.21	68.15	57.54	44.30	28.22	2606	2333	4231
0.95	5	50	85.60	73.12	64.81	54.38	41.49	2798	2525	4576
0.95	5	75	86.70	76.98	70.46	62.23	51.88	2957	2684	4863
0.95	5	100	87.25	78.91	73.29	66.17	57.14	3045	2772	5021
0.9091	10	20.9	80.67	55.68	39.31	19.00	—	2134	1861	3381
0.9091	10	35	83.99	67.38	56.43	42.74	26.47	2586	2313	4195
0.9091	10	50	85.45	72.58	64.02	53.28	40.31	2786	2513	4555
0.9091	10	75	86.59	76.62	69.93	61.51	51.13	2949	2676	4848
0.9091	10	100	87.17	78.63	72.88	65.61	56.62	3039	2766	5010
0.8	25	20.9	79.52	51.66	33.41	10.77	—	1984	1711	3111
0.8	25	35	83.30	64.98	52.88	37.78	20.25	2508	2235	4054
0.8	25	50	84.97	70.89	61.52	49.79	35.95	2736	2463	4465
0.8	25	75	86.28	75.47	68.24	59.13	48.23	2916	2643	4789
0.8	25	100	86.93	77.76	71.61	63.84	54.48	3015	2742	4967

The first three steps of Example 13-2 illustrate one of the many uses of the available heat concept -- sizing combustion equipment. Another important use of the available heat concept is for preliminary comparisons of a variety of ways to achieve better fuel efficiency and better flame temperature. Such comparisons might be between cold combustion air with various percents of excess air, or hot combustion air with various degrees of preheat, or oxygen-enriched air with any amount of oxygen enrichment, or any combination of these. See Tables 13.5a, b, and c.

Formula 13/3 can be used to derive % fuel savings for any pair of operating conditions, using the % available heat figures for that pair of conditions. The derivation of formula 13/3 is in Reference 13.m. Comparisons of fuel savings with preheated air and with oxygen enrichment are tabulated in Reference 13.n. *Fuel saved is only one of many factors that must be considered when comparing pàc and oec.*

Table 13.5a.   Comparisons of % available heat and adiabatic flame temperatures for an average natural gas* with selected amounts of excess air, air preheat, and oxygen enrichment.

% XS air	% XS O₂	phi = equiv ratio	p = preheat temp, F	p = preheat temp, C	e = % O₂ in a-O	% Available heat† 500F 260C	% Available heat† 1500F 816C	% Available heat† 2100F 1149C	Adiabatic flame temp‡ F	Adiabatic flame temp‡ C
5	1	0.95	60	16	20.9	81.03	56.95	41.16	3460	1905
5	1	0.95	60	16	35	84.21	68.15	57.54	4231	2333
5	1	0.95	60	16	100	87.25	78.91	73.29	5021	2772
5	1	0.95	1200	649	20.9		78.88	63.12	3946	2175
5	1	0.95	1600	871	20.9			71.48	4093	2257
5	1	0.95	2000	1093	20.9			80.07	4231	2333
10	2	0.909	60	16	20.9	80.67	55.68	39.31	3381	1861
10	2	0.909	60	16	35	83.99	67.38	56.43	4195	2313
10	2	0.909	60	16	100	87.17	78.63	72.88	5010	2766
10	2	0.909	1200	649	20.9		78.61	62.26	3905	2152
10	2	0.909	1600	871	20.9			71.01	4134	2279
10	2	0.909	2000	1093	20.9			79.98	4205	2318
25	4.55	0.80	60	16	20.9	79.52	51.66	33.41	3111	1711
25	4.55	0.80	60	16	35	83.30	64.98	52.88	4054	2235
25	4.55	0.80	60	16	100	86.93	77.76	71.61	4967	2742
25	4.55	0.80	1200	649	20.9		77.72	59.50	3748	2065
25	4.55	0.80	1600	871	20.9			69.44	3938	2170
25	4.55	0.80	2000	1063	20.9			79.64	4107	2264

* AVERAGE NATURAL GAS = 90% CH₄, 5% C₂H₆, 1% C₃H₈, 4% N₂; 1025 gross Btu/scf; 9.68 ft³ air/ft³ fuel, 10.71 ft³ poc/ft³ fuel (with air); 2.025 ft³ air/ft³ fuel, 3.06 ft³ poc/ft³ fuel (with oxygen).
† with an assumed furnace exit gas temperature of...
‡ Flame temperatures for #2 and #6 oils are calculated using an older program than used for natural gas. Temperatures from Tables 13.5b and 13.5c should not be compared with Table 13.5a. Comparisons within Table 13.5b and 13.5c are still useful.

Table 13.5b. Comparisons of % available heat and adiabatic flame temperatures for a #2 distillate fuel oil* with selected amounts of excess air, air preheat, and oxygen enrichment.

% XS air	% XS O_2	phi = equiv ratio	p = preheat temp, F	C	e = % O_2 in a-O	% Available heat† 500F 260C	1500F 816C	2100F 1149C	Adiabatic flame temp‡ F	C
5	1	0.95	60	16	20.9	84.2	59.9	43.9	3443	1895
5	1	0.95	60	16	35	87.7	69.9	61.0	4081	2249
5	1	0.95	60	16	100		83.0	77.6	4615	2546
5	1	0.95	1200	649	20.9		80.9	66.9	3859	2126
5	1	0.95	1800	982	20.9			77.3	4030	2221
10	2	0.909	60	16	20.9	86.2	58.5	41.9	3362	1850
10	2	0.909	60	16	35	87.4	70.8	59.8	4044	2229
10	2	0.909	60	16	100		82.7	77.1	4606	2541
10	2	0.909	1200	649	20.9		82.3	66.0	3815	2101
10	2	0.909	1800	982	20.9			79.8	3999	2204
25	4.40	0.80	60	16	20.9	82.6	54.3	35.7	3123	1717
25	4.40	0.80	60	16	35	86.7	68.2	56.1	3927	2164
25	4.40	0.80	60	16	100		82.8	75.8	4578	2526
25	4.40	0.80	1200	649	20.9		81.7	63.1	3681	2027
25	4.40	0.80	1800	982	20.9			75.6	3909	2154

Table 13.5c. Comparisons of % available heat and adiabatic flame temperatures for a #6 residual fuel oil* with selected amounts of excess air, air preheat, and oxygen enrichment.

% XS air	% XS O_2	phi = equiv ratio	p = preheat temp, F	C	e = % O_2 in a-O	% Available heat† 500F 260C	1500F 816C	2100F 1149C	Adiabatic flame temp‡ F	C
5	1	0.95	60	16	20.9	85.9	62.3	46.9	3544	1951
5	1	0.95	60	16	35	89.2	73.7	63.6	4156	2291
5	1	0.95	60	16	100		84.8	79.7	4676	2580
5	1	0.95	1200	649	20.9		84.8	69.3	3934	2168
5	1	0.95	1800	982	20.9			82.2	4094	2257
10	2	0.909	60	16	20.9	83.2	60.9	44.9	3468	1909
10	2	0.909	60	16	35	89.0	72.9	62.4	4123	2273
10	2	0.909	60	16	100		84.5	79.3	4668	2575
10	2	0.909	1200	649	20.9		84.5	68.4	3891	2144
10	2	0.909	1800	982	20.9			79.0	4065	2240
25	4.54	0.80	60	16	20.9	84.3	56.8	38.9	3231	1777
25	4.54	0.80	60	16	35	88.3	70.5	58.8	4012	2211
25	4.54	0.80	60	16	100		83.6	78.0	4643	2561
25	4.54	0.80	1200	649	20.9		83.5	65.6	3761	2072
25	4.54	0.80	1800	982	20.9			80.9	3977	2191

---

* per specifications on pages 16 and 17 of Volume I.

† with an assumed furnace exit gas temperature of...

‡ Flame temperatures for #2 and #6 oils are calculated using an older program than used for natural gas. Temperatures from Tables 13.5b and 13.5c should not be compared with Table 13.5a. Comparisons within Table 13.5b and 13.5c are still useful.

Flame Temperature. Adiabatic flame temperatures may be read from an available heat chart, such as Figure 13.4a, where the appropriate curve meets the zero available heat line (x-intercept). These are theoretical, calculated flame temperatures representing cases where the flame has not yet given off any heat to its surroundings (adiabatic conditions). These flame temperatures are therefore applicable *only for comparison* of various conditions, all determined by the same mathematical procedure. See Tables 13.5a, b, and c.

Actual measurement of flame temperatures is very difficult (a) because any intrusive sensor immediately becomes a radiator, cooling itself below the temperature of the flame it was to measure, and (b) because every segment of a flame immediately disperses heat to its surroundings, making even non-intrusive measurements variable with the nature of the surroundings.

Theoretical (adiabatic) flame temperature increases sharply with oxygen enrichment, and exceeds 5000 F (2760 C) with 100% pure oxygen (oxy-fuel) at stoichiometric ratio (phi = 1.0).

At low levels of oxygen enrichment, flame temperature increases as much as 100 degrees F (56 degrees C) with each 1% increase in oxygen concentration ("e"). At high concentrations (e), the rate of increase of flame temperature is diminished as dissociation of the combustion products consumes more energy. Similar changes occur with other fuels. Flame temperatures of a variety of fuels are compared in Table 1.10 of Volume I.

When oxygen enrichment is used in industrial burners designed for normal air combustion, the flame tends to become shorter, hotter, and more luminous as the flame temperature is increased and/or the flame momentum is reduced. Care must be exercised to prevent overheating of burner parts, tiles, and refractory walls.

Factors affecting flame stability are listed in detail in Table 1.10 in Volume I, including data for combustion of many fuels in air and in pure oxygen. The factors most specifically affecting burner flame stability* are discussed below.

1. *Minimum Ignition Temperature* is a measure of the minimum ignition energy level required to initiate a combustion reaction. Without a pilot flame or spark igniter to raise the incoming fuel and air to this level, a cold main burner or a cold combustion chamber cannot be started.

   After a minimum trial-for-ignition period (usually specified by insuring or safety authorities), the pilot or spark is timed out, and burner is expected to continue operating with its own continuous self-sustained re-ignition. This may be from heat built up in a refractory tile (quarl) or from the thermal and chemical energy of hot poc back-flowing to the root of the burner flame.

---

* Flame stability means reliability: easy lighting, staying continuously burning without a pilot and without pulsing or sputtering -- over the whole range of expected operating conditions.

Ignition temperature also has a bearing on flammability limits, as explained below. Table 13.6 lists minimum ignition temperatures for a few gaseous fuels in air and in oxygen.

Table 13.6. Ignition temperatures in air and oxygen for some gaseous fuels. (See also Table 1.10 in Volume I.)

IGNITION TEMPERATURE	In Air (e = 21% O_2)		In Oxygen (e = 100% O_2)	
Acetylene (C_2H_2)	612 F (q)	350 C (q)	565 F (q)	296 C (q)
Carbon monoxide (CO)	1128 F (d)	609 C (d)	1090 F (l)	588 C (l)
Ethane (C_2H_6)	882 F (q)	472 C (q)	842 F (q)	450 C (q)
Ethylene (C_2H_4)	914 F (q)	490 C (q)	905 F (q)	485 C (q)
Hydrogen (H_2)	1062 F (d)	572 C (d)	1040 F (l)	560 C (l)
Methane (CH_4)	1170 F (d)	632 C (d)	1033 F (l)	556 C (l)
Propane (C_3H_8)	919 F (l)	493 C (l)	874 F (l)	468 C (l)

Letters in parentheses relate to References at the end of Part 13.

2. *Flammability Limits.* With both premix and nozzle-mix (diffusion) flames, the mixture at the root of the flame must be within the flammability limits for a flame to be initiated. (Many nozzle-mix burners appear to operate beyond the flammability limits on a macro basis because extra air or oxygen is added downstream of the point where the flame root is established.) Table 13.7 lists flammability limits in air and in oxygen for a few fuels. Flammability limits are also explosive limits.*

Figure 13.9, later in this Part 13, compares the effects of O_2 enrichment and of air preheating on flammability limits of methane. The upper (rich) flammability limit increases substantially with oxygen enrichment, but the lower (lean) limit changes very little...because the excess oxygen acts as a heat sink, just as nitrogen does under the lean conditions.

---

* An explosion is a *detonation*; normal combustion or burning is termed *deflagration*. (See Glossary.)

Flammability limits really relate to minimum ignition temperature and the heat absorbing capability of inert molecules between "sets" of fuel and oxidant molecules that are stoichiometrically proportioned so that they are ready to burn. Higher concentrations of intervening inert material may absorb so much heat that they prevent transfer of sufficient heat to the next eligible "set" so it cannot be heated to its minimum ignition temperature. The inert materials may be:

(a) excess air or oxygen, unaccompanied by fuel molecules;
(b) excess fuel, unaccompanied by air or oxygen molecules;
(c) products of complete combustion, such as CO_2, H_2O; or
(d) intentional extinguishing (inert) gases such as nitrogen, steam, CO_2.

Flammability limits are measures of the magnitude of the heat absorbing capability of the intervening inert materials relative to the minimum ignition energies of the sets.

Flammability limits are also influenced by other factors such as the ease with which molecular bonds can be broken.

Table 13.7. **Flammability limits in air and oxygen for some gaseous fuels.** (See also Table 1.10 in Volume I.)

FLAMMABILITY LIMITS*	In Air (e = 21% O_2)		In Oxygen (e = 100% O_2)	
	lower (lean)	upper (rich)	lower (lean)	upper (rich)
Butane (n-C_4H_{10})	1.9 (b)	8.5 (b)	1.8 (f)	49 (f)
Carbon monoxide (CO)	12.5 (f)	74.2 (f)	15.5 (f)†	94 (f)
Hydrogen (H_2)	4.0 (o)	74.2 (o)	4 (g)	94 (g)
Methane (CH_4)	5.3 (b)	14 (b)	5.1 (f)	61 (f)
Propane (C_3H_8)	2.2 (b)	9.5 (b)	2.3 (f)	55 (f)

Letters in parentheses relate to references at the end of Part 13.

----

* % fuel in an air-fuel mix or in an oxygen-fuel mix. Example: For methane burning in air, the lower explosive limit (LEL) = 5.3%, or 94.7 volumes air/5.3 volumes gas = 17.87:1 air/gas ratio. From page 5 of Volume I, the stoichiometric air/gas ratio for methane is 9.53:1. Therefore excess air is 17.87 − 9.53 = 8.34 ft^3 air/ft^3 gas; so % excess air = 100% × 8.34/9.53 = 87.5% excess air; which is an equivalence ratio (ϕ) of 0.5333...Appendix Table C.12.

† The lower flammability limit (LEL) for CO with oxygen is significantly higher whereas LEL with oxygen is generally unchanged or slightly lower for other fuels. This is due to the catalytic effects of H_2O in CO combustion, which also affect the LEL of CO with humid air.

3. *Flame Speed* is also termed burning velocity, ignition velocity, flame propagation velocity. Most references list *maximum* burning velocity and specify the air/fuel ratio at which the maximum occurs (usually at stoichiometric or slightly fuel rich). The bell-shaped curves of flame speed variation with air/fuel ratio are discussed in Reference 13.r at the end of this Part 13. Table 13.8 lists velocity ranges and most probable flame speeds for a few fuels.

Table 13.8. Burning velocities in air and in oxygen for five gaseous fuels. (From Reference 13.d at the end of Part 13.)

Fuel	In Air (e = 21% O_2) Most probable, ft/sec	m/s	Range, % of probable, min	max	In Oxygen (e = 100% O_2) Most probable, ft/sec	m/s	Range, % under	% over
Hydrogen	9.19	2.80	0.89	1.30	38.55	11.75	0.76	1.01
Methane	1.24	0.38	0.87	1.16	12.96	3.95	0.82	1.22
Propane	1.41	0.43	0.93	1.09	12.30	3.75	0.96	1.07
Butane	1.35	0.41	0.92	1.12	11.65	3.55	0.94	1.00
Acetylene	5.25	1.60	0.69	1.13	37.08	11.30	0.84	1.13

Example 13-3. From Table 13.8, the range of burning velocities for hydrogen in air is [9.19 ft/sec × 0.89] to [9.19 × 1.30] = 8.18 to 11.9 ft/sec, or [2.80 m/s × 0.89] to [2.80 × 1.30] = 2.49 to 3.64 m/s.

Figure 13.9 gives a comparison of the effects of oxygen enrichment and air preheat on flame velocity and flammability limits for methane gas. (Most natural gases contain about 90% methane.) Flame velocity increases from about 1 to 11 ft/sec as the oxygen concentration increases from e = 20.9% to e = 100%.

For a flame to be stable, the feed speed of the oxidant-fuel mixture must equal the flame speed. If feed speed exceeds flame speed, the flame will be pushed away from the burner, causing a lift-off or blow-off, or moving the flame to a new "detached" position, i.e. not in contact with the burner nozzle, and therefore more subject to instability caused by furnace currents.

If flame speed exceeds feed speed, the flame will move upstream. This may cause a nozzle-mix flame to go out, or a premix flame to flash back, with possible damage. See Part 1, Volume I. Many burners are designed to minimize flashback by the quenching effect of the mass of a relatively cold nozzle and a steep velocity gradient in the mixture's boundary layer.

The combination of higher flame velocity and wider flammability limits with oxygen-enriched combustion improves flame stability and tends to create short intense flames. The stability of a premixed flame is usually measured in terms of the critical velocity gradients at flashback and blowoff limits. As much as 100-fold to 1000-fold increases in flashback and blowoff velocity gradients are measured with pure oxygen. With oxygen enrichment, the gradients are less, but substantially more than with air. *Extreme caution must be exercised* when oxygen enrichment is considered for premix combustion systems.

Figure 13.9. Effects of oxygen enrichment and air preheating on methane gas flame velocity and flammability limits. (From Reference 13.q at the end of Part 13.)

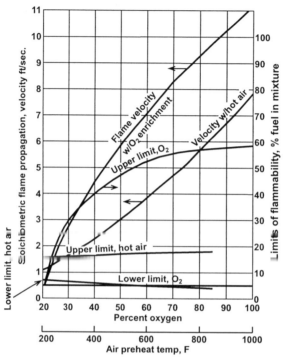

The flame speeds or velocities usually listed (including Table 13.8 and Figure 13.9) are for laminar flow, which exists in only a few industrial burners. Turbulent flame speeds vary with temperature, Reynolds Number, and flame configuration; but are estimated to be 6 to 8 times the listed laminar speeds. Even these higher turbulent flame speeds are not enough to satisfy industrial needs for high bulk throughput velocity across the nozzle. The feed speed = flame speed requirement is met by creating recirculating flows or low velocity boundary flows or by flames stretching out like long cones, thus satisfying the *feed speed = flame speed* requirement at right angles to the flame "surface." As input changes, or flame speed changes (due to temperature changes, for example), the cone length stretches or contracts.

High speed photography shows that most flame surfaces, particularly with large burners, are not smooth cones, but consist of myriad small spurs of flame (each a little cone) because of localized variable velocities, equivalence ratios, and temperatures within the unburned feed. Liquid fuel flames especially exhibit this characteristic because unvaporized fuel droplets behave as tiny projectiles. For all of these reasons, burner design aerodynamics can be very complex if a burner is to be flexible enough to operate with a variety of input rates (turndown ratios), equivalence ratios, oxidants, and fuels.

NOx emissions can be formed in combustion reactions above about 2000 F (1090 C). NOx means NO and NO_2, the NO transforming to NO_2 in the presence of ozone (O_3) and oxygen (O_2) and faster with sunlight and VOCs. NO_2 is a cause of smog and acid rain, and is therefore a criteria pollutant subject to control.

The amount of NO produced in combustion is greater at higher temperatures that break the bonds in O_2 and N_2 molecules, producing reactive atoms and radicals. Both oec (oxygen-enriched combustion) and pac (preheated air combustion) generate higher flame temperatures than cold air combustion; so they both tend to aggravate the NOx emission problem, but modern oec and pac burners incorporate features that reduce this effect.

The amount of NO also increases with higher concentrations of oxygen and nitrogen as found with excess air or oxygen enrichment.* The opposite effect (lower NOx emissions) can be achieved by lowering the oxygen and nitrogen concentrations through dilution with inert gases such as CO_2 and H_2O, which are readily available in the poc (products of combustion). External flue gas recirculation (FGR) and furnace poc recirculation with peripheral gas injectors are used to accomplish this effect, with concurrent lowering of flame temperature.

Conventional wisdom is that oxygen enrichment would increase NOx. Actually, oxygen enrichment, if applied properly, provides additional design flexibility for NOx reduction. The peak flame temperatures of some specifically designed low-flame-temperature oxy-fuel burners are lower than those of some low NOx air burners. Very good temperature uniformity has been demonstrated in full scale furnaces. Tests in a 2000 F furnace with 100% oxygen showed peak flame temperatures only 300 to 1000 degrees F above the furnace temperature. Flame momentum was maintained equivalent to that with air burners by use of small volumes of oxygen at high velocity. (See Reference 13.a.)

Also available to industry are low NOx oxy-fuel burners using staged combustion, or off-stoichiometric firing (fuel rich or oxygen rich), as discussed in Part 11.

------

* Theoretically, NO formation peaks in mid-range between air and pure oxygen (as encountered in some cases of oxygen enrichment), and NO formation falls to zero with 100/0 oxygen/nitrogen (as with oxy-fuel firing).

Dilute oxygen combustion burners avoid high flame temperatures by utilizing some of the same recirculation principles found in high velocity burners (or in direct gas injection), but using the high velocity of low volume, high pressure oxygen streams to induce mixing of inert poc with the oxygen before it mixes with the fuel. See Figure 13.10. If no air is used, and no tramp air has infiltrated, the induced inert poc (diluent) will contain zero or very low nitrogen; so NOx formation will be minimal.

Figure 13.10a. Conventional air-gas burner.

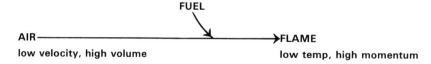

Figure 13.10b. Conventional oxygen-gas burner.

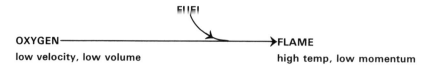

Figure 13.10c. Dilute oxygen combustion (doc) gas burner.

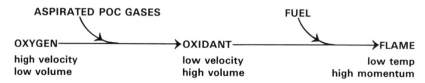

The ultimate design goal of the doc method is to react fuel with an oxidant stream containing the lowest possible oxygen concentration. Wet flue gas or furnace gases are recirculated so as to dilute pure oxygen or air and thereby provide an "oxidant" that is diluted with inert gases (preferably other than nitrogen).

As the recirculation (dilution) is increased, the oxygen concentration and the flame temperature are slowly lowered (from the theoretical adiabatic flame temperature with 100% oxygen). With further increase in poc recirculation, as O_2 concentration goes below about 40%, flame temperature falls off more and more rapidly. For example: 3100 F at 25% oxygen and 2100 F at 15% oxygen, for methane-oxygen combustion. Similar but (surprisingly) higher flame temperatures occur with methane/air combustion.

A minimum adiabatic flame temperature of 1800-2000 F is thought to be required for a stable flame in a practical furnace environment. That would say limit the lowest practical doc to 12-13% oxygen for methane-oxygen combustion, and 10-11% oxygen for methane-air. However, preheating the oxidant overcomes this limitation.

At low oxygen concentrations, the increases in flame temperature become close to the increases in oxidant temperature. At 5% O_2, adiabatic flame temperatures are only 300-500 degrees F above the preheat temperatures; so the use of a very dilute preheated oxygen stream offers potential for stable low NOx combustion.

The volume of oxidant (oxygen-inert mix) required is approximately inversely proportional to the required concentration of oxygen in the oxidant. A huge amount of oxidant is required for dilute oxygen combustion. High fuel and oxygen jet momentums are important. Small high velocity jets can be used to entrain large volumes of furnace gas rapidly.

NOx emission figures presented in this book are burner-specific and furnace-specific. They are offered only as general indications of the levels of NOx that might be found in a process, and to show the magnitude of change that various factors and modifications might have on these levels. Consideration must be given to the specific furnace's operating characteristics, such as:

type of burner and heat release pattern;
heat release per unit of chamber volume;
retention time for the poc;
arrangement of the load relative to the flame;
type of refractory construction; and
air leakage through and around car, doors, seals, and ports.

METHODS FOR COMBUSTION WITH OXYGEN

Techniques available for applying oxygen combustion in industrial furnaces include oxygen enrichment, oxygen lancing, and oxy-fuel firing. They have different retrofit requirements and offer different flexibilities in changing the thermal conditions in a furnace.

The overall furnace productivity and fuel efficiency improvements are largely determined by the amount of equivalent pure oxygen used, and are little influenced by the methods for combustion with oxygen.

Direct oxygen enrichment of combustion air is the simplest technique for applying oxygen using burners designed for normal air combustion. Oxygen is typically injected through a special sparger into combustion air in the combustion air supply pipe near the burner. The sparger should be designed for rapid dispersion and uniform mixing with the air. It should be made of an oxygen-compatible material; and direct impingement of oxygen jets on the combustion air pipe must be avoided to prevent possible oxygen fire in the steel pipes. In addition, it is recommended that dirt and combustible materials be cleaned from the combustion air pipes and from burner parts prior to oxygen service, especially for high oxygen enrichment levels. See Reference 13.s.

O_2-air upstream-mixing, shown in Figure 13.11, mixes oxygen with the combustion air just upstream of the burner. GREAT CARE MUST BE TAKEN to assure that, in the event of blower air flow reduction, an interlock will prevent oxygen from backing out through the blower where it might contact grease or oil.

Flame temperature increases sharply with oxygen enrichment, as discussed in the third sub-topic of the preceding section on "Effects of Oxygen on Combustion." If the fuel input is kept the same after addition of oxygen enrichment, the volume flow rate of the enriched air-oxygen mixture will be reduced compared with the base case of air combustion; so the momentum of the flame will be reduced. Both of these factors (higher flame temperature and lower flame momentum) tend to increase the temperature near the burner port. The highest enrichment level (e) is often limited to less than 30% oxygen by the maximum service temperatures of the burner block and the refractory walls near the flame.

The principal advantage of this direct enrichment method is the relatively simple modification required to retrofit existing burners. It is well suited to small increases in net furnace heat input (approximately 10-20%), which can be accomplished by a small percent increase in oxygen concentration. This method must be applied with caution for furnaces in which uniform temperature distribution is critical.

O_2-air nozzle-mixing is a relatively simple way to add oxygen to an existing burner, piping it through the burner in the manner of nozzle-mixing. This is most conveniently accomplished with a gas burner for which a dual-fuel retrofit is available as a standard add-on atomizer for oil. The oil tube or the atomizing air tube may be used for delivery of oxygen to the center of the flame if (a) the % oxygen in the total oxidant through the burner is very low, e.g. 30-35%, and if (b) the atomizer has never been used with oil, is thoroughly cleaned, and complies with all material requirements of Reference 13.s.

Figure 13.11. O_2-air upstream-mixing is a simple and popular method for oxygen-enriching for higher flame temperature and improved fuel efficiency, but it is not very flexible and the enrichment level is limited. Some sort of directional air flow sensor should be used upstream of the mixing point as a safety interlock.

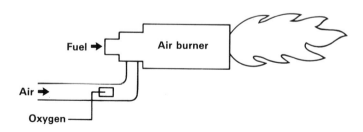

Oxygen lancing is usually installed through a furnace wall near a conventional burner, shooting oxygen angularly into a side of the conventional flame, as shown in Figure 13.12. The extent of flame modification by this technique is strongly dependent on the location, direction, and momentum of the oxygen jet relative to the fuel and air streams of the main burner.

Figure 13.12.   Undershot lancing of oxygen beneath a conventional burner flame.

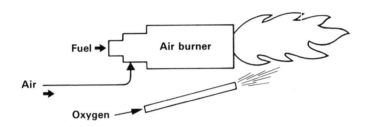

In the so-called "undershot" method, an oxygen jet is injected upward toward a main burner flame to create a hotter oxygen flame on the lower side of the main flame. The higher temperature of the lower side of the flame and its closer proximity to the load enhance local heat transfer, while minimizing overheating of the furnace roof. This arrangement has been popular in reverberatory melting furnaces.

Reference 13.h reports the following results of a study of oxygen injected parallel to an oil burner:

a. Much higher peak temperature with oxygen enrichment.
b. Peak temperature zone shifted from the axis of the main burner to the axis of the oxygen jet.
c. Significant change in flame radiation because of changes in carbon concentration and in gas temperature profile.
d. Local heat flux rates were modified by the above changes.
e. Only small differences were observed in overall fuel efficiency by different oxygen injection methods tested.

The principal advantages of the oxygen lancing method are the low cost of retrofit, and some flexibility in modifying the flame characteristics of the main air burner.

Oxy-fuel burners may be used to supplement conventional air burners, or to replace some of the air burners. In the supplemental case, additional oxy-fuel burners are placed where more heat is desirable. In the replacement case, one or more air burners are replaced with oxy-fuel burners to increase the overall available heat (efficiency) for the furnace.

The main advantage of oxy-fuel burners is the high flexibility in influencing heat flux distribution within the furnace. Auxiliary oxy-fuel burners can be positioned to improve the temperature distribution within the furnace, or high temperature oxygen flames can be directly applied to furnace load, as in glass furnaces to accelerate melting.

Traditional oxy-fuel burners have used intense mixing of oxygen and fuel to create high temperature flames. Many specialized oxy-fuel burners have been designed for specific industrial applications. New burners are available with adjustable flame temperatures.

Low NOx emissions have been achieved by using staged combustion or flue gas recirculation (with in-furnace aspirating of furnace gases) to reduce peak flame temperature.

Impinging flames are sometimes used to achieve very high convective heat transfer rates, resulting from the high concentration of dissociated species in oxy-fuel flames.

In the past, water cooling and high temperature alloys have been used to protect burners from the high heat flux of oxy fuel flames. However, a number of modern burners use gaseous cooling instead of liquid cooling.

Suggested Modes of Operation. For optimum performance, oxygen use should be limited to those parts of the heating cycle where (a) the gross input to the furnace is the greatest, and (b) the potential for generating NOx is greatest (high furnace temperature).

Dual air-or-oxygen burners are available. They can be operated on either air or oxygen, minimizing oxygen costs while conserving fuel and limiting NOx emissions.

For applications where greater productivity is desired, there is limited gain after soak temperature is reached; so oxygen is often used during the heat-up period. With the temperature low, NOx generation will be low. At low temperatures, radiation heat transfer is low; so there is greater dependence on convection heating. By operating with air instead of oxygen, the mass and volume of hot gases circulating within the furnace will be higher, thereby aiding convection heat transfer and uniformity.

After a furnace reaches temperatures above 1400 F (760 C), the major mode of heat transfer is radiation, and also NOx formation is more probable. At this point oxygen operation will eliminate much nitrogen from the furnace atmosphere.

Elimination of nitrogen reduces the potential for NOx formation and decreases the mass of stack gas, reducing stack loss, or increasing the % available heat; so less fuel is required. Less fuel consumed and elimination of nitrogen result in much less volume flow of poc through the furnace. The retention time of the poc will therefore increase. That assures more complete combustion and increases heat transfer time, improving overall efficiency.

At furnace temperatures above about 1400 F (760 C), a great part of the heat transfer comes from gas radiation and also by reradiation from the walls and roof. The poc from a standard air-gas fired system contain about 26% triatomic molecules, principally CO_2 and H_2O vapor. The energy radiated by these molecules varies with their concentration, thickness of the gas blanket, and temperature. Figure 13.13, specifically for the CO_2 and H_2O concentration from a typical natural gas burned with air, shows how the gas radiation varies with temperature and blanket thickness.

Figure 13.13. Gas radiation heat flux rates from the poc of a typical natural gas can be calculated from the formulas at the top of this graph if the gas temperature and blanket thickness are known.

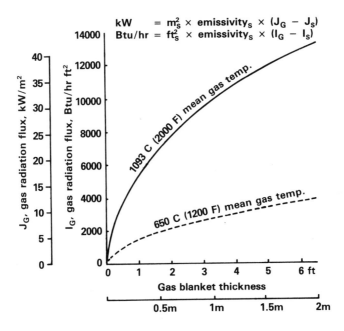

When burners are switched from air-gas operation to oxy-fuel operation, the concentration of triatomic molecules increases from 26% to near 100%, thus increasing the gas radiation heat transfer. This rise in gas radiation helps heat transfer. For real furnace conditions with constant heat input, however, the gas temperature tends to drop slightly. The added gas radiation apparently about balances the lower temperature, luminosity, and convection. Depending on the type of furnace, the gas radiation could aid temperature uniformity and provide a modest increase in production.

APPLICATIONS AND ECONOMICS

Productivity improvement, fuel savings, and reduced pollutant emissions are the main benefits of oxygen-enriched combustion (oec). Because of the additional cost of oxygen, oec was traditionally considered only for special cases, to overcome the deficiencies of conventional air-fuel combustion. The most common reason for oec was to increase productivity of a furnace by raising heat flux to the furnace load. Specialized oxy-fuel burners and oxygen lancing techniques have been applied for high temperature furnaces such as electric arc steel scrap melting furnaces and glass melting furnaces.

Recent advancement in air separation technology, especially in PSA and membrane technology, and the escalation of fuel costs have justified the use of oxygen more than before. In the early 1980s, the complete oxy-fuel conversion of certain high temperature furnaces was shown to be economical for fuel savings alone. New oxy-fuel burners with flame characteristics equivalent to conventional air burners have been developed (References 13.a, e, and l) and adopted for some steel reheating furnaces and for aluminum melting furnaces.

Tighter air pollution regulations, especially for NOx emissions, have provided additional economic benefits to 100% oxy-fuel combustion. Many glass melting furnaces and hazardous waste incinerators have been converted to oxy-fuel firing to reduce emissions and to save capital costs for pollution abatement and heat recovery. The economics of these applications are complex and site-specific. Capital costs for furnace conversion and air pollution control equipment, plus operating costs for fuel and oxygen must be analyzed together with potential productivity improvement.

In the following sections, general considerations are discussed for oec applications for industrial furnaces. More detailed discussions and some examples for economic analyses are found in Reference 13.j.

Productivity improvement has been successful by use of oxygen enrichment in a broad range of industrial furnaces as listed in Table 13.14. In most furnaces, throughput increases of 10% to 20% are typically possible with a few percentage points increase in oxygen concentration. For example, at 23% oxygen concentration (i.e. 2% above that of air), the total amount of oxygen present in a unit volume of enriched air is 10% greater than it was with air. Consequently, the fuel input can be increased by 10%. The available heat from the combustion with extra oxygen is much greater than that with air; so the total input of available heat to the furnace is increased by 10% to 20%, depending on the furnace temperature. From the overall energy balance, it is expected that about 10% to 20% increase in productivity will be possible with only 2% oxygen enrichment above the air level.

Table 13.14. Industrial furnaces and kilns for applications of oxygen enrichment

Industry	Furnaces/Kilns	Primary benefits*
Aluminum	Remelting	1, 2
	Coke calcining	1
Cement	Calcining	1
Chemical	Incineration	1, 2, 3, 4
Clay	Brick firing	1, 2, 3
Copper	Smelting	1, 2, 3
	Anode	2
Glass	Regenerative melters	1, 2, 4
	Unit melters	1, 2, 4
	Day tanks	1, 2, 4
Iron and Steel	Soaking pits	2, 1
	Reheat furnaces	2, 1
	Ladle preheat	1
	Electric arc melters	1, 2
	Forging furnaces	1, 2
Petroleum	FCC Regenerator	1
	Claus sulfur	1
Pulp and paper	Lime kilns	1, 2, 3
	Black liquor	1, 2

The extent of productivity improvement possible varies with the nature of the furnace limitations. The most common limitations relate to the air supply system capacity and the flue gas handling system capacity. Oxygen enrichment is very effective in overcoming these limitations because of the lesser volume of oxidant and flue gas for the same fuel input and higher available heat to the furnace. Productivity increases of more than 40% have been reported with oxygen enrichment in glass melting, steel reheating, and aluminum remelting furnaces. Dust carry-over problems in glass melters and in rotary drum incinerators, calciners, and kilns (cement, lime) have also been effectively alleviated by the smaller flue gas volumes resulting from oxygen enrichment.

Although not common, heat transfer from flame can be a limiting factor in some applications. High temperature oxygen-enriched or oxy-fuel flames have been successfully applied to certain glass melters and kilns to increase heat transfer to strategic areas in the furnaces. In most furnaces, however, uniform temperature distribution is a critical requirement. High temperature flames also cause serious concerns for furnace refractory walls and roofs.

---

* Benefits of oxygen: 1 = productivity improvement, 2 = energy saving, 3 = quality improvement, 4 = emissions reduction.

Higher flame temperature is not the only way to achieve a higher heat transfer rate. Gas radiation from hot combustion products to the surrounding refractory walls and re-radiation to the load constitute the primary modes of heat transfer in many high temperature furnaces. The intensity of gas radiation is a function of the gas temperature and the concentrations of CO_2, H_2O, and soot. With oxygen enrichment, reduced concentration of N_2 in the poc means the concentrations of triatomic gases will be higher; so the gas radiation will be stronger.

In a radiation-dominant furnace with uniform temperature requirement, the preferred method for productivity improvement is to increase the bulk gas temperature in the furnace; not the localized flame temperature. Various oxygen enrichment techniques are available to create intense high temperature flames or low temperature high momentum flames, depending on the process requirement.

Fuel Savings. Both oxygen-enriched combustion (oec) and preheated air combustion (pac) improve fuel efficiency. In Figure 13.15, fuel required to provide one million Btu of available heat to a furnace is plotted as a function of furnace flue gas exit temperature for ambient air combustion and three different levels of pac and oec. As the flue gas temperature increases, more fuel is required to provide the same amount of available heat to the furnace. With ambient air combustion, the fuel requirement increases sharply at high temperatures. With 100% oxygen, the fuel requirement is much smaller than with ambient air, and increases only slightly with flue gas temperature because of the small flue gas mass and resulting small flue heat loss.

Figure 13.15.   Fuel requirement to provide 1 000 000 Btu of available heat.

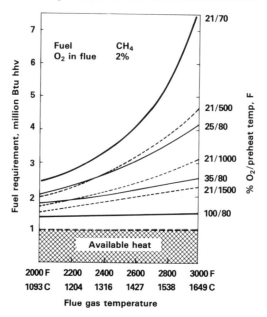

Higher fuel savings are achieved as oec approaches 100% oxygen, or with higher temperature preheated air. State-of-the-art direct-fired integral burner/regenerators can produce air preheat temperatures within 400 F (204 C) of the furnace's flue gas exit temperature. The thermal efficiencies (%s available heat) of such pac systems can exceed those of 100% oec at flue temperatures above 2100 F (1150 C). The choice between oec and pac depends on the overall economics and other process requirements, which will be discussed in the next section.

When considering oec for fuel savings, an important economic parameter is the ratio of fuel saved to oxygen required, or specific fuel saving. Figure 13.16 shows that % fuel saved increases sharply at low enrichment levels and increases at a lesser rate at high enrichment levels. Substantial fuel savings are obtained at the modest enrichment levels achievable with membrane air separation systems (28 to 35% oxygen). Very little fuel is saved by increasing oxgyen enrichment above about 80% purity. Thus the high purity oxygen from a cryogenic separation system offers little fuel-saving advantage over the 80 to 95% purity of a PSA separation system.

Figure 13.16.   Fuel savings and specific fuel savings from oxygen enrichment, based on natural gas, with 2400 F flue gas temperature and 2% excess oxygen in the flue gas.

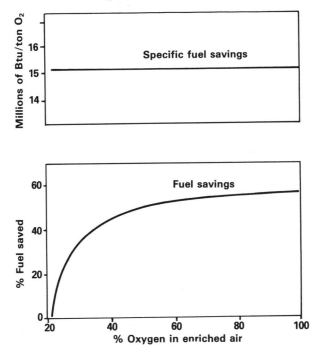

Although Figure 13.16 suggests diminishing benefits with higher oxygen enrichment, the incremental fuel saving per unit amount of additional equivalent pure oxygen (Figure 13.17) actually remains constant for any oxygen enrichment level, when the furnace flue gas exit temperature, excess O_2, and heat load are kept constant.

Figure 13.17 shows specific fuel savings in millions of Btu/ton of oxygen consumed, for methane as the fuel, applicable at any level of oec. Each curve represents the fuel saving over a base air case with a different combustion air preheat temperature. For example, with 2400 F flue gas temperature, the specific fuel saving is 15.2 million Btu per ton of oxygen over ambient air, or 5.1 million Btu/ton compared with 1000 F air. The x-intercept of each curve (i.e. at zero savings) represents the flue gas temperature at which the thermal efficiencies of oec and pac become the same.

Figure 10.17. Specific fuel savings with 100% oxygen compared with various air preheat temperatures. (From Reference 13.i)

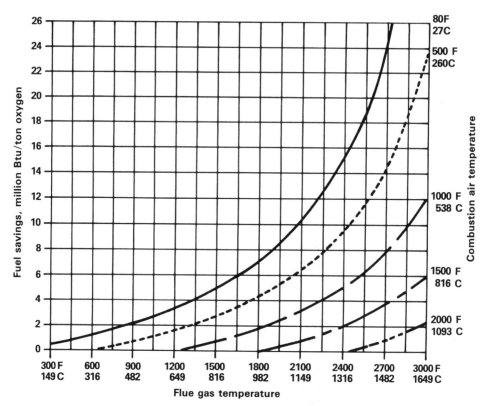

If the cost of fuel and oxygen are $3 per million Btu and $30 per ton of equivalent pure oxygen, the break-even specific fuel saving becomes

$$\frac{\$30/\text{ton oxygen}}{\$3/\text{million Btu}} = 10 \text{ million Btu/ton oxygen.}$$

In the above example, oec resulted in a specific fuel saving of 15.2 million Btu/ton O_2, or $3 \times 15.2 = \$45.6$ saved/ton O_2. Subtracting the $30 cost of oxygen, shows a net saving of $15.6 compared with ambient air. Compared with 1000 F preheated air, (3×5.1) $-$ $30 = -\$14.7$, a net loss if oxygen were used. Graphs are available in Reference 13.i for similarly checking the economics of oec for a variety of fuels and combustion conditions.

Economics. Oxygen-enriched combustion (oec) offers a fast, low-capital retrofit without major changes in existing furnace and combustion equipment. Although a significant reduction in specific energy consumption is often realized through productivity increases, fuel savings alone may not pay the cost of oxygen required. The economics of oec depend primarily on the value of the additional products produced through the furnace; so the optimum arrangement uses just enough oxygen enrichment to achieve the desired productivity increase.

Oec is not usually economical, fuel-wise, for boilers and other low temperature processes, or those with good heat recovery devices. It may be economical for some high temperature furnaces, particularly if the existing heat recovery system is not efficient. Economic evaluation of oec for fuel savings compared with other heat recovery options is usually very complex and site-specific, because so many factors influence the retrofit costs and the cost of oxygen. The true operating cost of oec is the cost of electric power for the oxygen source system. The capital cost is included in the "oxygen cost."

The capital costs of oec and pac heat recovery systems increase with furnace temperature. (With oxy-fuel firing, the same equipment is generally used regardless of furnace temperature.) Maintenance and replacement costs increase with more corrosive flue gases. The costs of flue gas handling and cleaning systems (including stacks) are less with oec because of the reduced flue gas volume.

In general, oec has a low capital cost and high operating cost as compared with pac. Economically, oec is more favorable at high temperatures and for flue gases laden with particulates, corrosive gases, and high NOx emissions. Glass melting is an example where many furnaces have been converted to 100% oec in recent years to attain the combined benefits of reduced NOx and other emissions, furnace rebuild costs, and fuel consumption. Further information on the economic aspects can be found in Reference 13.i.

References 13.m and 13.n predict fuel savings with combustion air preheat (cap) and oxygen-enriched combustion (oec). These are based on 10% excess air (2% oxygen in the flue gas) with natural gas as the fuel; but savings will be about the same with fuel oils. (Both references are derived from the same enthalpy and dissociation formulas; so they give comparable savings. Use of tables from different sources may give misleading results.) *Fuel saved is only one of many factors to be considered when comparing pac and oec.*

NOTE: Oxygen is usually sold in units of 100 cubic feet measured at 70 F; sometimes by the ton (2000 pounds = 909.1 kilograms). One ton of pure oxygen occupies 24 180 cubic feet measured at 70 F or 21 C, and 14.696 psia or one atmosphere. To convert an oxygen price from $ per ton to $/ft³, divide $/ton by 24 180 ft³/ton.

Example 13-4: Convert an oxygen price from $80/ton, to $/hundred cubic feet:

$$\frac{\$80/\text{ton}}{24\ 180\ \text{ft}^3/\text{ton}} = \$0.00331/\text{ft}^3, \quad \text{or } \$0.331/\text{hundred ft}^3.$$

SUMMARY

Oxygen-enriched combustion and oxy-fuel firing have been found to be economically viable in some applications. Engineers must make thorough studies of all aspects of their processes and of state-of-the-art combustion and control equipment.

Both oxygen-enriched combustion (oec) and preheated air combustion (pac) improve fuel efficiency and raise flame temperature, which improves radiation heat transfer. Both improve burner stability by broadening flammability limits and increasing flame velocity. However, both oec and pac require special burners, piping or ducting, valves, and controls; and both tend to aggravate NOx by their high flame temperatures. Lower NOx burners are continually being developed for both oec and pac. Low flame temperature oxy-fuel firing can minimize NOx, especially when little nitrogen enters the flame from the fuel, load, or furnace infiltration.

The volume, and sometimes the velocity or momentum, of circulating poc gases within a furnace is reduced by any effort (oec or pac) to lower fuel and air consumption. Reduced poc volume means less space and expense for post-combustion cleanup equipment (scrubbers, baghouses, ID fans), and less chance of particulate carryover loading. Oxygen-enriched combustion tends to reduce poc volume more than pac does.

The above benefits must be weighed against the fact that less circulation may mean poorer temperature uniformity, and therefore longer fuel use to regain the uniformity. Oxygen enrichment tends to lower the circulation volume more than does use of pac. The recirculation can sometimes be maintained by keeping the momentum constant.

When oec or pac (preheated air combustion) is being considered for application to a furnace, the effects on heat transfer and temperature uniformity must be carefully evaluated. The selected burner type has a strong effect on both heat transfer and NOx emissions; so careful consultation with burner experts is advised. The number of burners and their placement may have to be changed. Flame momentum may have to be increased to maintain the same furnace gas recirculation ratio.

A very thorough analysis should be made comparing *all* costs. Safety costs are difficult to evaluate, but must be considered. Handling oxygen may create some safety hazards; so it is important to utilize the judgment of an engineer experienced in selecting oxygen valve/control trains.

REFERENCES

13.a **Anderson, J. E.:** "A Low NOx, Low Temperature Oxygen Burner", 1986 Symposium on Industrial Combustion Technologies, Gas Research Institute, Chicago, IL, 1986.

13.b **Bodurtha, F. T.:** "Industrial Explosion Prevention and Protection", pg. 21, McGraw-Hill Inc., New York, NY; 1980.

13.c **Bomelburg, H. J.:** "Efficiency Evaluation of Oxygen Enrichment in Energy Conversion Processes", Report No. PNL-4917, for U.S. Department of Energy, Washington, D.C., December 1983.

13.d **Booker:** Document F21/ca/50, International Flame Research Foundation, IJmuiden, Netherlands, 1981.

13.e **Browning, R. A. et al:** "Recent Advances in Oxygen Combustion Technology", CIM Conference of Metallurgists, Winnipeg, 1987.

13.f **Coward and Jones:** "Limits of Flammability of Gases & Vapors", pg. 131, Bulletin 503, U.S. Bureau of Mines, 1952

13.g **Glassman, I.:** "Combustion", pg. 80, Academic Press, NY, NY, 1977.

13.h **Kissel, R. R. and Michand, M.:** "International Flame Research Foundation: First Experiments at IJmuiden on the Combustion of Oil Using Oxygen," pp. 109-120, *J. Inst. of Fuel*, March, 1962.

13.i **Kobayashi, H.:** "Oxygen Enriched Combustion System Performance Study, Volume I = Technical and Economic Analysis", U.S. Dept. of Energy, Idaho Operations; 1987; DOE/ID/12597.

13.j **Kobayashi, H.:** "Oxygen Enriched Combustion System Performance Study, Volume II = Market Assessment", U.S. Dept. of Energy, Idaho Operations; 1988; DOE/ID/12597-3.

13.k **Kobayashi, H. et al:** "Oxygen Enriched Combustion System Performance Study, Volume III = Burner Tests and Combustion Modeling", U.S. Dept. of Energy, Idaho Operations; 1988.

13.l **Kobayashi, H. and Du, Z.:** "Dilute Oxygen Combustion", American Flame Research Foundation, Cambridge, Oct., 1992.

13.m **North American Mfg. Co..** "Fuel Savings from Preheated Air", (Handbook Supplement 155b), 1993.

13.n **North American Mfg. Co.:** "Fuel Savings from Oxygen Enrichment or Oxy-Fuel", (Handbook Supplement 276), 1994.

13.o **Reed, R. J.:** "Combustion Handbook", Volume I; 3rd edition, pp. 10,12; North American Mfg. Co., Cleveland, OH; 1986.

13.p **Spiers, H. M. (ed.):** "Technical Data on Fuel", 6th ed., pp. 260-265, British National Committee, World Power Conference, London, England, 1962.

13.q **Turin, J. J. and Huebler, J. H.:** Gas-Air-Oxygen Combustion Studies, *Report to Committee on Industrial and Commercial Gas Research*, American Gas Association (AGA), Project I.G.R.-61, pp. 1-28, 1951.

13.r **Vandaveer, F. E. and Segeler, C. G.:** "Combustion" in Segeler, C. G. (ed.): "Gas Engineers Handbook", pp. 2/1-2/148, 2/78, The Industrial Press, New York, NY, 1967.

13.s **Werley, B. L. (ed.):** "Flammability and Sensitivity of Materials in Oxygen-Enriched Atmospheres". ASTM Special Technical Publication 812, (ASTM PCN 04-812000-17), 1983.

13.t **Zabetakis:** "U.S. Bureau of Mines Bulletin 627", 1965, as reported on page 253 of Rose and Cooper, "Technical Data on Fuel", 1977.

ADDITIONAL SOURCES of information relative to oxygen

Bonnekamp, H., et al: "Extension of Possibilities of Utilization of Fuels by Addition of Oxygen", 1978, Stahl & Eisen, 98:141-149.

Brame, J. S. S. and King, J. G: "Fuel: Solid, Liquid, and Gaseous", 6th ed., St. Martin's Press, New York, NY, 1967. (Out of print.)

Compressed Gas Association: "Cleaning Equipment for Oxygen Service", CGA Pamphlet G-4.1, 1985.

Compressed Gas Association: "Industrial Practices for Gaseous Oxygen Transmission/Piping", CGA Pamphlet G-4.4, 1980.

National Fire Protection Association: "Bulk Oxygen Systems at Consumer Sites", NFPA JN-50, Quincy, MA, 1990.

National Fire Protection Association: "Fire Hazards in Oyxgen-Enriched Atmospheres", NFPA JF-53, Quincy, MA, 1994.

Puri, I. K. (ed.): "Environmental Implications of Combustion Processes", CRC Press, Boca Raton, FL, 1993.

Strahlel, W. C.: "An Introduction to Combustion", Gordon and Breach Science Publishers, 1993.

APPENDIX
List of tables and charts

APPENDIX
List of tables and charts

Table A.1 US. Properties of air at elevated temperatures

Temperature, F	G, gas gravity	Density, lb/ft³	Volume expansion ratio†	Volume of 1 lb dry air, ft³	Heat content of dry air		%Volume of water vapor in saturated air at 1 atm.	Absolute viscosity, lb/ft · sec	Thermal conductivity Btu ft/ft² hr °F
					Btu/lb	Btu/ft³ stp air			
60	1.000 0	0.076 5	1.000	13.07	0.00	0.000	1.744	0.000 012 06	0.014 6
80	0.963 4	0.073 7	1.038	13.57	4.86	0.372	3.452	0.000 012 42	0.015 1
100	0.928 1	0.071 0	1.078	14.08	9.72	0.744	6.467	0.000 012 76	0.015 5
120	0.896 7	0.068 6	1.115	14.58	14.6	1.12	11.53	0.000 013 10	0.016 0
140	0.866 7	0.066 3	1.154	15.08	19.4	1.49	19.68	0.000 013 45	0.016 5
160	0.839 2	0.064 2	1.192	15.58	24.3	1.86	32.29	0.000 013 79	0.017 0
180	0.813 1	0.062 2	1.230	16.08	29.2	2.23	51.14	0.000 014 11	0.017 4
200	0.788 2	0.060 3	1.269	16.58	34.0	2.60	78.45	0.000 014 43	0.017 8
250	0.732 0	0.056 0	1.366	17.86	46.2	3.54		0.000 015 22	0.018 9
300	0.683 7	0.052 3	1.463	19.12	58.4	4.47		0.000 015 97	0.020 0
400	0.605 2	0.046 3	1.652	21.60	79.7	6.1		0.000 017 40	0.022 0
500	0.541 2	0.041 4	1.848	24.15	106	8.1		0.000 018 76	0.024 0
600	0.490 2	0.037 5	2.040	26.67	129	9.9		0.000 020 04	0.026 0
700	0.448 4	0.034 3	2.230	29.15	154	11.8		0.000 021 26	0.027 8
800	0.413 1	0.031 6	2.421	31.65	179	13.7		0.000 022 41	0.029 6
900	0.381 7	0.029 2	2.620	34.25	205	15.7		0.000 023 52	0.031 4
1000	0.355 6	0.027 2	2.812	36.76	233	17.8		0.000 024 58	0.033 2
1100	0.334 7	0.025 6	3.000	39.22	260	19.9		0.000 025 59	0.035 0
1200	0.313 7	0.024 0	3.188	41.67	288	22.0		0.000 026 59	0.036 7
1300	0.296 7	0.022 7	3.382	44.20	315	24.1		0.000 027 54	0.038 2
1400	0.279 7	0.021 4	3.575	46.73	342	26.2		0.000 028 49	0.039 7
1500	0.266 0	0.020 4	3.769	49.27	370	28.3		0.000 029 45	0.041 1

(continued)

† Volume of one cubic foot of dry stp air at the listed temperature, = 1/G.

Table A.1 US (concluded)

Temperature, F	G, gas gravity	Density, lb/ft³	Volume expansion ratio†	Volume of 1 lb dry air, ft³	Heat content of dry air Btu/lb	Heat content of dry air Btu/ft³ stp air	Absolute viscosity, lb/ft · sec	Thermal conductivity Btu ft / ft² hr °F
1600	0.2523	0.0193	3.964	51.81	397	30.1	0.000 030 41	0.0425 5
1700	0.2412	0.0185	4.155	54.32	425	32.5	0.000 031 17	0.0438 8
1800	0.2301	0.0176	4.346	56.82	452	34.5	0.000 031 93	0.0451 1
1900	0.2210	0.0169	4.534	59.28	481	36.3	0.000 032 74	0.0462 2
2000	0.2118	0.0162	4.721	61.73	510	39.0	0.000 033 54	0.0475 5
2100	0.2037	0.0156	4.918	64.30	538	41.1	0.000 034 31	0.0487 7
2200	0.1955	0.0150	5.115	66.87	568	43.5	0.000 035 08	0.0499 9
2300	0.1886	0.0145	5.309	69.41	597	45.1	0.000 035 82	
2400	0.1817	0.0139	5.504	71.94	626	47.1	0.000 036 55	
2500	0.1758	0.0135	5.695	74.43	655	50.1	0.000 037 26	
2600	0.1699	0.0130	5.886	76.92	685	52.1	0.000 037 97	
2700	0.1647	0.0126	6.078	79.45	714	54.1	0.000 038 66	
2800	0.1595	0.0122	6.270	81.97	742	56.1	0.000 039 34	
2900	0.1549	0.0119	6.462	84.47	771	59.1		
3000	0.1503	0.0115	6.653	86.96	801	61.1		
3100	0.1464	0.0112	6.835	89.35	830	63.1		
3200	0.1425	0.0109	7.018	91.74	859	65.1		
3300	0.1386	0.0106	7.223	94.42	887	67.1		
3400	0.1346	0.0103	7.429	97.09	918	70.1		
3500	0.1314	0.0101	7.618	99.57	946	72.1		
3600	0.1281	0.0098	7.806	102.04	975	74.1		

† Volume of one cubic foot of dry stp air at the listed temperature, = 1/G.

Table A.2 Metric. Properties of air at elevated temperatures

Temperature, C	G, gas gravity	Density, kg/m³	Volume expansion ratio†	Volume of 1 kg dry air, m³	Heat content of dry air kcal/kg	Heat content of dry air kcal/m³	% Volume of water vapor in saturated air at 1 atm.	Absolute viscosity, centipoise	Thermal conductivity, W·m/m²°K
20	0.985 6	1.208	1.015	0.827 9	1.1	1.3	2.308	0.018 13	0.025 71
40	0.922 9	1.131	1.084	0.884 2	5.9	7.2	7.287	0.019 07	0.027 25
60	0.866 7	1.062	1.154	0.941 5	10.8	13.2	19.68	0.019 99	0.028 77
80	0.818 4	1.003	1.222	0.997 1	15.6	19.1	46.77	0.020 88	0.030 26
100	0.773 9	0.948 4	1.292	1.054	20.5	25.0	100.0	0.021 73	0.031 73
125	0.725 5	0.889 1	1.378	1.125	26.6	32.5		0.022 75	0.033 52
150	0.682 3	0.836 2	1.466	1.196	32.7	39.9		0.023 78	0.035 27
200	0.610 4	0.748 1	1.638	1.337	43.3	52.9		0.025 70	0.038 66
250	0.552 9	0.677 6	1.809	1.476	56.3	68.7		0.027 53	0.041 89
300	0.504 6	0.618 4	1.982	1.617	68.1	83.2		0.029 28	0.045 00
350	0.464 1	0.568 7	2.155	1.758	80.3	98.0		0.030 92	0.047 98
400	0.428 8	0.525 5	2.332	1.903	92.8	113		0.032 49	0.050 88
450	0.400 0	0.490 2	2.500	2.040	106	129		0.034 01	0.053 63
500	0.373 9	0.458 2	2.675	2.182	119	145		0.035 49	0.056 33
600	0.330 7	0.405 3	3.024	2.467	146	178		0.038 26	0.061 45
700	0.296 7	0.363 6	3.370	2.750	174	212		0.040 86	0.066 30
800	0.269 3	0.330 0	3.713	3.030	201	245		0.043 32	
900	0.245 8	0.301 2	4.068	3.320	229	280		0.045 64	
1000	0.227 4	0.278 7	4.398	3.588	256	313		0.047 88	
1200	0.196 1	0.240 3	5.099	4.161	314	383		0.052 07	
1400	0.172 5	0.211 5	5.797	4.729	373	456		0.055 95	
1600	0.154 2	0.189 0	6.485	5.291	431	526		0.059 60	
1800	0.139 9	0.171 4	7.148	5.834	489	597			
2000	0.126 8	0.155 4	7.886	6.435	548	669			

† Volume of one cubic foot of dry stp air at the listed temperature, = 1/G.

Table A.3 US. Factors for correcting gas volumes for pressure

Listed below are multipliers for correcting the measured volume of any perfect gas (including air) from the pressure at which it was measured to a base pressure of atmospheric pressure (zero base) or to a base pressure of 6 ounces per square inch. These correction factors are based on an atmospheric pressure of 29.92 inches of mercury. Use of this table is illustrated in Example 2-7. When applied to air, the factors in the zero base column represent the gravity of the air relative to air at standard atmospheric pressure.

Gauge pressure	Factor Zero base	Factor 6 osi base	Gauge pressure	Factor Zero base	Factor 6 osi base
20 in. Hg (vac)	0.331 7	0.323 5	1 psi	1.068 0	1.041 4
19 in. Hg (vac)	0.365 2	0.356 1	2 psi	1.136 0	1.107 8
18 in. Hg (vac)	0.398 6	0.388 6	3 psi	1.204 1	1.174 1
17 in. Hg (vac)	0.432 0	0.421 2	4 psi	1.272 1	1.240 5
16 in. Hg (vac)	0.465 6	0.453 8	5 psi	1.340 2	1 306 8
15 in. Hg (vac)	0.498 8	0.486 4	6 psi	1.408 2	1.373 2
14 in. Hg (vac)	0.532 2	0.519 0	7 psi	1.476 3	1.439 5
13 in. Hg (vac)	0.565 6	0.551 5	8 psi	1.544 3	1.505 9
12 in. Hg (vac)	0.599 0	0.584 1	9 psi	1.612 4	1.572 2
11 in. Hg (vac)	0.632 4	0.610 7	10 psi	1.680 4	1.638 6
10 in. Hg (vac)	0.665 8	0.649 3	12 psi	1.816 5	1.771 3
9 in. Hg (vac)	0.699 3	0.681 9	14 psi	1.952 6	1.904 0
8 in. Hg (vac)	0.732 7	0.714 4	16 psi	2.088 7	2.036 7
7 in. Hg (vac)	0.766 1	0.747 0	18 psi	2.224 8	2.169 4
6 in. Hg (vac)	0.799 5	0.779 6	20 psi	2.360 9	2.302 1
5 in. Hg (vac)	0.832 9	0.812 2	22 psi	2.497 0	2.434 8
4 in. Hg (vac)	0.866 3	0.844 8	24 psi	2.633 0	2.567 5
3 in. Hg (vac)	0.899 7	0.877 3	26 psi	2.769 1	2.700 2
2 in. Hg (vac)	0.933 1	0.909 9	28 psi	2.905 2	2.832 9
1 in. Hg (vac)	0.966 5	0.942 5	30 psi	3.041 3	2.965 6
0.8 in. Hg (vac)	0.973 2	0.949 0	32 psi	3.177 4	3.098 4
0.6 in. Hg (vac)	0.979 9	0.955 5	34 psi	3.313 5	3.231 1
0.4 in. Hg (vac)	0.986 6	0.962 1	36 psi	3.449 6	3.363 8
0.2 in. Hg (vac)	0.993 3	0.968 6	38 psi	3.585 7	3.496 5
0 in. Hg	1.000 0	0.975 1	40 psi	3.721 8	3.629 2
1 osi	1.004 2	0.979 2	42 psi	3.857 9	3.761 9
2 osi	1.008 5	0.983 4	44 psi	3.994 0	3.894 6
3 osi	1.012 7	0.987 5	46 psi	4.130 1	4.027 3
4 osi	1.017 0	0.991 7	48 psi	4.266 1	4.160 0
5 osi	1.021 2	0.995 8	50 psi	4.402 2	4.292 7
6 osi	1.025 5	1.000 0	52 psi	4.538 3	4.425 4
7 osi	1.029 8	1.004 1	54 psi	4.674 4	4.558 1
8 osi	1.034 0	1.008 2	56 psi	4.810 5	4.690 8
9 osi	1.038 3	1.012 4	58 psi	4.946 6	4.823 5
10 osi	1.042 5	1.016 5	60 psi	5.082 7	4.956 2
11 osi	1.046 8	1.020 7	62 psi	5.218 8	5.088 9
12 osi	1.051 0	1.024 8	64 psi	5.354 9	5.221 6
13 osi	1.055 3	1.029 7	66 psi	5.491 0	5.354 3
14 osi	1.059 5	1.033 1	68 psi	5.627 1	5.487 0
15 osi	1.063 8	1.037 3	70 psi	5.763 2	5.619 7
16 osi	1.068 0	1.041 4	72 psi	5.899 2	5.752 5
17 osi	1.072 3	1.045 6	74 psi	6.035 2	5.885 2
18 osi	1.076 5	1.049 7	76 psi	6.171 4	6.017 9
19 osi	1.080 8	1.053 9	78 psi	6.307 5	6.150 6
20 osi	1.085 0	1.058 0	80 psi	6.443 6	6.283 0
22 osi	1.093 5	1.066 3	84 osi	6.715 8	6.548 7
24 osi	1.102 0	1.074 6	88 osi	6.988 0	6.814 1
26 osi	1.110 5	1.082 9	92 psi	7.260 2	7.079 5
28 osi	1.119 0	1.091 2	96 psi	7.532 3	7.344 9
30 osi	1.127 5	1.099 5	100 psi	7.804 5	7.610 3

Figure A.4 US. **High temperature psychrometric chart.** Example: Find the air required to dry 100#/hr of water from granular material that cannot be exposed to >250 F. The air input to the once-through dryer is at 80 F DB (dry bulb temp) and 80 RH (% relative humidity) and is heated to 250 F. Moisture added by burning H_2 from the fuel is 0.0037#H_2O/#da (dry air). All air and flue gas exhausts at 220 F.

Solution: Plot the fresh air input, point 1 at 80 DB and 80 RH. (Table below lists data from the chart.) Add H_2O from combustion and input air, 0.0037 + 0.0175 = 0.0212. Plot point 2 at this AH_2 (absolute humidity) and DB_2 = 250 F. Assume the drying process is at constant total heat; so find point 3, exhaust condition, at TH_3 = TH_2 = 87½ and at DB_3 = 220 F; there read AH_3 = 0.0280. The moisture pickup is 0.0280 − 0.0212 = 0.0068#H_2O/#da; so to remove 100#H_2O/hr requires 100 ÷ 0.0068 = 14 700#da/hr, or 14 700 × 14 ft^3/#da = 205 800 cfh air.

point	DB, F	RH, %	AH, #H₂O/#da	TH, Btu/#da	ft³/#da	
1. air inlet	**80**	**80**	0.0175	38	14	**Bold = given data.**
2. after heating	**250**	2½	**0.0212**	87½	18.5	Standard = answers for problem from chart. *Italics = other readings*
3. exhaust	**220**	4	0.0280	**87½**	17.9	*from chart.*

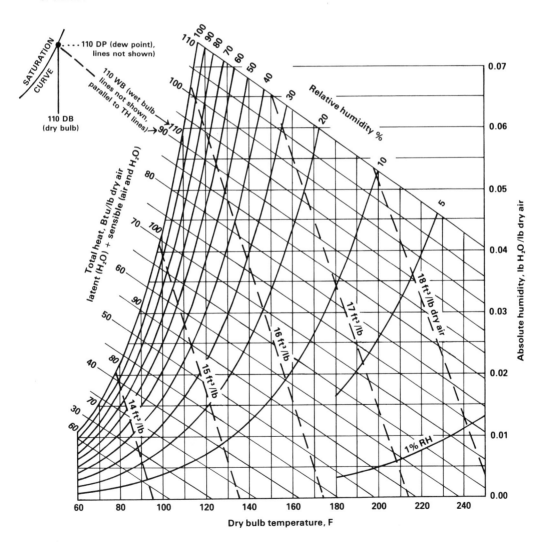

Figure A.5 Metric. High temperature psychrometric chart. Example: Find the air required to dry 50 kg/h of water from granular material that cannot be exposed to >120 C. The air input to the once-through dryer is at 25 C DB (dry bulb temp) and 80 RH (% relative humidity) and is heated to 120 C. Moisture added by burning H_2 from the fuel is 0.0037 kg H_2O/kg da (dry air). All air and flue gas exhausts at 105 C.

Solution: Plot the fresh air input, point 1 at 25 C DB and 80 RH. (Table below lists data from the chart.) Add H_2O from combustion and input air, 0.0037 + 0.0160 = 0.0197. Plot point 2 at this AH_2 (absolute humidity) and DB_2 = 120 C. Assume the drying process is at constant total heat; so find point 3, exhaust condition, at $TH_3 = TH_2 = 46\frac{1}{2}$ and at DB_3 = 105 C; there read AH_3 = 0.0255. The moisture pickup is 0.0255 – 0.0197 = 0.0058 kg H_2O/kg da; so to remove 50 kg H_2O/h requires 50 ÷ 0.0058 = 8621 kg da/h, or 8621 × 2.30 m³/kg da = 19 800 m³/h air.

point	DB, C	RH, %	AH, kgH₂O/kgda	TH, kcal/kgda	m³/kgda	Bold = given data.
1. air inlet	**25**	**80**	0.0160	*20*	*2.30*	Standard = answers for problem from chart.
2. after heating	**120**	*2½*	**0.0197**	*46½*	*3.05*	*Italics = other readings*
3. exhaust	**105**	*4*	0.0255	*46½*	*2.97*	*from chart.*

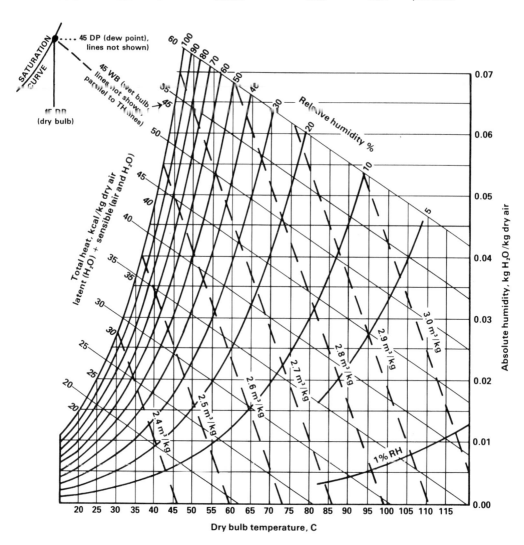

Table A.6 US. Properties of saturated air-water vapor mixtures. Temperature Range = 80 to 211 F; Pressure = 29.921"Hg.
(Source = Zimmerman and Lavine, Psychrometric Tables and Charts.)

Temp., F	Saturation pressure psi	Saturation pressure "Hg	Saturation humidity, weight of water vapor/lb dry air lb/lb	Saturation humidity, weight of water vapor/lb dry air grains/lb	Saturation moisture content, weight of water vapor/cf saturated mixture pounds per cf	Saturation moisture content, weight of water vapor/cf saturated mixture grains per cf	Saturation density, weight of air plus water vapor/cf saturated mixture pounds per cf	Saturation density, weight of air plus water vapor/cf saturated mixture grains per cf	Specific volume dry air, cf/lb	Specific volume saturated mixture, cf/lb dry air	Heat content, above 32 F dry air, Btu/lb	Heat content, above 32 F saturated mixture, Btu/lb dry air
80	0.50689	1.0320	0.02231	156.1	0.001583	11.08	0.07257	508.0	13.601	14.088	11.533	35.985
85	0.59588	1.2132	0.02639	184.7	0.001844	12.91	0.07173	502.1	13.728	14.309	12.735	41.718
90	0.69816	1.4215	0.03115	218.0	0.002141	14.99	0.07088	496.2	13.854	14.547	13.938	48.212
95	0.81537	1.6601	0.03668	256.8	0.002478	17.34	0.07003	490.2	13.980	14.804	15.139	55.586
100	0.94926	1.9327	0.04312	301.8	0.002859	20.01	0.06916	484.1	14.107	15.083	16.341	63.980
105	1.1018	2.2432	0.05061	354.3	0.003289	23.02	0.06827	477.9	14.233	15.389	17.544	73.563
110	1.2750	2.5959	0.05932	415.3	0.003772	26.41	0.06737	470.6	14.359	15.725	18.747	84.535
115	1.4711	2.9952	0.06946	486.2	0.004315	30.20	0.06643	465.0	14.486	16.098	19.950	97.128
120	1.6927	3.4463	0.08128	569.0	0.004922	34.45	0.06547	458.3	14.612	16.515	21.153	111.65
125	1.9423	3.9544	0.09511	665.8	0.005600	39.20	0.06448	451.4	14.738	16.983	22.356	128.44
130	2.2227	4.5255	0.1113	779.4	0.006357	44.50	0.06345	441.2	14.864	17.514	23.560	147.99
135	2.5373	5.1659	0.1304	913.0	0.007199	50.39	0.06239	436.7	14.991	18.119	24.764	170.79
140	2.8890	5.8821	0.1530	1071	0.008131	56.92	0.06128	428.9	15.117	18.816	25.968	197.59
145	3.2814	6.6809	0.1798	1259	0.009163	64.14	0.06012	420.8	15.243	19.626	27.172	229.26
150	3.7182	7.5703	0.2120	1484	0.01030	72.13	0.05890	412.3	15.370	20.576	28.376	267.06
155	4.2034	8.5581	0.2509	1756	0.01156	80.92	0.05763	403.4	15.496	21.704	29.581	312.55
160	4.7412	9.6531	0.2985	2089	0.01294	90.59	0.05630	394.1	15.622	23.063	30.786	368.13
165	5.3358	10.864	0.3575	2502	0.01446	101.2	0.05490	384.3	15.748	24.725	31.991	436.61
170	5.9923	12.200	0.4320	3024	0.01612	112.8	0.05343	374.0	15.875	26.804	33.196	523.06
175	6.7156	13.673	0.5284	3699	0.01793	125.5	0.05187	363.1	16.001	29.465	34.402	634.63

(continued)

NORTH AMERICAN COMBUSTION HANDBOOK

Table A.6 US. (concluded)

Temp., F	Saturation pressure		Saturation humidity, weight of water vapor/lb dry air		Saturation moisture content, weight of water vapor/cf saturated mixture		Saturation density, weight of air plus water vapor/cf saturated mixture		Specific volume		Heat content, above 32 F	
	psi	"Hg	lb/lb	grains/lb	pounds per cf	grains per cf	pounds per cf	grains per cf	dry air, cf/lb	saturated mixture, cf/lb dry air	dry air, Btu/lb	saturated mixture, Btu/lb dry air
180	7.5109	15.292	0.6569	4598	0.01992	139.4	0.05023	351.6	16.127	32.984	35.607	783.08
185	8.3836	17.069	0.8352	5847	0.02207	154.5	0.04850	339.5	16.253	37.839	36.813	988.8
190	9.3392	19.015	1.097	7681	0.02442	170.9	0.04667	326.7	16.380	44.935	38.019	1291.0
195	10.385	21.143	1.517	10629	0.02697	188.8	0.04474	313.2	16.506	56.265	39.226	1775.0
200	11.526	23.466	2.292	16046	0.02973	208.1	0.04270	298.9	16.632	77.102	40.433	2667.2
205	12.769	25.998	4.181	29269	0.03272	229.0	0.04054	283.8	16.758	127.80	41.640	4840.5
210	14.122	28.753	15.54	108773	0.03595	251.6	0.03826	267.8	16.885	432.25	42.849	17906

Table A.7 Metric. Properties of saturated air-water vapor mixtures. Temperature range = 26.7 to 99.4 C; pressure = 1.0 atmosphere. (Source = Zimmerman and Lavine, "Psychrometric Tables and Charts".) Abbreviations: atm = atmospheres, C = Celsius, da = dry air, kg = kilograms, kPa = kilopascals, m³ = cubic metres, sm = saturated mixture of water vapor and air, wv = water vapor.

Temp., C	Saturation pressure,		Saturation humidity, kg wv/kg da	Saturation moisture content, kg wv/m³ sm	Saturation density, (kg da + kg wv)/m³ sm	Specific volume		Heat content above 0 C		Temp., C
	atm	kPa				m³ da/kg da	m³ sm/kg da	kcal in da/kg da	kcal in sm/kg da	
26.7	0.034 49	3.494	0.022 31	0.025 36	1.162	0.8493	0.8798	6.409	20.00	26.7
27.5	0.036 22	3.670	0.023 48	0.026 66	1.158	0.8518	0.8839	6.809	20.91	27.5
30.0	0.041 86	4.241	0.027 28	0.030 45	1.146	0.8587	0.8963	7.210	23.87	30.0
40.0	0.072 79	7.377	0.049 02	0.051 22	1.096	0.8872	0.9569	9.616	39.76	40.0
50.0	0.1217 7	12.33	0.086 55	0.083 02	1.042	0.9155	1.042	12.02	65.61	50.0
60.0	0.196 6	19.92	0.153 0	0.130 2	0.9811	0.9439	1.175	14.43	109.8	60.0
70.0	0.307 5	31.16	0.278 2	0.198 1	0.9100	0.9723	1.404	16.84	191.4	70.0
80.0	0.467 4	47.35	0.551 1	0.293 4	0.8253	1.0007	1.879	19.25	367.3	80.0
90.0	0.691 9	70.10	1.414	0.423 5	0.7227	1.0291	3.349	21.66	920.4	90.0
99.4	0.980 3	99.32	31.49	0.587 4	0.6052	1.056	53.70	23.94	20150	99.4

Table A.8 US. Properties of saturated steam*

Saturation temperature F	Saturation pressure, inches of mercury absolute	Specific volume of the vapor ft³/lb (v_g)	Latent heat of vaporization Btu/lb (h_{fg})	Heat content of the vapor Btu/lb (h_g)
40	0.2477	2445	1070.9	1078.9
50	0.3625	1704.2	1065.2	1083.3
60	0.5218	1206.9	1059.6	1087.7
70	0.7395	867.7	1054.0	1092.0
80	1.0329	632.8	1048.3	1096.4
90	1.4228	467.7	1042.7	1100.7
100	1.9349	350.0	1037.0	1105.0
110	2.5986	265.1	1031.3	1109.3
120	3.4501	203.0	1025.5	1113.5
130	4.5302	157.17	1019.8	1117.8
140	5.8883	122.88	1014.0	1121.9
150	7.5782	96.99	1008.1	1126.1
160	9.6611	77.23	1002.2	1130.1
170	12.208	62.02	996.2	1134.2
180	15.301	50.20	990.2	1138.2
190	19.023	40.95	984.1	1142.1
200	23.474	33.63	977.9	1145.9
210	28.759	27.82	971.6	1149.7
212	29.926	26.80	970.3	1150.5
220	34.996	23.15	965.3	1153.5

Saturation temperature F	Saturation pressure psig†	Specific volume of the vapor ft³/lb (v_g)	Latent heat of vaporization Btu/lb (h_{fg})	Heat content of the vapor Btu/lb (h_g)
239.4	10.00	16.50	952.7	1160.5
250.0	15.12	13.83	945.6	1164.2
258.8	20.00	12.00	939.7	1166.2
286.7	40.00	7.830	920.1	1176.2
300.0	52.26	6.472	910.4	1180.2
307.4	60.00	5.840	905.0	1182.3
323.9	80.00	4.668	892.3	1186.8
337.9	100.00	3.893	881.2	1190.3
350.0	119.8	3.346	871.3	1193.1
365.9	150.0	2.759	857.8	1196.4
387.8	200.0	2.136	838.2	1200.3
400.0	232.4	1.866	826.8	1202.0
448.2	400.0	1.121	777.4	1205.5
488.9	600.0	0.7513	729.1	1203.8
500.0	665.3	0.6761	714.8	1202.5
520.5	800.0	0.5581	686.6	1199.1
546.5	1000.0	0.4389	647.2	1191.7
568.9	1200.0	0.3573	609.6	1180.9
600.0	1526.3	0.2677	549.7	1166.4
700.0	3075.3	0.0744	167.5	990.2

* Adapted by permission from Steam Tables by J. H. Keenan, Keyes, Hill, and Moore, published by John Wiley & Sons, Inc, New York, 1969.

† Pressure in psi absolute is pressure in psig plus 14.696.

Table A.9. Properties of saturated steam*

Saturation temperature C	Saturation pressure, mm Hg or *atm*	Specific volume of the vapor, m³/kg (v_g)	Latent heat of vaporization, kcal/kg (h_{f_g})	Heat content of the vapor, kcal/kg (h_g)
10	9.2077	106.39	591.19	601.23
15	12.7904	77.94	588.41	603.40
20	17.5383	57.80	585.58	605.62
25	23.7705	43.36	582.75	607.78
30	31.8492	32.90	579.92	609.95
35	42.2139	25.22	577.09	612.11
40	55.3818	19.53	574.26	614.22
45	71.9580	15.26	571.43	616.38
50	92.6305	12.03	568.54	618.49
55	118.180	9.569	565.66	620.60
60	149.574	7.671	562.77	622.66
65	187.744	6.197	559.83	624.71
70	233.930	5.043	556.89	626.76
75	289.373	4.132	553.89	628.82
80	355.472	3.407	550.89	630.81
85	433.776	2.828	547.84	632.76
90	526.096	2.3611	544.79	634.70
95	634.346	1.9828	541.68	636.64
100	760.181	1.6731	538.52	638.53
105	906.393	1.4197	535.41	640.42
115	*1.67*	1.0369	528.9	644.0
125	*2.29*	0.7708	522.2	647.5
150	*4.70*	0.3928	504.5	655.3
175	*8.80*	0.2169	485.0	661.8
200	*15.33*	0.1274	463.1	666.5
250	*39.21*	0.0509	409.5	668.5
300	*84.70*	0.0217	335.2	655.9
350	*162.97*	0.0088	213.1	611.7

* Calculated from Steam Tables by J. H. Keenan, Hill, and Moore, published by John Wiley & Sons, Inc., New York, NY, 1969.

Table A.10 US. Properties of superheated steam*

Specific volume ft³/lb
Heat content Btu/lb

For each pressure the upper line is specific volume (ft³/lb) and the lower line is heat content (Btu/lb).

Pressure psia	Property	200	250	300	350	400	450	500	550	600	700	800	900	1000	1200	1400	1600
1	v	392.5	422.4	452.3	482.1	511.9	541.8	571.5	601.3	631.1	690.7	750.3	809.9	869.5	988.6	1107.7	1226.9
1	h	1150.1	1172.8	1195.7	1218.6	1241.8	1265.1	1288.5	1312.2	1336.1	1384.5	1433.7	1483.8	1534.8	1639.6	1748.1	1860.4
5	v	78.15	84.21	90.24	96.25	102.24	108.25	114.20	120.18	125.15	138.08	150.01	161.94	173.86	197.70	221.54	245.4
5	h	1148.6	1171.7	1194.8	1218.0	1241.2	1264.6	1288.2	1311.9	1335.8	1384.3	1433.5	1483.7	1534.7	1639.5	1748.1	1860.3
10	v	38.85	41.93	44.99	48.02	51.03	54.04	57.04	60.04	63.03	69.01	74.98	80.95	86.91	98.84	110.76	122.68
10	h	1146.6	1170.2	1193.7	1217.1	1240.5	1264.1	1287.7	1311.5	1335.5	1384.0	1433.3	1483.5	1534.6	1639.4	1748.0	1860.3
14.396	v	—	28.42	30.52	32.60	34.67	36.72	38.77	40.82	42.86	46.93	51.00	55.07	59.13	67.25	75.36	83.47
14.396	h	—	1168.8	1192.6	1216.3	1239.9	1263.6	1287.3	1311.2	1335.2	1383.8	1433.1	1483.4	1534.5	1639.3	1747.9	1860.2
25	v	—	16.589	17.860	19.106	20.34	21.56	22.78	23.99	25.20	27.61	30.01	32.40	34.80	39.59	44.37	49.15
25	h	—	1165.7	1190.3	1214.5	1238.5	1262.4	1286.3	1310.3	1334.5	1383.3	1432.7	1483.0	1534.2	1639.2	1747.8	1860.1
50	v	—	—	8.772	9.425	10.061	10.687	11.305	11.919	12.529	13.742	14.949	16.152	17.352	19.747	22.138	24.53
50	h	—	—	1184.4	1210.0	1235.0	1259.5	1284.0	1308.4	1332.8	1381.9	1431.7	1482.2	1533.5	1638.7	1747.4	1859.8
75	v	—	—	—	6.206	6.646	7.074	7.495	7.910	8.322	9.139	9.949	10.755	11.558	13.159	14.810	15.457
75	h	—	—	—	1205.3	1231.4	1256.6	1281.6	1306.3	1331.0	1380.6	1430.7	1481.4	1532.8	1637.7	1747.1	1859.6
100	v	—	—	—	4.592	4.934	5.265	5.587	5.903	6.216	6.834	7.445	8.053	8.657	9.861	11.060	12.257
100	h	—	—	—	1200.8	1227.8	1253.9	1279.3	1304.3	1329.4	1379.4	1429.7	1480.5	1532.1	1637.1	1746.7	1859.3
150	v	—	—	—	—	3.221	3.455	3.679	3.897	4.111	4.531	4.944	5.353	5.759	6.566	7.368	8.167
150	h	—	—	—	—	1219.5	1247.4	1274.1	1300.1	1325.7	1376.6	1427.5	1478.8	1530.7	1636.7	1746.0	1858.7
200	v	—	—	—	—	2.361	2.548	2.724	2.893	3.058	3.379	3.693	4.003	4.310	4.918	5.521	6.123
200	h	—	—	—	—	1210.8	1240.7	1268.8	1295.7	1322.1	1373.8	1425.3	1477.1	1529.3	1635.7	1745.3	1858.2
300	v	—	—	—	—	—	1.6361	1.7662	1.8878	2.004	2.227	2.442	2.653	2.860	3.270	3.675	4.078
300	h	—	—	—	—	—	1226.2	1257.5	1286.7	1314.5	1368.3	1421.0	1473.6	1526.5	1633.8	1743.8	1857.0
400	v	—	—	—	—	—	1.1745	1.2843	1.3833	1.4760	1.6503	1.8163	1.9776	2.136	2.446	2.752	3.055
400	h	—	—	—	—	—	1209.6	1245.2	1277.0	1306.6	1362.5	1416.6	1470.1	1523.6	1631.8	1742.4	1855.9
500	v	—	—	—	—	—	—	0.9924	1.0792	1.1583	1.3040	1.4407	1.5723	1.7008	1.9518	2.198	2.442
500	h	—	—	—	—	—	—	1231.5	1266.6	1298.3	1356.7	1412.1	1466.5	1520.7	1629.8	1741.0	1854.8
600	v	—	—	—	—	—	—	0.7947	0.8749	0.9456	1.0727	1.1900	1.3021	1.4108	1.622	1.8289	2.033
600	h	—	—	—	—	—	—	1216.2	1255.4	1289.5	1350.6	1407.6	1462.9	1517.8	1627.8	1739.5	1853.7
700	v	—	—	—	—	—	—	—	0.7275	0.7929	0.9073	1.0109	1.1089	1.2036	1.3868	1.5652	1.7409
700	h	—	—	—	—	—	—	—	1243.2	1280.2	1344.4	1402.9	1459.3	1514.9	1625.8	1738.1	1852.6
800	v	—	—	—	—	—	—	—	0.6154	0.6776	0.7829	0.8764	0.9640	1.0482	1.2102	1.3674	1.5218
800	h	—	—	—	—	—	—	—	1229.9	1270.4	1338.0	1398.2	1455.6	1511.9	1623.8	1736.6	1851.6
900	v	—	—	—	—	—	—	—	0.5265	0.5871	0.6859	0.7717	0.8513	0.9273	1.0729	1.2135	1.3515
900	h	—	—	—	—	—	—	—	1215.2	1260.0	1331.4	1393.4	1451.9	1508.9	1621.7	1735.1	1850.4
1000	v	—	—	—	—	—	—	—	0.4534	0.5140	0.6080	0.6878	0.7610	0.8305	0.9630	1.0905	1.2152
1000	h	—	—	—	—	—	—	—	1198.7	1248.8	1324.6	1388.5	1448.1	1505.9	1619.7	1733.7	1849.3

Temperature, F

Pressure, psia

* From Steam Tables by J. H. Keenan, Keyes, Hill, and Moore, published by John Wiley & Sons, Inc., New York, NY, 1969.

Table A.11 Metric. Properties of superheated steam*

Specific volume — m³/kg
Heat content — kcal/kg

Pressure, atmospheres		100	150	200	250	300	350	400	450	500	550	600	650	700	750	800	850
0.1	m³/kg	17.13	19.44	21.74	24.05	26.34	28.79	30.95	33.24	35.55	37.85	40.15	42.49	44.74	47.05	49.34	51.64
	kcal/kg	641.3	664.0	687.2	710.4	734.1	758.1	782.6	806.8	832.6	858.1	884.2	910.6	937.5	964.8	992.5	1021
0.5	m³/kg	3.377	3.842	4.303	4.758	5.220	5.677	6.134	6.591	7.048	7.504	7.960	8.417	8.873	9.329	9.785	10.24
	kcal/kg	640.1	663.3	686.7	710.1	733.8	757.9	782.4	807.2	832.4	858.0	884.1	910.5	937.4	964.7	992.5	1021
1.0	m³/kg	1.674	1.910	2.144	2.374	2.604	2.833	3.062	3.291	3.519	3.747	3.976	4.203	4.431	4.659	4.887	5.115
	kcal/kg	639.0	662.5	686.0	709.7	733.5	757.7	782.2	807.0	832.3	858.0	884.0	910.4	937.4	964.7	992.3	1021
1.5	m³/kg	—	1.269	1.424	1.572	1.725	1.877	2.029	2.181	2.345	2.497	2.650	2.802	2.959	3.107	3.259	3.411
	kcal/kg	—	664.2	685.4	709.2	733.2	757.5	782.0	806.9	832.2	857.9	884.0	910.4	937.3	964.6	995.4	1020
2.0	m³/kg	—	0.948	1.067	1.184	1.300	1.415	1.578	1.645	1.760	1.875	1.989	2.103	2.217	2.332	2.446	2.560
	kcal/kg	—	660.6	684.9	708.9	732.9	757.2	781.8	806.8	832.1	857.8	883.8	910.3	937.2	964.5	992.2	1020
3.0	m³/kg	—	0.625	0.707	0.786	0.864	0.941	1.018	1.095	1.171	1.248	1.324	1.400	1.477	1.553	1.629	1.704
	kcal/kg	—	658.7	683.7	708.0	732.3	756.7	786.9	806.4	831.8	857.5	883.7	910.1	937.1	964.4	992.1	1020
4.0	m³/kg	—	0.465	0.535	0.587	0.646	0.710	0.763	0.820	0.878	0.935	0.993	1.049	1.083	1.167	1.221	1.279
	kcal/kg	—	656.7	681.9	707.2	731.7	756.3	781.1	806.1	831.5	857.4	883.5	910.0	937.0	964.3	992.1	1020
5.0	m³/kg	—	—	0.419	0.467	0.511	0.563	0.609	0.656	0.702	0.748	0.794	0.840	0.885	0.924	0.977	1.023
	kcal/kg	—	—	681.2	709.2	728.3	755.8	780.7	805.8	831.3	857.1	883.3	909.8	936.8	964.2	992.0	1020
10.0	m³/kg	—	—	0.203	0.230	0.255	0.279	0.294	0.327	0.362	0.386	0.410	0.434	0.458	0.481	0.505	0.529
	kcal/kg	—	—	674.6	701.9	727.9	753.4	784.4	804.3	830.1	856.1	882.5	909.1	936.2	963.6	991.4	1020
15.0	m³/kg	—	—	0.130	0.150	0.167	0.184	0.200	0.216	0.232	0.232	0.263	0.279	0.294	0.310	0.325	0.340
	kcal/kg	—	—	667.0	697.3	724.7	750.9	776.8	802.6	828.6	854.9	881.4	908.3	935.4	963.0	990.8	1007

Temperature, C

* Calculated from Steam Tables by J. H. Keenan, Keyes, Hill, and Moore, published by John Wiley and Sons, Inc., New York, NY, 1969.

Table A.12 US.   Thermal properties of metals

Substance	Composition	Density lb/ft³	Mean specific heat, 60 F to melting point Btu/lb °F	Latent heat of fusion Btu/lb	Mean specific heat of liquid Btu/lb °F	Melting point °F	Average pouring temp F	Ht content of solid at melting point Btu/lb	Ht content of liquid at melting temp Btu/lb	Ht content of liquid at pouring temp Btu/lb
Aluminum	Al	166.7	0.248	169.0	0.26	1215	1380	286.0	455.0	497.0
Babbitt, lead base	75 Pb, 15 Sb, 10 Sn	—	0.039	26.2	0.038	462	625	15.8	42.0	48.0
Babbitt, tin base	83.3 Sn, 8.4 Sb, 8.3 Cu	462	0.071	34.1	0.063	464	916	28.6	67.7	91.0
Bismuth	Bi	612	0.033	18.5	0.035	518	620	15.1	33.6	37.2
Brass, Muntz metal	60 Cu, 40 Zn	524	0.105	69.0	0.125	1630	1850	165.0	234.0	261.0
Brass, red	90 Cu, 10 Zn	546	0.104	86.5	0.115	1952	2250	197.0	283.5	317.8
Brass, yellow	67 Cu, 33 Zn	528	0.105	71.0	0.123	1688	1950	171.0	242.0	274.2
Bronze, aluminum	90 Cu, 10 Al	510	0.126	98.6	0.125	1922	2200	235.0	333.6	368.0
Bronze, bearing	80 Cu, 10 Pb, 10 Sn	556	0.095	79.9	0.109	1832	2050	168.3	248.2	272.0
Bronze, bell metal	78 Cu, 22 Sn	540	0.100	76.3	0.119	1634	1900	157.4	233.7	265.4
Bronze, gun metal	90 Cu, 10 Sn	550	0.107	84.2	0.106	1830	2100	191.5	275.7	302.0
Bronze, tobin	60 Cu, 39.2 Zn, 0.8 Sn	525	0.107	73.5	0.124	1625	1850	167.5	241.0	268.9
Copper	Cu	559	0.104	91.0	0.111	1942	2200	200.0	291.0	315.0
Die casting metal	92 Al, 8 Cu	176	0.236	163.0	0.241	1110	1400	257.3	420.3	481.0
Die casting metal	80 Pb, 10 Sn, 10 Sb	—	0.038	17.5	0.037	610	820	20.5	38.0	146.0
Die casting metal	90 Sn, 4.5 Cu, 5.5 Sb	—	0.070	30.2	0.062	450	650	27.6	57.8	70.0
Die casting metal	87.3 Zn, 8.1 Sn, 4.1 Cu, 0.5 A	—	0.103	48.0	0.138	780	980	74.0	122.0	150.0
German silver	60 Cu, 25 Zn, 15 Ni	—	0.109	86.2	0.123	1850	2100	194.0	280.2	311.0
Gold	Au	1205	0.033	28.5	0.034	1945	2150	62.2	90.7	97.7
Iron, pure	Fe	491	0.168	117	0.150	2802	3100	451	568	626
Iron, cast, gray	94 Fe, 3.5 C, 2.5 Si	—	0.190	41.4	—	2246	2800	415	456	583
Iron, cast, white	97 Fe, 3 C	—	0.180	60.3	—	2102	2900	368	428	612
Iron, pig	93 Fe, 4.22 C, 1.48 Si, 0.73 Mn, 0.12 P, 0.03 S	—	0.153	83.6	—	2012	2300	299	384	450
Lead	Pb	708	0.032	10.0	0.034	621	720	18.0	28.0	31.0
Linotype	86 Pb, 11 Sb, 3 Sn	—	0.036	21.5	0.036	486	620	15.3	36.8	41.6
Magnesium	Mg	108.6	0.272	83.7	0.266	1204	1380	311.2	394.9	441.7
Manganese	Mn	464	0.171	66.0	0.192	2246	2400	374.0	440.0	469.0
Monel metal	67 Ni, 28 Cu; Fe, Mn, Si	550	0.129	117.4	0.139	2415	2750	304.0	421.4	468.0
Nickel 60 to 2644 F	Ni	556	0.134	131.5	0.133	2644	2850	346.0	477.5	505.0
Silver	Ag	655	0.063	46.8	0.070	1762	1950	107.0	153.8	167.0
Solder, bismuth	40 Pb, 20 Sn, 40 Bi	—	0.040	16.4	0.039	232	330	9.3	25.7	29.5
Solder, plumbers	50 Pb, 50 Sn	580	0.051	23.0	0.049	414	500	18.0	41.0	45.0
Tin	Sn	455	0.069	25.0	0.0637	450	650	27.0	52.0	64.0
Zinc	Zn	445	0.107	48.0	0.146	786	900	77.8	125.8	142.0

Table A.13 Metric.   Thermal properties of metals

Substance	Composition	Density kg/m³	Mean specific heat, 15.6 C to melting point cal/g °C	Latent heat of fusion cal/g	Mean specific heat of liquid cal/g °C	Melting point C	Average pouring temp C	Ht content of solid at melting point kcal/kg	Ht content of liquid at melting temp kcal/kg	Ht content of liquid at pouring temp kcal/kg
Aluminum	Al	2 670	0.248	93.9	0.26	657.2	749	158.9	252.8	276.1
Babbitt, lead base	75 Pb, 15 Sb, 10 Sn	—	0.039	14.6	0.038	239	329	8.8	23.3	26.7
Babbitt, tin base	83.3 Sn, 8.4 Sb, 8.3 Cu	7 400	0.071	18.9	0.063	240	491	15.9	37.6	50.6
Bismuth	Bi	9 800	0.033	10.3	0.035	270	327	8.4	18.7	20.7
Brass, Muntz metal	60 Cu, 40 Zn	8 390	0.105	38.3	0.125	887.8	1010	91.7	130.0	145.0
Brass, red	90 Cu, 10 Zn	8 750	0.104	48.1	0.115	1067	1232	109.5	157.5	176.6
Brass, yellow	67 Cu, 33 Zn	8 460	0.105	39.4	0.123	920	1066	95.0	134.5	152.3
Bronze, aluminum	90 Cu, 10 Al	8 170	0.126	54.8	0.125	1050	1204	130.6	185.3	204.5
Bronze, bearing	80 Cu, 10 Pb,10 Sn	8 910	0.095	44.4	0.109	1000	1121	93.5	137.9	151.1
Bronze, bell metal	78 Cu, 22 Sn	8 650	0.100	42.4	0.119	890	1038	87.5	129.8	147.5
Bronze, gun metal	90 Cu, 10 Sn	8 810	0.107	46.8	0.106	1010	1149	106.4	153.2	167.8
Bronze, tobin	60 Cu, 39.2 Zn, 0.8 Sn	8 410	0.107	40.8	0.124	885	1010	93.1	133.9	149.4
Copper	Cu	8 960	0.104	50.6	0.111	1083	1204	111.1	161.7	175.0
Die casting metal	92 Al, 8 Cu	2 820	0.236	90.6	0.241	621	760	143.0	233.5	267.2
Die casting metal	80 Pb, 10 Sn, 10 Sb	—	0.038	9.7	0.037	316	438	11.4	21.1	81.1
Die casting metal	90 Sn, 4.5 Cu, 5.5 Sb	—	0.070	16.8	0.062	232	343	15.3	32.1	38.9
Die casting metal	87.3 Zn, 8.1 Sn, 4.1 Cu, 0.5 A	—	0.103	26.7	0.138	416	527	41.1	67.8	83.3
German silver	60 Cu, 25 Zn, 15 Ni	—	0.109	47.9	0.123	1010	1149	107.8	155.7	172.8
Gold	Au	19 300	0.033	15.8	0.034	1063	1177	34.6	50.4	54.3
Iron, pure	Fe	7 870	0.168	65.0	0.150	1539	1704	251	316	348
Iron, cast, gray	94 Fe, 3.5 C, 2.5 Si	—	0.190	23.2	—	1230	1538	231	253	324
Iron, cast, white	97 Fe, 3 C	—	0.180	33.5	—	1150	1593	204	238	340
Iron, pig	93 Fe, 4.22 C, 1.48 Si, 0.73 Mn, 0.12 P, 0.03 S	—	0.153	47.0	—	1100	1260	166	213	250
Lead	Pb	11 300	0.032	5.6	0.034	327	382	10.0	15.6	17.2
Linotype	86 Pb, 11 Sb, 3 Sn	—	0.036	11.9	0.036	252	327	8.5	20.4	23.1
Magnesium	Mg	1 740	0.272	46.5	0.266	651	749	172.9	219.4	245.4
Manganese	Mn	7 430	0.171	36.7	0.192	1230	1316	207.8	244.5	260.6
Monel metal	67 Ni, 28 Cu; Fe, Mn, Si	8 810	0.129	65.2	0.139	1329	1510	168.9	234.1	260.0
Nickel 16 to 1450 C	Ni	8 910	0.134	73.1	0.133	1451	1566	192.2	265.3	280.6
Silver	Ag	10 500	0.063	26.0	0.070	961	1066	59.4	85.5	92.8
Solder, bismuth	40 Pb, 20 Sn, 40 Bi	—	0.040	9.1	0.039	111	166	5.2	14.3	16.4
Solder, plumbers	50 Pb, 50 Sn	9 290	0.051	12.8	0.049	212	260	10.0	22.8	25.0
Tin	Sn	7 290	0.069	13.9	0.0637	232	343	15.0	28.9	35.6
Zinc	Zn	7 130	0.107	26.7	0.146	419	482	43.2	69.9	78.9

Figure A.14. Heat contents of common metals at various temperatures

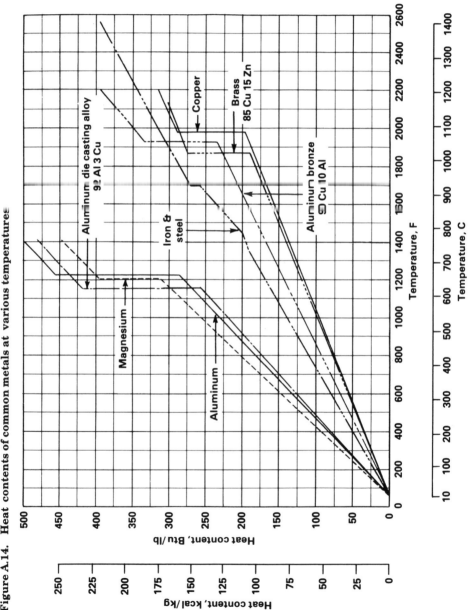

Figure A.15. Heat contents of lead, tin, zinc, and their alloys

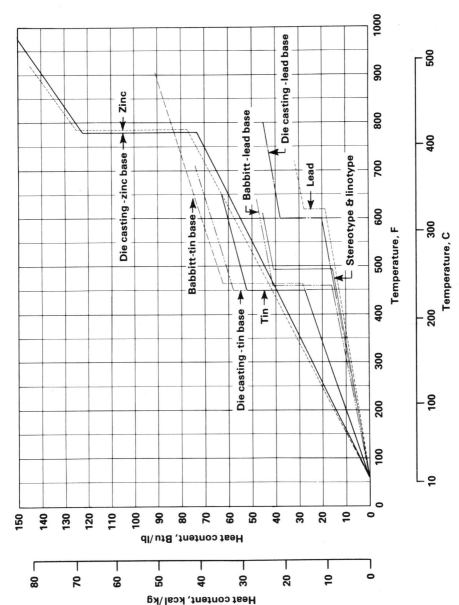

Table A.16 US. Densities and thermal properties of various substances. (See Part 2, Vol. I) for more information on fuels; Part 4 (Vol. I) for insulations and refractories; Tables and Figures A.12-A.15 for metals and alloys; Tables A.19-A.20 for heat transfer fluids.)

Name – Description	Other[a] classification	Normal state	Density lb/ft³	Specific heat in Btu/lb °F — Solid state Specific heat	Solid state Temp range F	Liquid state Specific heat	Liquid state Temp range F	Gaseous state Specific heat	Gaseous state Temp range F	Melting (fusion) point F	Boiling point, F (at std barometric pressure)	Latent heat of fusion Btu/lb	Latent heat of vaporization Btu/lb
Acetic acid (CH₃COOH)	O	Liquid	65.8	0.487	32	0.51	32-212	–	–	62.6	244.4	80.5	174
Acetone (CH₃COCH₃)	O	Liquid	49.75	–	–	0.544	32-212	0.346	79-230	-138.2	128-134	42	239
Acetylene (C₂H₂)	O, F	Gas	0.0691	–	–	–	–	0.64	59	-113.8	-118.8	–	–
Air (see also Tables A.2 and A.3)	–	Gas	0.0763	–	–	–	–	0.239 / 0.246 / 0.256	-22-+50 / 68-824 / 68-1472	–	-311.0	–	–
Alcohol, ethyl (C₂H₅OH)	O, F	Liquid	49.3	–	–	0.648	104	0.453	226-428	-173.2	172.4	46	369
Alcohol, ethyl (90%) and water	S	Liquid	51.4	–	–	0.718	–	–	–	–	–	–	–
Alcohol, ethyl (50%) and water	S	Liquid	57.3	–	–	0.923	–	–	–	–	–	–	–
Alcohol, ethyl (10%) and water	S	Liquid	61.4	–	–	0.99	–	–	–	–	–	–	–
Alcohol, methyl (CH₃OH)	O, F	Liquid	49.6	–	–	0.601	59-68	0.33	260-440	142.6	150.8	29.5	480.6
Alcohol, methyl (90%) and water	S	Liquid	51.4	–	–	0.643	–	–	–	–	–	–	–
Alcohol, methyl (50%) and water	S	Liquid	57.3	–	–	0.846	–	–	–	–	–	–	–
Alcohol, methyl (10%) and water	S	Liquid	61.4	–	–	0.986	–	–	–	–	–	–	–
Alumina – see Aluminum Oxide. Alumina (fused) refractory, high-alumina refractory	–	–	–	–	–	–	–	–	–	–	–	–	–
Alumina (fused) refractory (see also high-alumina refractory)	R	Solid	153-181	0.20	60-1200	–	–	–	–	3390+	–	–	–
Aluminum (see also Table A.12 and Figure A.8)	M, E	Solid	166.7	0.225	61-579	–	–	–	–	1220	3272	167.5	360.0
Aluminum foil	I	Solid	–	0.24	290	–	–	–	–	–	–	–	–
Aluminum oxide (alumina)	–	Solid	243.5	0.183	32-212	–	–	–	–	–	–	–	–
Ammonia (NH₃)	–	Gas	0.046	–	–	1.08	-8.4-+5	0.525	70-220	-103	-28.3	150	589
Ammonium chloride (10%) and water	S	Liquid	64.4	–	–	0.788	–	–	–	–	–	–	–
Ammonium sulfate(NH₄)₂SO₄	–	Solid	110	0.283	–	–	–	–	–	956[b]	–	–	–

[a] A = alloy, E = element, F = fuel or fuel component, I = insulation, M = metal, O = organic compound, R = refractory, S = solution.
[b] Decomposes.

Table A.16 US. Densities and thermal properties of various substances *(continued)*

Name – Description	Other classi-fication[a]	Normal state	Density lb/ft³	Solid state Specific heat	Solid state Temp range F	Liquid state Specific heat	Liquid state Temp range F	Gaseous state Specific heat	Gaseous state Temp range F	Melting (fusion) point F	Boiling point, F (at std barometric pressure)	Latent heat of fusion Btu/lb	Latent heat of vaporization Btu/lb
Andalusite (AL_2SiO_3)	—	Solid	199.8	0.228	—	—	—	—	—	3290	—	—	—
Aniline (C_6H_5N)	O, F	Liquid	63.9	—	—	0.514	59	—	—	17.6	363	38	198.0
Antimony (Sb)	M, E	Solid	422	0.052	392	—	—	—	—	1166	2624	68.9	703
Argon (A)	E	Gas	0.105	—	—	—	—	0.1233	68-194	-306.4	-303	12	68
Arsenic, gray	E	Solid	358	0.0622	12-212	—	—	—	—	c	c	—	—
Asbestos	I	Solid	124-174	0.20	12-212	—	—	—	—	—	—	—	—
Ashes	—	Solid	—	0.20	12-212	—	—	—	—	—	—	—	—
Asphalt, Bermudez	—	Solid	67.4	0.55	—	—	—	—	—	180	—	—	—
Gilsonite	—	Solid	64.9	0.55	—	—	—	—	—	300	—	—	—
Oil	—	Solid	61.7-64.9	0.55	—	—	—	—	—	140-180	—	—	—
Trinidad	—	Solid	87.4	0.55	—	—	—	—	—	190	—	—	—
Babbitt, lead base	A	Solid	—	0.039	60-462	0.038	—	—	—	462	—	26.2	—
tin base	A	Solid	465	0.071	60-464	0.063	—	—	—	464	—	34.1	—
(see also Table A.12)													
Bakelite (see Resin, phenol)	E	Solid	—	—	—	—	—	—	—	—	—	—	—
Barium (Ba)	E	Solid	218	0.068	-121+68	—	—	—	—	1562	2080	—	1120
Basalt (see lava)	—	—	—	—	—	—	—	—	—	—	—	—	—
Beeswax	—	Solid	59.9	0.82	60-144	—	—	—	—	143.6	—	76.2	—
Benzene (C_6H_6)	O, F	Liquid	54.9	—	—	0.423	—	0.3325	95-356	41.8	176.3	55.6	169.4
Benzoic acid ($C_7H_6O_2$)	O	Solid	81.2	0.287	60	—	—	—	—	249.8	480.2	61.0	—
Benzol (C_6H_6) (water white)	O, F	Liquid	55	—	—	0.423	104	—	—	41.7	176.3	55.08	169.4
Beryllium (Be)	E, M	Solid	113.6	0.52	—	0.425	—	—	—	2732	5020	570	—
Bismuth (Bi) (see also Table A.12)	E, M	Solid	612	0.0302	68-212	0.036	535-725	—	—	519.8	2606	22.4	395
Bismuth (63.8)–Tin (36.2) alloy	A	Solid	—	0.040	68-210	—	—	—	—	—	—	—	—
Blast furnace gas	F	Gas	0.0778	—	—	—	—	0.2277	—	—	—	—	—
Boron (B)	E	Solid	152	0.307	32-212	—	—	—	—	3992-4532	4600	—	—
Borax ($Na_2B_4O_7$ - 10 H_2O)	—	Solid	107	0.385	95	—	—	—	—	1366	—	—	—
Brass,													
Muntz metal (60Cu, 40Zn)	A	Solid	524	0.105	60-1630	—	—	—	—	1630	—	69.0	—
Red (85Cu, 15Zn)	A	Solid	546	0.104	60-1952	—	—	—	—	1952	—	86.5	—
Yellow (67Cu, 33Zn)	A	Solid	528	0.105	60-1688	—	—	—	—	1688	—	71.0	—
(see also Table A.12)													
Brick, red	—	Solid	118	0.22	32-212	—	—	—	—	—	—	—	—
Britania metal (90Sn, 10Pb)	A	Solid	—	—	—	—	—	—	—	—	—	50.4	—
Bromine (Br)	E	Liquid	193	0.0843	-108-4	0.107	34-89	0.0555	181-442	18.86	141.8	29.16	82.1

[a] A = alloy, E = element, F = fuel or fuel component, I = insulation, M = metal, O = organic compound, R = refractory, S = solution.

c Sublimes at 1038 F, melts at 1562 F under 36 atmospheres of pressure.

Table A.16 US. Densities and thermal properties of various substances (continued)

Name – Description	Other[a] classification	Normal state	Density lb/ft³	Solid state Specific heat	Solid state Temp range F	Liquid state Specific heat	Liquid state Temp range F	Gaseous state Specific heat	Gaseous state Temp range F	Melting (fusion) point F	Boiling point, F (at std barometric pressure)	Latent heat of fusion Btu/lb	Latent heat of vaporization Btu/lb
Bronze (80 Cu, 20 Sn)	A	Solid	—	0.0862	57-208	—	—	—	—	—	—	—	—
Aluminum	A	Solid	510	0.126	60-1922	—	—	—	—	1922	—	98.6	—
Bearing (see also Table A.12)	A	Solid	556	0.095	60-1832	—	—	—	—	1832	—	79.9	—
Bell metal	A	Solid	540	0.100	60-1634	—	—	—	—	1634	—	76.3	—
Gun metal	A	Solid	550	0.107	60-1850	—	—	—	—	1850	—	84.2	—
Tobin	A	Solid	525	0.107	60-1625	—	—	—	—	1625	—	73.5	—
Butane (C_4H_{10})	O, F	Gas	0.149	—	—	0.55	60	0.456	60	-210	30.9	—	165.5
Cadmium (Cd)	E, M	Solid	540	0.057	212	—	—	—	—	609.6	1412	23.8	409
Calcium (Ca)	E	Solid	96.6	0.170	32-358	—	—	—	—	1560	2625	—	—
Calcium carbonate ($CaCO_3$)	—	Solid	168-184	0.210	32-212	—	—	—	—	1517[b]	—	—	—
Calcium chlor.de ($CaCl_2$)	—	Solid	134	0.292	60	—	—	—	—	1425.2	>2910	—	—
Calcium chlor.de (30%) and water	S	Liquid	78.7	—	—	0.676	104	—	—	—	—	—	—
Camphor ($C_{10}H_6O$)	O	Solid	62.4	0.44	68-353	0.61	353-410	—	—	353	408.2	19.4	—
Carbon (C) (graphite)	E, F	Solid	138	0.160	52	—	—	—	—	6332[c]	8730	—	—
Carbon bisulfide (see carbon disulfide)	—	—	—	—	—	—	—	—	—	—	—	—	—
Carbon dioxide (CO_2)	—	Gas	0.117	—	—	0.232	60	0.216	52-417	—	-109[h]	—	—
Carbon disulfide (CS_2)	—	Liquid	79.3	0.467	1789	0.0615	-82.6--73.8	0.1596	187-374	-166	115.0	—	150.8
Carbon monoxide (CO)	F	Gas	0.0741	—	—	—	—	0.2425	79-388	-340	-312[c]	—	—
Carbon tetrachloride (CCl_4)	—	Liquid	98.8	—	—	0.215	122	—	—	-9	170	—	83.5
Castor oil	—	Liquid	60.1	—	—	0.434	—	—	—	—	—	—	—
Cellulose	—	Solid	94.97	0.32	32-212	—	—	—	—	—	—	—	—
Cerium (Ce)	E	Solid	430	0.0448	32-212	—	—	—	—	1184	2540	—	—
Cesium (Cs)	E	Solid	118.6	0.0482	32-79	—	—	—	—	83	1238	6.77	—
Chalk	F	Solid	—	0.215	32-212	—	—	—	—	—	—	—	—
Charcoal	E	Solid	18-38	0.165-0.25	75	—	—	—	—	—	—	—	—
Chlorine (Cl)	O	Gas	0.190	—	—	0.229	-82	0.112	61-649	-150.7	-28.5	41.3	121
Chloroform ($CHCl_3$)	R	Liquid	95.5	—	—	0.23	32-212	0.148	72-172	-85	142.1	—	105.3
Chrome refractory, burned	R	Solid	188	0.20	60-1200	—	—	—	—	3580+	—	—	—
Chrome refractory, unburned	R	Solid	193	0.21	60-1200	—	—	—	—	3580+	—	—	—
Chromite (chrome ore) ($FeCr_2O_4$)	R	Solid	281	0.22	—	—	—	—	—	3956	—	—	—

[a] A = alloy, E = element, F = fuel or fuel component, I = insulation, M = metal, O = organic compound, R = refractory, S = solution.
[b] Decomposes.
[c] Sublimes at -109 F, melts at -49.2 F under 5.2 atmospheres of pressure.
[h] Sublimes.

Table A.16 US. Densities and thermal properties of various substances (continued)

Name – Description	Other[a] classi-fication	Normal state	Density lb/ft³	Specific heat — Solid state: Specific heat	Solid state: Temp range F	Liquid state: Specific heat	Liquid state: Temp range F	Gaseous state: Specific heat	Gaseous state: Temp range F	Melting (fusion) point F	Boiling point, F (at std barometric pressure)	Latent heat of fusion Btu/lb	Latent heat of vaporization Btu/lb
Chromium (Cr)	E, M	Solid	449	0.1039 0.1121 0.1872	32 212 1112	–	–	–	–	2929	4500	–	–
Cinders	–	Solid	–	–	–	–	–	–	–	–	–	–	–
Clay	–	Solid	112-162	0.18	32-212	–	–	–	–	3160	–	–	–
Coal	F	Solid	–	0.224	68-208	–	–	–	–	–	–	–	–
Coal tar oil	F	Liquid	–	0.3	32-212	0.34	–	–	–	–	390-910	–	–
Cobalt (Co)	E	Solid	556	0.1542 0.204	932 1832	–	–	–	–	2723	5250	115.2	–
Coke	F	Solid	–	0.203 0.376	32-212 100-2200	–	–	–	–	–	–	–	–
Columbium (Cb)	E	Solid	535	0.065	–	–	–	–	–	4380	–	–	–
Concrete	–	Solid	137	0.156-0.27	32-212	–	–	–	–	–	–	–	–
Constantan	A	Solid	–	0.098	32-212	–	–	–	–	–	–	–	–
Copper (Cu)	M, E	Solid	559	0.0951 0.1259	59-450 1652	–	–	–	–	1981 ±5	4700	75.6	–
Copper sulfate (CuSO₄) (16.7%) and water	S	Liquid	73.7	0.848	53.6-59	–	–	–	–	–	–	–	–
Cork, natural	I	Solid	15	0.419	77	–	–	–	–	–	–	–	–
Cork, granulated	I	Solid	5.4-7.3	0.43	77	–	–	–	–	–	–	–	–
Corkboard	I	Solid	6.9-20.7	0.204 0.417	-115 +150	–	–	–	–	–	–	–	–
Corundum (Al₂O₃)	–	Solid	250	0.1976	42-208	–	–	–	–	3722	–	–	–
Cottonseed oil	–	Liquid	58	–	–	0.474	–	–	–	32	–	–	–
Cream	–	Liquid	–	0.780	–	–	–	–	–	–	–	–	–
Cupric oxide (CuO)	–	Solid	374-405	0.227	68-212	–	–	–	–	1944	–	–	–
Cuprous oxide (Cu₂O)	–	Solid	375	0.111	32-212	–	–	–	–	2254	–	–	–
D'Arcet's metal (50 Bi, 25 Pb, 25 Sn)	A	Solid	–	0.050	32-212	–	–	–	–	–	–	10.4	–
Decane (C₁₀H₂₂)	O, F	Liquid	45.3	–	–	0.42	60	–	–	-21.5	346	–	110
Diatomaceous earth	R	Solid	12.5-25	0.21	77	–	–	–	–	–	–	–	–
Die casting metal (see also Table A.12) Aluminum base	A	Solid	176	0.236	60-1150	–	–	–	–	1150	–	163.1	–
Lead base	A	Solid	–	0.038	60-600	–	–	–	–	600	–	17.4	–
Tin base	A	Solid	–	0.070	60-450	–	–	–	–	450	–	30.3	–
Zinc base	A	Solid	–	0.103	60-780	–	–	–	–	780	–	48.3	–
Diphenyl (C₆H₅C₆H₅)	O	Liquid	62	0.693	104	0.481	159-492	–	–	159	492	47	136.5

[a] A = alloy, E = element, F = fuel or fuel component, I = insulation, M = metal, O = organic compound, R = refractory, S = solution.

Table A.16 US. Densities and thermal properties of various substances (continued)

Name – Description	Other[a] classification	Normal state	Density lb/ft³	Specific heat in Btu/lb — Solid state		Liquid state		Gaseous state		Melting (fusion) point F	Boiling point, F (at std barometric pressure)	Latent heat of fusion Btu/lb	Latent heat of vaporization Btu/lb
				Specific heat	Temp range F	Specific heat	Temp range F	Specific heat	Temp range F				
Diphenylamine ($C_6H_5NHC_6H_5$)	O	Liquid	72.3	0.443	133	–	–	–	–	129	576	45.4	–
Dolomite	–	Solid	181	0.222	68-208	–	–	–	–	–	–	–	–
Dowtherm A	–	Liquid	58.8	0.53	–	0.41	122	–	–	180	500	–	–
Earth (see also humus)	–	Solid	–	0.44	32-212	–	–	–	–	–	–	–	–
Ebonite	–	Solid	–	0.33	32-212	–	–	–	–	–	–	–	–
Ether, ethyl ($C_4H_{10}O$)	O	Liquid	45.9	–	–	0.529	32	0.48	156-435	-180.4	94.3	–	159.1
Ethyl acetate ($CH_3CO_2CH_2CH_3$)	O	Liquid	55.8	–	–	0.478	32-212	–	–	-118.3	–	–	–
Ethyl bromide (CH_3CH_2Br)	O	Liquid	90.5	–	–	0.21	60	–	–	-182	100.7	–	108.7
Ethyl iodide (CH_3CH_2I)	O	Liquid	120	–	–	0.25	122	–	–	-163	159.8	–	84.6
Ethylene glycol ($C_2H_6O_2$)	O	Liquid	68.6	–	–	0.602	32-212	–	–	–	387	–	–
Fiberglas board	I	Solid	2-6	0.129	-117	–	–	–	–	–	–	–	–
				0.192	-22								
				0.236	111								
Firebrick, fireclay insulating (2600 F)	R	Solid	137-150	0.243	60-2195	–	–	–	–	2900-3200	–	–	–
	R	Solid	38.4	0.22	60-1200	–	–	–	–	2980-3000	–	–	–
silica	R	Solid	144-162	0.258	60-2195	–	–	–	–	2100+	–	–	–
Flourine (F_2)	E	Gas	0.10	–	–	–	–	0.32	80	-369.4	-304.6	–	–
Forsterite refractory	R	Solid	153	0.25	60-1200	–	–	–	–	3430	–	–	–
Fuel oil (see Tables 2.5 and 2.10, Vol I)	F	Liquid	52-61	–	–	0.50	–	–	–	–	–	–	145-150
Fusel oil	–	Liquid	–	–	–	0.56	32-212	–	–	–	–	–	–
Galena (PbS)	–	Solid	467	0.0466	32-212	–	–	–	–	2050	–	–	–
Gallium (Ga)	E	Solid	367	0.080	54-235	–	–	–	–	86.18	3090	–	–
Gasoline (commercial)	F	Liquid	45.7	–	–	0.5135	–	–	–	–	158-194	–	126-146
Germanium (Ge)	E	Solid	335	0.0737	32-212	–	–	–	–	1756.4	–	–	–
German silver (see also Table A.12)	A	Solid	–	0.0946	32-212	0.123	–	–	–	1850	–	86.2	–
Glass	–	Solid	144-187	0.109	60-1850	–	–	–	–	–	–	–	–
				0.15-0.23	32-212								
Glass block, expanded, foamglas	–	Solid	10.6	0.132	-117	–	–	–	–	2190±	–	–	–
				0.157	-20								
Glass wool	I	Solid	1.4-4.8	0.179	112	–	–	–	–	–	–	–	–
				0.210	150								
Glycerine ($C_3H_8O_3$) (glycerol)	O	Liquid	78.7	0.279	624	0.576	59-212	–	–	-68	554	76.5	–
Gneiss	–	Solid	–	0.196	63-210	–	–	–	–	–	–	–	–

[a] A = alloy, E = element, F = fuel or fuel component, I = insulation, M = metal, O = organic compound, R = refractory, S = solution.

Table A.16 US.   Densities and thermal properties of various substances *(continued)*

| Name – Description | Other[a] classi-fication | Normal state | Density lb/ft³ | Specific heat in Btu/lb °F | | | | | | Melting (fusion) point F | Boiling point, F (at std barometric pressure) | Latent heat of fusion Btu/lb | Latent heat of vapor-ization Btu/lb |
| | | | | Solid state | | Liquid state | | Gaseous state | | | | | |
				Specific heat	Temp range F	Specific heat	Temp range F	Specific heat	Temp range F				
Gold (Au) (see also Table A.12)	E, M	Solid	1205	0.0316	32-212	0.034	–	–	–	1945.4	5380	28.7	29
Granite	–	Solid	162-175	0.192	54-212	–	–	–	–	–	–	–	–
Graphite	–	Solid	138.3	0.201 / 0.38	32-212 / 70-2200	–	–	–	–	6300	8720	–	–
Gypsum	–	Solid	145	0.259	50-212	–	–	–	–	2480	–	–	–
Hairfelt	I	Solid	–	0.334	148	–	–	–	–	–	–	–	–
Helium (He)	E	Gas	0.01043	–	–	–	–	1.25	456	<-456	-448.6	–	–
Hematite (Fe₂O₃)	–	Solid	–	0.1645	59-210	–	–	–	–	–	–	–	–
Heptane (C₇H₁₆)	O, F	Liquid	42.4	–	–	0.55	122	–	–	-130	194	–	140.0
Hexane (C₆H₁₄)	O, F	Liquid	40.8	–	–	0.600	68-212	–	–	-138	158	–	142.6
High-alumina refractory	R	Solid	128	0.23	60-1200	–	–	–	–	3290	–	–	–
Hornblende	–	Solid	–	0.195	32-212	–	–	–	–	–	–	–	–
Humus (soil) (see also earth)	–	Solid	–	0.44	32-212	–	–	–	–	–	–	–	–
Hydrochloric acid (HCl) (45.2%) + H₂O	–	Liquid	92	–	–	–	–	–	–	4.5	–	–	–
Hydrofluoric acid (HF) (35.35%) + H₂O	–	Liquid	72	–	–	–	–	–	–	-28	248	–	–
Hydrogen (H₂)	E, F	Gas	0.0053	–	–	1.75-2.33	-111.3 -108.0	3.410	70-212	-434.2	-422.6	27	194
Hydrogen chloride (HCl)	–	Gas	0.0967	–	–	–	–	0.1940	55-212	-168.3	-117	–	–
Hydrogen fluoride (HF)	–	Gas	0.0754	–	–	–	–	–	–	-134.14	-33.8	–	–
Hydrogen sulfide (H₂S)	–	Gas	0.0907	–	–	–	–	0.2451	68-403	-125	-81	–	–
Ice (H₂O)	–	Solid	55.8-57.4	0.463	-103-0	–	–	–	–	32	–	144	–
India rubber (Para) (see also rubber)	–	Solid	–	0.27-0.48	32-212	–	–	–	–	–	–	–	–
Indium (In)	E	Solid	456	0.0570	32-212	–	–	–	–	313.5	2630	–	–
Insulation, high temp block type	I	Solid	14-24	0.203 / 0.269	149 / 710	–	–	–	–	–	–	–	–
Iodine (I)	E	Solid	308	0.0541	48-206	–	–	–	–	236.3	364	28.4	42.3
Iridium (Ir)	E	Solid	1400	0.0323	54-212	–	–	–	–	4449	9600	–	–
Iron, (see also Figure A.14 and Table A.12)	E, M	Solid	491	0.11	68-212	–	–	–	–	2802	5430	117	–
gray cast	M	Solid	443	0.119	68-212	–	–	–	–	2330	–	41.4	–
white cast	M	Solid	480	0.119	68-212	–	–	–	–	2000	–	59.5	–
wrought	M	Solid	487-493	0.115	59-212	–	–	–	–	–	–	–	–

[a] A = alloy, E = element, F = fuel or fuel component, I = insulation, M = metal, O = organic compound, R = refractory, S = solution.

Table A.16 US. Densities and thermal properties of various substances (continued)

Name – Description	Other[a] classification	Normal state	Density lb/ft³	Solid state: Specific heat	Solid state: Temp range F	Liquid state: Specific heat	Liquid state: Temp range F	Gaseous state: Specific heat	Gaseous state: Temp range F	Melting (fusion) point F	Boiling point, F (at std barometric pressure)	Latent heat of fusion Btu/lb	Latent heat of vaporization Btu/lb
Kaolin	R	Solid	131	0.224	68-206	-	-	-	-	3200	-	-	-
				0.22	60-1200								
Kapok fiber	I	Solid	0.9	0.320	65	-	-	-	-	-	-	-	-
Kerosene	F	Liquid	48-49	-	-	0.50	32-212	-	-	-	-	-	105-110
Krypton (Kr)	E	Gas	0.215	0.1069	-	-	-	-	-	-272.2	-241	-	-
Lanthanum (La)	E	Solid	384	0.0448	32-212	-	-	-	-	1490	3270	-	-
Lava	-	Solid	-	0.197	77-212	-	-	-	-	-	-	-	-
Lead (Pb) (see also Table A.12)	M, E	Solid	706	0.0319	61-493	0.041	590-680	-	-	620 ± 4	2950	9.9	-
Lead-antimony alloy (62.9 Pb, 37.1 Sb)	A	Solid	-	0.0388	50-206	-	-	-	-	-	-	-	-
Lead-bismuth alloy (40 Pb, 60 Bi)	A	Solid	-	0.0317	32-212	-	-	-	-	-	-	-	-
Lead oxide (PbO)	-	Solid	574-593	0.049	60-212	-	-	-	-	1749	-	-	-
Lead slag wool	I	Solid	-	0.178	150	-	-	-	-	-	-	-	-
				0.235	718								
Light oil	F	Liquid	50	-	-	0.50	-	-	-	-	-	-	145-150
Limestone	-	Solid	168-175	0.216	59-212	-	-	-	-	-	-	-	-
Linotype (see also Table A.12)	A	Solid	-	0.036	60-486	-	-	-	-	486	-	21.5	-
Linseed oil	-	Liquid	58	-	-	0.441	-	-	-	-5	600	-	-
Lipowitz's metal (Pb26, Sn 13, Cd 10, Bi 51)	A	Solid	-	0.041	60-140	-	-	-	-	140	-	17.2	-
Lithium (Li)	E	Solid	33	1.0407	212	-	-	-	-	366.8	2552	286	-
Lodestone (magnetite)	-	Solid	322	0.156	32-212	-	-	-	-	-	-	-	-
Machine oil	-	Liquid	-	-	-	0.40	32-212	-	-	-	-	-	-
Magnesia, (85%)	I	Solid	11-13	0.276	150	-	-	-	-	-	-	-	-
				0.283	279								
Magnesite refractory	R	Solid	171	0.27	60-1200	-	-	-	-	3580+	-	-	-
Magnesite refractory (unburned)	R	Solid	183	0.26	60-1200	-	-	-	-	3580+	-	-	-
Magnesite refractory (fused)	R	Solid	179	0.27	60-1200	-	-	-	-	3580+	-	-	-
Magnesium (Mg) (see also Table A.12)	M, E	Solid	108.6	0.2492	68-212	0.266	-	-	-	1203.8	2048	160	-
Magnesium ox-de	-	Solid	228	0.234	86-100	-	-	-	-	5070	6580	-	-
				0.295	86-1800								

[a] A = alloy, E = element, F = fuel or fuel component, I = insulation, M = metal, O = organic compound, R = refractory, S = solution.

Table A.16 US. Densities and thermal properties of various substances (continued)

Name – Description	Other[a] classification	Normal state	Density lb/ft³	Solid state — Specific heat	Solid state — Temp range F	Liquid state — Specific heat	Liquid state — Temp range F	Gaseous state — Specific heat	Gaseous state — Temp range F	Melting (fusion) point F	Boiling point F (at std barometric pressure)	Latent heat of fusion Btu/lb	Latent heat of vaporization Btu/lb
Magnetite (Lodestone)	–	Solid	322	0.156	32-212	–	–	–	–	–	–	–	–
Manganese (Mn) (see also Table A.12)	E, M	Solid	464	0.1211	68-212	–	–	–	–	2246.0	3452	115	–
Marble	–	Solid	162-175	0.210	32-212	–	–	–	–	–	–	–	–
Mercuric chloride ($HgCl_2$)	–	Solid	339	–	–	–	–	–	–	539.6	581	–	–
Mercurous chloride (Hg_2Cl_2)	–	Solid	446	0.05	68	–	–	–	–	576	722	–	–
Mercury (Hg)	E, M	Liquid	847	0.032	-121-+68	0.033	32-212	–	–	-37.97	674.6	5.07	117.0
Methane (CH_4)	O, F	Gas	0.04243	–	–	0.992	-172	0.5929	64-406	–	-260	–	–
Methanol (CH_3OH, Methyl Alcohol)	F, O	Liquid	49.5	–	–	0.6	–	–	–	–	148	–	470
Methyl chloride (CH_3Cl)	O	Gas	0.135	–	–	0.385	60	–	–	-144	-11	–	–
Mica	I	Solid	–	0.10	68	–	–	–	–	–	–	–	–
Mica, expanded (vermiculite)	I	Solid	–	0.205, 0.236	149, 290	–	–	–	–	–	–	–	–
Milk	I	Liquid	64.2	0.095	-120	0.847	–	–	–	–	–	–	–
Mineral wool board with binder	I	Solid	14.3	0.247	150	–	–	–	–	–	–	–	–
Molasses	–	Liquid	87.3	–	–	0.60	–	–	–	–	–	–	–
Molybdenum (Mo)	E, M	Solid	637	0.0647	68-212	–	–	–	–	4595	6680	126	–
Monel metal (see also Table A.12)	A	Solid	550	0.129	60-2415	0.139	–	–	–	2415	–	117.4	–
Naphtha	–	Liquid	41.2	–	–	0.493	–	–	–	–	306	–	184
Naphthalene ($C_6H_4C_4H_4$)	O	Solid	71.8	0.325	68-140	0.427	262	–	–	175.8	424.2	64.1	135.7
Neat's-foot oil	E	Liquid	–	0.457	68-86	–	–	–	–	32	–	–	–
Neodymium (Nd)	E	Solid	431	–	–	–	–	–	–	1544	–	–	–
Neon (Ne)	A	Gas	0.0514	–	–	–	–	0.443	any	-423?	-398.2	–	–
Nichrome	A	Solid	517	0.111	–	–	–	–	–	–	–	–	–
Nickel (Ni) (see also Table A.12)	E, M	Solid	556	0.109, 0.1608	64-212, 1832	–	–	–	–	2645	4950	133	2670
Nickel steel	A	Solid	–	0.109	32-212	–	–	–	–	–	–	–	–
Nitric acid (HNO_3)	A	Liquid	96.1	–	–	0.445	–	–	–	-43.6	186.8	–	207.2
Nitric acid (10%) & water	S	Liquid	–	–	–	0.768	–	–	–	–	–	–	–
Nitric acid (2%) & water	S	Liquid	–	–	–	0.930	–	–	–	–	–	–	–
Nitric acid (1%) & water	S	Liquid	–	–	–	0.963	–	–	–	–	–	–	–
Nitric oxide (NO)	–	Gas	0.079	–	–	0.580	-249, -252	0.2317	55-342	–	-240	–	–
Nitrobenzene ($C_6H_5O_2N$)	O	Liquid	74.7	–	–	0.38	122	–	–	41	411.8	–	–

ª A = alloy, E = element, F = fuel or fuel component, I = insulation, M = metal, O = organic compound, R = refractory, S = solution.

Table A.16 US. Densities and thermal properties of various substances *(continued)*

Name – Description	Other[a] classi-fication	Normal state	Density lb/ft³	Solid state Specific heat	Solid state Temp range F	Liquid state Specific heat	Liquid state Temp range F	Gaseous state Specific heat	Gaseous state Temp range F	Melting (fusion) point F	Boiling point, F (at std barometric pressure)	Latent heat of fusion Btu/lb	Latent heat of vapor-ization Btu/lb
Nitrobenzole	–	Liquid	–	–	–	0.35	57.2	–	–	–	–	–	–
Nitrogen (N₂)	–	Gas	0.0741	–	–	0.475	57.2 -322- -344	0.2415	68-824	-347.8	-319	11.1	85.6
Nitrous oxide (N₂O)	–	Gas	0.117	–	–	–	–	–	–	–	-129	–	–
Octane (C₈H₁₈)	O, F	Liquid	43.6	–	–	0.52	60	0.226	61-405	–	266.0	–	126.0
Oil (see castor oil, coal tar oil, cottonseed oil, fuel oil, light oil, linseed oil, machine oil, oil of citron, oil of juniper, oil of orange, oil of turpentine, olive oil, paraffin oil, petroleum, and Part 2, Vol. I.)													
of citron	–	Liquid	53.2	–	–	0.438	42	–	–	–	–	–	–
of juniper	–	Liquid	–	–	–	0.477	–	–	–	–	–	–	–
of orange	–	Liquid	–	–	–	0.489	–	–	–	–	–	–	–
of turpentine	–	Liquid	53.7	–	–	0.411	32	–	–	14	320	–	184
Olive oil	–	Liquid	57.4	–	–	0.471	44	–	–	68±	572±	–	–
Osmium (Os)	E	Solid	1402	0.0311	68-208	–	–	–	–	4892?	9900	–	–
Oxalic acid (C₂H₂O₄ · 2H₂O)	O	Solid	103.8	0.338	0	–	–	–	–	372	302h	–	–
Oxygen (O₂)	E	Gas	0.0847	0.416	100	0.398	-88.0- -79.2	0.2178	55-405	-360.4	-296.9	5.98	91.6
Palladium (Pd)	E	Solid	749	0.0714	32-2309	0.0714	–	–	–	2820 ± 9	3992	64.6	–
Paper, expanding blanket ("Kimsul")	I	Solid	–	0.349	148	–	–	–	–	–	–	–	–
Parafin	–	Solid	54-57	0.622	95-104	0.712	140-145	–	–	100-133	662-806	63-70	–
Parafin oil	–	Liquid	–	–	–	0.52	32-212	–	–	–	–	–	–
Pentane n-C₅H₁₂	O, F	Liquid	38.8	–	–	–	–	–	–	–	97	–	154.4
Petroleum	F	Liquid	47-55	–	–	0.511	69.8-136.4	–	–	–	–	–	–
Phenol (C₆H₆O) (see aldo resin)	–	Solid	66.8	–	–	–	–	–	–	105.6	360	45.0	–
Phosphorus (P)	E	Solid	113.8	0.1829	32-124	–	–	–	–	111.6	550.4	9.05	–
Pitch (coal tar)	F	Solid	62-81	0.45	60-212	0.35-0.45	–	–	–	86-302	325	–	–
Plaster	–	Solid	90	0.20	–	–	–	–	–	–	–	–	–
Platinum (Pt)	E, M	Solid	1335	0.0359	68-2372	–	–	–	–	3191 ± 9	7750	48.96	–
Porcelain	R	Solid	143-156	0.26	59-1742	–	–	–	–	2140-3000	–	–	–
Porcelain, refractory	R	Solid	–	0.23	60-1200	–	–	–	–	–	–	–	–
Potassium (K)	E	Solid	53.6	0.170	-301- +68	–	–	–	–	144.1	1400	26.2	–
Potassium chlorate (KClO₃)	–	Solid	145	0.205	122	–	–	–	–	701.6	–	–	–

[a] A = alloy, E = element, F = fuel or fuel component, I = insulation, M = metal, O = organic compound, R = refractory, S = solution.
h Sublimes.

Table A.16 US. Densities and thermal properties of various substances (continued)

Name – Description	Other[a] classification	Normal state	Density lb/ft³	Solid state Specific heat	Solid state Temp range F	Liquid state Specific heat	Liquid state Temp range F	Gaseous state Specific heat	Gaseous state Temp range F	Melting (fusion) point F	Boiling point, F (at std barometric pressure)	Latent heat of fusion Btu/lb	Latent heat of vaporization Btu/lb
Potassium hydroxide (KOH + 30 H₂O)	–	Liquid	–	–	–	0.876	64.4	–	–	–	–	–	–
Potassium nitrate (KNO₃)	E	Solid	129.2	0.19	59-212	–	–	–	–	646	–	88	–
Praseodymium (Pr)	E	Solid	–	–	–	–	–	–	–	1724	–	–	–
Propane, commercial	F, O	Gas	0.1170	–	–	0.588	60	0.404	60	–	-51	–	185
Propane, pure C₃H₈	F, O	Gas	0.1162	–	–	0.592	60	0.388	60	–	-43.7	–	183
Pyrex	–	Solid	–	0.196	68-212	–	–	–	–	–	–	–	–
Pyrites, copper	–	Solid	–	0.1291	59-210	–	–	–	–	–	–	–	–
Quartz	–	Solid	165	0.17-0.28	32-212	–	–	–	–	–	–	–	–
Quicklime	–	Solid	–	0.217	32-212	–	–	–	–	–	–	–	–
Radium (Ra)	E	Solid	312	–	–	–	–	–	–	1742	2080	–	–
Redwood bark, shredded ("Palco Bark")	I	Solid	4.0	0.172 / 0.246	-127 / 109	–	–	–	–	–	–	–	–
Resin, phenol, pure[d]	–	Solid	75-81	0.33-0.37	–	–	–	–	–	167-212[f]	–	–	–
Resin, phenol, wood flour filler[d]	–	Solid	81-87	0.30-0.36	–	–	–	–	–	257-266[f]	–	–	–
Resin, phenol, asbestos filler[d]	–	Solid	112-125	0.38-0.40	–	–	–	–	–	266-302[f]	–	–	–
Resin, copals[e]	–	Solid	65-71	0.38-0.40	–	–	–	–	–	300-680	–	–	–
Rhodium (Rh)	E	Solid	777	0.058	50-207	–	–	–	–	3542	>4580	–	–
Rock wool	–	Solid	7-12	0.201 / 0.250	149 / 652	–	–	–	–	–	–	–	–
Rose's metal (25 Pb, 25 Sn, 50 Bi)	A	Solid	–	0.043	60-230	0.041	–	–	–	230	–	18.3	–
Rosin	–	Solid	68	0.525	68-450	–	–	–	–	170-212	–	–	–
Rubber	–	Solid	62-125	0.481	60-212	–	–	–	–	248	–	–	–
Rubber board, expanded ("Rubatex")	–	Solid	4.9	0.152 / 0.273	-125 / 111	–	–	–	–	–	–	–	–
Rubidium (Rb)	E	Solid	95.5	0.0802	32	–	–	–	–	100.4	1284.8	–	–
Ruthenium (Ru), black	E	Solid	537	0.0611	32-212	–	–	–	–	3530	–	–	–
Ruthenium (Ru), gray	–	Solid	760	–	–	–	–	–	–	4442	–	–	–
Salt, rock	–	Solid	135	0.219	55-113	–	–	–	–	1495	–	–	–
Samarium (Sm)	E	Solid	481	–	–	–	–	–	–	2372-2552	–	–	–
Sand	–	Solid	162	0.195	59-212	–	–	–	–	–	–	–	–
Sea water	S	Liquid	64	–	–	0.938	63.5	–	–	–	–	–	–

[a] A = alloy, E = element, F = fuel or fuel component, I = insulation, M = metal, O = organic compound, R = refractory, S = solution.
[d] Manufactured under trade names: Bakelite, Redmanol, Condensite, etc.
[e] Resins used in varnish making: Kauric, Congo, Zanzibar, and Manila copals.
[f] Softening point under load.

Table A.16 US. Densities and thermal properties of various substances (continued)

Name – Description	Other[a] classi-fication	Normal state	Density lb/ft³	Specific heat in Btu/lb °F — Solid state: Specific heat	Solid state: Temp range F	Liquid state: Specific heat	Liquid state: Temp range F	Gaseous state: Specific heat	Gaseous state: Temp range F	Melting (fusion) point F	Boiling point, F (at std barometric pressure)	Latent heat of fusion Btu/lb	Latent heat of vaporization Btu/lb
Selenium (Se)	E	Solid	300	0.068	-306-+64	–	–	–	–	422.6-428.0	1274	–	–
Serpentine	–	Solid	–	0.25	32-212	–	–	–	–	–	–	–	–
Shellac (Lac)	–	Solid	75-76	0.40	60-212	–	–	–	–	170-180	–	–	–
Silica (SiO₂)	–	Solid	180	0.1910	32-212	–	–	–	–	3182	–	–	–
Silica aerogel ("Santocel")	I	Solid	5.3	0.205	147	–	–	–	–	–	–	–	–
				0.274	630								
Silica refractory	R	Solid	111	0.23	60-1200	–	–	–	–	3060-3090	–	–	–
Silicon (Si)	E	Solid	145	0.1833	135	–	–	–	–	2588	4149	607	–
				0.2029	450								
Silicon carbide (SiC)	–	Solid	199	0.23	60-950	–	–	–	–	4082	–	–	–
Silicon carbide (clay-bonded) refractory	R	Solid	136-159	0.20	60-1200	–	–	–	–	3390	–	–	–
Silk, raw	–	Solid	81-87	0.33	60-212	–	–	–	–	–	–	–	–
Sillimanite (mullite) refr.	R	Solid	145-202	0.23	60-1200	–	–	–	–	3310-3340	–	–	–
Silver (Ag) (see also Table A.12)	E, M	Solid	655	0.05987	63-945	–	–	–	–	1760.9	3551	45.1	–
Slag, blast furnace (powdered)	–	Solid	22.5	–	–	–	–	–	–	–	–	90.0	–
Slag wool	I	Solid	9.4-18.7	0.17	77	–	–	–	–	–	–	–	–
Soda, baking	–	Solid	137	0.231	32-212	–	–	–	–	–	–	–	–
Sodium (Na)	E	Solid	60.6	0.253	-301-+68	–	–	–	–	207.5	1614	49.3	–
Sodium carbonate (Na₂CO₃)	–	Solid	151.5	0.306	–	–	–	–	–	1565.6	–	–	–
Sodium carbonate (2%) and water	S	Liquid	–	–	–	0.896	–	–	–	–	–	–	–
Sodium chloride (NaCl) (see also rock salt)	–	Solid	135	–	–	–	–	–	–	1472	–	–	–
Sodium chloride (10%) and water (see also sea water)	S	Liquid	67	–	–	0.791	64.4	–	–	–	–	–	–
Sodium hydroxide (2% NaOH) and water	S	Liquid	63.8	–	–	0.942	64.4	–	–	–	–	–	–
Sodium nitrate (NaNO₃)	–	Solid	140.5	0.231	–	–	–	–	–	597	716	116.8	–
Sodium sulfate (Na₂SO₄)	–	Solid	–	–	–	–	–	–	–	1623.2	–	–	–
Solder (Pb and Sn)	A	Solid	580	0.040-0.051	–	–	–	–	–	361-594	–	11.6-30.6	–
Spermaceti (whale oil)	F	Solid	–	0.051	–	–	–	–	–	113±	–	66.56	–

[a] A = alloy, E = element, F = fuel or fuel component, I = insulation, M = metal, O = organic compound, R = refractory, S = solution.

Table A.16 US. Densities and thermal properties of various substances (continued)

Name – Description	Other[a] classi-fication	Normal state	Density lb/ft³	Specific heat in Btu/lb °F						Melting (fusion) point F	Boiling point, F (at std barometric pressure)	Latent heat of fusion Btu/lb	Latent heat of vapor-ization Btu/lb
				Solid state		Liquid state		Gaseous state					
				Specific heat	Temp range F	Specific heat	Temp range F	Specific heat	Temp range F				
Steel (see also Figure A.14, iron, nickel steel)	A	Solid	490	0.165[k]	60-2900	–	–	–	–	–	–	–	–
Sterotype (see Figure A.15)	A	Solid	670	0.036	–	0.036	–	–	–	500	–	26.2	–
Stones, all kinds (see also marble, granite, limestone, sandstone)	–	Solid	168	0.18-0.23	54-212	–	–	–	–	–	–	–	–
Sugar, cane, amorphous	–	Solid	–	0.342	68	–	–	–	–	–	–	–	–
Sugar, cane, crystaline	–	Solid	102	0.301	68	–	–	–	–	320	–	–	–
Sugar, cane, (4%) and water	S	Liquid	–	0.7558	–	–	–	–	–	–	–	–	–
Sulfur (S)	E	Solid	119-130	0.190	59-130	0.2337	235-840	–	–	239	832.5	16.87[g]	651.6
Sulfur dioxide (SO₂)	–	Gas	0.1733	–	–	0.36	122	0.1544	61-396	-104.8	14	–	–
Sulfuric acid (H₂SO₄)	–	Liquid	115.8	–	–	0.336	32-212	–	–	50.9	640.4[b]	–	–
Talc	–	Solid	–	–	–	0.2092	68-208	–	–	–	–	–	–
Tallow, beef	–	Solid	56.8-60.5	–	–	0.79	79-108	–	–	80-100	–	–	–
						0.54	151-216						
Tantalum (Ta)	E	Solid	1035	0.043	2552	–	–	–	–	5252	–	–	–
Tartaric acid (C₄H₆O₆)	–	Solid	104	0.287	97	–	–	–	–	338	–	–	–
Tellurium (Te)	E	Solid	389	0.0483	59-212	–	–	–	–	845.6	2534	13.1	–
Thallium (Tl)	E	Solid	740	0.0326	68-212	–	–	–	–	575.6	3000	–	–
Thorium (Th)	E	Solid	699	0.0276	32-212	–	–	–	–	3350	>5432	–	–
Tile, hollow	–	Solid	75	0.15	–	–	–	–	–	–	–	–	–
Tin (Sn) (see also Table A.12)	E, M	Solid	455	0.0551	70-228	0.05799	482	–	–	449 ± 0.4	4118	–	–
						0.0758	2012						
Titanium (Ti)	E	Solid	283	0.1125	32-212	–	–	–	–	3263	–	–	–
Toluene (C₇H₈)	O	Liquid	53.6	–	–	0.40	32-212	–	–	-133.6	230.5	–	150.3
Toluol (C₆H₈)	O	Liquid	–	–	–	0.490	149	–	–	–	230.5	–	154.8
Tufa	–	Solid	–	0.33	32-212	–	–	–	–	–	–	–	–
Tungsten (W)	E, M	Solid	1202	0.0336	32-212	–	–	–	–	6152	10526	79	–
				0.0337	1832								
Turpentine	–	Liquid	53.6	0.42	32-212	–	–	–	–	–	318.8	–	133.3
Type metal	A	Solid	–	0.0388	32-212	–	–	–	–	–	–	–	–
Uranium (U)	E	Solid	1167	0.028	32-208	–	–	–	–	<3344	6330	–	–
Vanadium (V)	E	Solid	375	0.1153	32-212	–	–	–	–	3128	5430	–	–

[a] A = alloy, E = element, F = fuel or fuel component, I = insulation, M = metal, O = organic compound, R = refractory, S = solution.
[b] Decomposes.
[g] Transformation from rhombic to monoclinic absorbs 5.06 Btu/lb.
[k] Specific heat steel is about 0.11 in the 60-600 F range.

Table A.16 US. Densities and thermal properties of various substances (concluded)

Name – Description	Other[a] classi-fication	Normal state	Density lb/ft³	Specific heat in Btu/lb°F — Solid state — Specific heat	Solid state Temp range F	Liquid state Specific heat	Liquid state Temp range F	Gaseous state Specific heat	Gaseous state Temp range F	Melting (fusion) point F	Boiling point, F (at std barometric pressure)	Latent heat of fusion Btu/lb	Latent heat of vapor-ization Btu/lb
Varnish (see resins)	–	–	–	–	–	–	–	–	–	–	–	–	–
Vegetable fiberboard ("Celotex")	I	Solid	14.4	0.171 / 0.279	-116 / 109	–	–	–	–	–	–	–	–
Vermiculite (see mica)	I	–	–	–	–	–	–	–	–	–	–	–	–
Vulcanite	–	Solid	–	0.3312	68-212	–	–	–	–	–	–	–	–
Water (H₂O) (see also sea water) (see Tables A.8 thru A.11)	–	Liquid	62.37	0.480	<32	1.00	60	–	–	32	212	144	970.2
Wood (see redwood bark)	–	Solid	19-56	0.33-0.67	–	–	–	–	–	–	–	–	–
Wood fiber blanket ("Balsam Wool")	I	Solid	2.6	0.330	150	–	–	–	–	–	–	–	–
Wood fiberboard	I	Solid	12-19	0.341	148	–	–	–	–	–	–	–	–
Wood, oak	–	Solid	48	0.57	32-212	–	–	–	–	–	–	–	–
Wood, pine	–	Solid	30	0.67	32-212	–	–	–	–	–	–	–	–
Wood's metal (26Pb, 13Sn, 12Cd, 49Bi)	A	Solid	–	0.041	60-158	0.042	–	–	–	158	–	17.2	–
Wool (see also glass wool, mineral wool, rock wool, lead slag wool, slag wool, etc.)	–	Solid	80-83	0.325	–	–	–	–	–	–	–	–	–
Xenon (Xe)	E	Gas	0.346	–	–	–	–	–	–	-220	-164.4	–	–
Xylene	–	Liquid	54.3	0.42	122	–	–	–	–	-18	288	–	147
Yttrium (Y)	E	Solid	343	–	–	–	–	–	–	2714	1663	–	–
Zinc (Zn) (see Table A.12)	E, M	Solid	445	0.0931 / 0.1040	68-212 / 572	–	–	–	–	786.9	–	46.8	758
Zinc chloride (ZnCl₂)	–	Solid	181.5	–	–	–	–	–	–	689	1350	–	–
Zinc oxide (ZnO)	–	Solid	350	0.125	32-212	–	–	–	–	>3240	–	–	–
Zinc sulfate (ZnSO₄)	S	Liquid	234	–	–	0.174	106	–	–	1330[b]	–	–	–
Zircon	–	Solid	293	0.132	–	–	–	–	–	4622	–	–	–
Zirconium (Zr)	E	Solid	405	0.0660	32-212	–	–	–	–	3092	9122	–	–

[a] A = alloy, E = element, F = fuel or fuel component, I = insulation, M = metal, O = organic compound, R = refractory, S = solution.
[b] Decomposes.

Table A.17 Metric. Densities and thermal properties of various substances. (See Part 2, Vol. I) for more information on fuels; Part 4 (Vol. I) for insulations and refractories; Tables and Figures A.12-A.15 for metals and alloys; Tables A.19-A.20 for heat transfer fluids.)

Name – Description	Other[a] classification	Normal state	Density kg/m³	Specific heat in cal/g °C						Melting (fusion) point C	Boiling point, C (at std barometric pressure)	Latent heat of fusion cal/g	Latent heat of vaporization cal/g
				Solid state		Liquid state		Gaseous state					
				Specific heat	Temp range C	Specific heat	Temp range C	Specific heat	Temp range C				
Acetic acid (CH₃COOH)	O	Liquid	1054	0.487	0	0.51	0-100	–	–	17	118	44.8	96.7
Acetone (CH₃COCH₃)	O	Liquid	797.0	–	–	0.544	0-100	0.3468	21-110	-95	53-57	23	133
Acetylene (C₂H₂)	O, F	Gas	1.1070	–	–	–	–	0.64	15	-81	-84	–	–
Air (see also Tables A.2 and A.3)	–	Gas	1.2223	–	–	–	–	0.2394 / 0.2469 / 0.2562	-30-+10 / 20-440 / 20-800	–	-191	–	–
Alcohol, ethyl (C₂H₅OH)	O, F	Liquid	790	–	–	0.648	40	0.4534	108-220	-114	78	26	205
Alcohol, ethyl (90%) and water	S	Liquid	823	–	–	0.718	–	–	–	–	–	–	–
Alcohol, ethyl (50%) and water	S	Liquid	918	–	–	0.923	–	–	–	–	–	–	–
Alcohol, ethyl (10%) and water	S	Liquid	984	–	–	0.99	–	–	–	–	–	–	–
Alcohol, methyl (CH₃OH)	O, F	Liquid	795	–	–	0.601	15-20	0.33	127-227	-97	66	16.4	267.2
Alcohol, methyl (90%) and water	S	Liquid	823	–	–	0.643	–	–	–	–	–	–	–
Alcohol, methyl (50%) and water	S	Liquid	918	–	–	0.846	–	–	–	–	–	–	–
Alcohol, methyl (10%) and water	S	Liquid	984	–	–	0.986	–	–	–	–	–	–	–
Alumina – see Aluminum Oxide. Alumina (fused) refractory, high-alumina refractory	–	–											
Alumina (fused) refractory (see also high-alumina refractory)	R	Solid	2450-2900	0.20	16-649	–	–	–	–	1866+	–	–	–
Aluminum (see also Table A.12 and Figure A.8)	M, E	Solid	2670	0.225	16-304	–	–	–	–	660	1800	93.13	200
Aluminum foil	I	Solid	–	0.24	143	–	–	–	–	–	–	–	–
Aluminum oxide (alumina)	–	Solid	3901	0.183	0-100	–	–	–	–	2020	–	–	–
Ammonia (NH₃)	–	Gas	0.737	–	–	1.08	-22.4--15	0.525	21-104	-75	-34	83.4	327
Ammonium chloride (10%) and water	S	Liquid	1032	–	–	0.788	–	–	–	–	–	–	–
Ammonium sulfate[(NH₄)₂SO₄]	–	Solid	176	0.283	–	–	–	–	–	513[b]	–	–	–

[a] A = alloy, E = element, F = fuel or fuel component, I = insulation, M = metal, O = organic compound, R = refractory, S = solution.
[b] Decomposes.

Table A.17 Metric. Densities and thermal properties of various substances (continued)

Name – Description	Other[a] classification	Normal state	Density kg/m³	Specific heat in cal/g — Solid state Specific heat	Solid state Temp range C	Liquid state Specific heat	Liquid state Temp range C	Gaseous state Specific heat	Gaseous state Temp range C	Melting (fusion) point C	Boiling point, C (at std barometric pressure)	Latent heat of fusion cal/g	Latent heat of vaporization cal/g
Andalusite (AL₂SiO₃)	—	Solid	3201	0.228	—	—	—	—	—	1810	—	—	—
Aniline (C₆H₅N)	O, F	Liquid	1024	—	—	0.514	15	—	—	−8	184	21	110.0
Antimony (Sb)	M, E	Solid	6760	0.052	200	—	—	—	—	630	1440	38.3	391
Argon (A)	E	Gas	1.682	—	—	—	—	0.1233	20-90	−188	−186	6.7	38
Arsenic, gray	E	Solid	5735	0.0622	0-100	—	—	—	—	c	c	—	—
Asbestos	I	Solid	1986-2787	0.20	0-100	—	—	—	—	—	—	—	—
Ashes	—	—	—	0.20	0-100	—	—	—	—	—	—	—	—
Asphalt, Bermudez	—	Solid	1080	0.55	—	—	—	—	—	82	—	—	—
Gilsonite	—	Solid	1040	0.55	—	—	—	—	—	149	—	—	—
Oil	—	Solid	1000-1040	0.55	—	—	—	—	—	60-82	—	—	—
Trinidad	—	Solid	1400	0.55	—	—	—	—	—	88	—	—	—
Babbitt, lead base	A	Solid	—	0.039	16-239	0.038	—	—	—	239	—	14.6	—
tin base (see also Table A.12)	A	Solid	7449	0.071	16-240	0.063	—	—	—	240	—	19	—
Bakelite (see Resin, phenol)	—	—	—	—	—	—	—	—	—	—	—	—	—
Barium (Ba)	E	Solid	3492	0.068	−85+20	—	—	—	—	850	1138	—	622.7
Basalt (see lava)	—	—	—	—	—	—	—	—	—	—	—	—	—
Beeswax	—	Solid	960	0.82	16-62	—	—	—	—	62	—	42.4	—
Benzene (C₆H₆)	O, F	Liquid	879	—	—	0.423	—	0.3325	35-180	5.4	80.2	30.9	94.19
Benzoic acid (C₇H₆O₂)	O	Solid	1301	0.287	16	—	—	—	—	121	249	33.9	—
Benzol (C₆H₆) (water white)	O, F	Liquid	811	—	—	0.423	—	—	—	5.4	80.2	30.62	94.19
Beryllium (Be)	E, M	Solid	1820	0.52	—	0.425	40	—	—	1500	2771	317	—
Bismuth (Bi) (see also Table A.12)	E, M	Solid	9804	0.0302	20-100	0.036	279-385	—	—	271	1430	12.5	220
Bismuth (63.8)–Tin (36.2) alloy	A	Solid	—	0.040	20-99	—	—	—	—	—	—	—	—
Blast furnace gas	F	Gas	1.25	—	—	—	—	0.2277	—	—	—	—	—
Boron (B)	E	Solid	2435	0.307	0-100	—	—	—	—	2200-2500	2538	—	—
Borax (Na₂B₄O₇ - 10 H₂O)	—	Solid	1714	0.385	35	—	—	—	—	741	—	—	—
Brass,	—		—	—	—	—	—	—	—	—	—	—	—
Muntz metal :60Cu, 40Zn)	A	Solid	8394	0.105	16-888	—	—	—	—	888	—	38.4	—
Red (85Cu, 15Zn)	A	Solid	8747	0.104	16-1067	—	—	—	—	1067	—	48.1	—
Yellow (67Cu, 33Zn) (see also Table A.12)	A	Solid	8459	0.105	16-920	—	—	—	—	920	—	39.5	—
Brick, red	—	Solid	1890	0.22	0-100	—	—	—	—	—	—	—	—
Britania metal (90Sn, 10Pb)	A	Solid	—	—	—	—	—	—	—	—	—	28.0	—
Bromine (Br)	E	Liquid	3092	0.0843	−78 − −20	0.107	1-32	0.0555	83-228	−7.3	61	16.21	45.6

[a] A = alloy, E = element, F = fuel or fuel component, I = insulation, M = metal, O = organic compound, R = refractory, S = solution.
c Sublimes at 559 C, melts at 850 C under 36 atmospheres of pressure.

Table A.17 Metric. Densities and thermal properties of various substances (continued)

(Specific heat in cal/g °C spans the Solid state, Liquid state, and Gaseous state column groups.)

Name – Description	Other[a] classification	Normal state	Density kg/m³	Solid state		Liquid state		Gaseous state		Melting (fusion) point C	Boiling point, C (at std barometric pressure)	Latent heat of fusion cal/g	Latent heat of vaporization cal/g
				Specific heat	Temp range C	Specific heat	Temp range C	Specific heat	Temp range C				
Bronze (80 Cu, 20 Sn)	A	Solid	–	0.0862	14-98	–	–	–	–	–	–	–	–
Aluminum (see also Table A.12)	A	Solid	8170	0.126	16-1050	–	–	–	–	1050	–	54.8	–
Bearing	A	Solid	8907	0.095	16-1000	–	–	–	–	1000	–	44.4	–
Bell metal	A	Solid	8651	0.100	16-890	–	–	–	–	890	–	42.4	–
Gun metal	A	Solid	8811	0.107	16-1010	–	–	–	–	1010	–	46.8	–
Tobin	A	Solid	8411	0.107	16-885	–	–	–	–	885	–	40.9	–
Butane (C_4H_{10})	O, F	Gas	2.387	–	–	0.55	16	0.458	16	-134	0.61	13.2	92.02
Cadmium (Cd)	E, M	Solid	8651	0.057	100	–	–	–	–	320.9	767	–	227
Calcium (Ca)	E	Solid	1548	0.170	0-181	–	–	–	–	849	1441	–	–
Calcium carbonate ($CaCO_3$)	–	Solid	2691-2948	0.210	0-100	–	–	–	–	825[b]	–	–	–
Calcium chloride ($CaCl_2$)	–	Solid	2147	0.292	16	–	–	–	–	774	>1599	–	–
Calcium chloride (30%) and water	S	Liquid	1261	–	–	0.676	40	–	–	–	–	–	–
Camphor ($C_{10}H_6O$)	O	Solid	1000	0.44	20-178	0.61	178-210	–	–	178	209	10.8	–
Carbon (C) (graphite)	E, F	Solid	2211	0.160	11	–	–	–	–	3500[h]	4832	–	–
Carbon bisulfide (see carbon disulfide)													
Carbon dioxide (CO_2)	–	Gas	1.87	–	–	–	–	0.2169	11-214	–	[h]	–	–
Carbon disulfide (CS_2)	–	Liquid	1270	0.467	976	0.232	16	0.1596	86-190	-110	46.1	–	83.84
Carbon monoxide (CO)	F	Gas	1.1871	–	–	0.0615	-63.7 to -58.8	0.2426	26-198	-207	-191[c]	–	–
Carbon tetrachloride (CCl_4)	–	Liquid	1583	–	–	0.215	50	–	–	-22.7	77	–	46.4
Castor oil	–	Liquid	963	–	–	0.434	–	–	–	–	–	–	–
Cellulose	–	Solid	1521	0.32	0-100	–	–	–	–	–	–	–	–
Cerium (Ce)	E	Solid	6889	0.0448	0-100	–	–	–	–	640	1393	–	–
Cesium (Cs)	E	Solid	1900	0.0482	0-16	–	–	–	–	28.3	670	3.76	–
Chalk	–	Solid	–	0.215	0-100	–	–	–	–	–	–	–	–
Charcoal	F	Solid	288-609	0.165-0.25	24	–	–	–	–	–	–	–	–
Chlorine (Cl)	E	Gas	3.044	–	–	0.229	-63.3	0.1125	16-343	-101.5	-33.6	23	67.3
Chloroform ($CHCl_3$)	O	Liquid	1520	–	–	0.23	0-100	0.1489	22-78	-65	61.2	–	58.55
Chrome refractory, burned	R	Solid	3012	0.20	16-649	–	–	–	–	1971+	–	–	–
Chrome refractory, unburned	R	Solid	3092	0.21	16-649	–	–	–	–	1971+	–	–	–
Chromite (chrome ore) ($FeCr_2O_4$)	R	Solid	4502	0.22	–	–	–	–	–	2180	–	–	–

[a] A = alloy, E = element, F = fuel or fuel component, I = insulation, M = metal, O = organic compound, R = refractory, S = solution.
[b] Decomposes.
[c] Sublimes at -78.3 C, melts at 56.5 C under 5.2 atmospheres of pressure.
[h] Sublimes.

Table A.17 Metric. Densities and thermal properties of various substances (continued)

Name – Description	Other[a] classification	Normal state	Density kg/m³	Specific heat in cal/g °C Solid state — Specific heat	Solid state — Temp range C	Liquid state — Specific heat	Liquid state — Temp range C	Gaseous state — Specific heat	Gaseous state — Temp range C	Melting (fusion) point C	Boiling point, C (at std barometric pressure)	Latent heat of fusion cal/g	Latent heat of vaporization cal/g
Chromium (Cr)	E, M	Solid	7193	0.1039	0	–	–	–	–	1609	2482	–	–
				0.1121	100								
				0.1872	600								
Cinders	–	Solid	–	–	–	–	–	–	–	–	–	–	–
Clay	–	Solid	1794-2595	0.18	0-100	–	–	–	–	1738	–	–	–
Coal	F	Solid	–	0.224	20-98	–	–	–	–	–	–	–	–
Coal tar oil	F	Liquid	–	0.3	0-100	0.34	–	–	–	–	199-488	–	–
Cobalt (Co)	E	Solid	8907	0.1542	500	–	–	–	–	1495	2899	64.05	–
				0.204	1000								
Coke	F	Solid	–	0.203	0-100	–	–	–	–	–	–	–	–
				0.376	38-1204								
Columbium (Cb)	E	Solid	8571	0.065	–	–	–	–	–	2416	–	–	–
Concrete	–	Solid	2195	0.156-0.27	0-100	–	–	–	–	–	–	–	–
Constantan	A	Solid	–	0.098	0-100	–	–	–	–	–	–	–	–
Copper (Cu)	M, E	Solid	8955	0.0951	15-232	–	–	–	–	1083 ±2.8	2593	42.0	–
				0.1259	900								
Copper sulfate ($CuSO_4$) (16.7%) and water	S	Liquid	1181	0.848	12-15	–	–	–	–	–	–	–	–
Cork, natural	I	Solid	240	0.419	25	–	–	–	–	–	–	–	–
Cork, granulated	I	Solid	87-117	0.43	25	–	–	–	–	–	–	–	–
Corkboard	I	Solid	111-332	0.204	-82	–	–	–	–	–	–	–	–
				0.417	+66								
Corundum (Al_2O_3)	–	Solid	4005	0.1976	6-98	–	–	–	–	2050	–	–	–
Cottonseed oil	–	Liquid	929	–	–	0.474	–	–	–	0	–	–	–
Cream	–	Liquid	–	0.780	–	–	–	–	–	–	–	–	–
Cupric oxide (CuO)	–	Solid	5991-6488	0.227	20-100	–	–	–	–	1062	–	–	–
Cuprous oxide (Cu_2O)	–	Solid	6008	0.111	0-100	–	–	–	–	1234	–	–	–
D'Arcet's metal (50 Bi, 25 Pb, 25 Sn)	A	Solid	–	0.050	0-100	–	–	–	–	–	–	5.78	–
Decane ($C_{10}H_{22}$)	O, F	Liquid	726	–	–	0.42	16	–	–	-29.7	174	–	61.2
Diatomaceous earth	R	Solid	200-401	0.21	25	–	–	–	–	–	–	–	–
Die casting metal: Aluminum base (see also Table A.12)	A	Solid	2820	0.236	16-621	0.241	–	–	–	621	–	90.68	–
Lead base	A	Solid	–	0.038	16-316	0.037	–	–	–	316	–	9.67	–
Tin base	A	Solid	–	0.070	16-232	0.062	–	–	–	232	–	16.8	–
Zinc base	A	Solid	–	0.103	16-416	0.138	–	–	–	416	–	26.9	–
Diphenyl ($C_6H_5C_6H_5$)	O	Liquid	993	0.693	40	0.481	71-256	–	–	71	256	26	75.89

[a] A = alloy, E = element, F = fuel or fuel component, I = insulation, M = metal, O = organic compound, R = refractory, S = solution.

Table A.17 Metric. Densities and thermal properties of various substances (continued)

Name – Description	Other[a] classification	Normal state	Density kg/m³	Specific heat in cal/g °C — Solid state: Specific heat	Solid state: Temp range °C	Liquid state: Specific heat	Liquid state: Temp range °C	Gaseous state: Specific heat	Gaseous state: Temp range °C	Melting (fusion) point °C	Boiling point, C (at std barometric pressure)	Latent heat of fusion cal\g	Latent heat of vaporization cal\g
Diphenylamine $(C_6H_5NHC_6H_5)$	O	Liquid	1158	0.443	56	—	—	—	—	54	302	25.2	—
Dolomite	—	Solid	2900	0.222	20-98	—	—	—	—	—	—	—	—
Dowtherm A	—	Liquid	942	0.53	—	0.41	50	—	—	82	260	—	—
Earth (see also humus)	—	Solid	—	0.44	0-100	—	—	—	—	—	—	—	—
Ebonite	—	Solid	—	0.33	0-100	—	—	—	—	—	—	—	—
Ether, ethyl $(C_4H_{10}O)$	O	Liquid	735	—	—	0.529	0	0.48	69-224	−118	34.6	—	88.46
Ethyl acetate $(CH_3CO_2CH_2CH_3)$	O	Liquid	894	—	—	0.478	0-100	—	—	−83.5	—	—	—
Ethyl bromide (CH_3CH_2Br)	O	Liquid	1450	—	—	0.21	16	—	—	−119	38.2	—	60.44
Ethyl iodide (CH_3CH_2I)	O	Liquid	1922	—	—	0.25	50	—	—	−108	71	—	47.0
Ethylene glycol $(C_2H_6O_2)$	O	Liquid	1099	—	—	0.602	0-100	—	—	—	197	—	—
Fiberglas board	I	Solid	32.96	0.129 / 0.192 / 0.236	−82.7 / −30 / 44	—	—	—	—	—	—	—	—
Firebrick, fireclay	R	Solid	2195-2403	0.243	16-1202	—	—	—	—	1593-1760	—	—	—
insulating (2600 F)	R	Solid	615	0.22	16-649	—	—	—	—	1638-1649	—	—	—
silica	R	Solid	2307-2595	0.258	16-1202	—	—	—	—	1149+	—	—	—
Flourine (F_2)	E	Gas	1.60	—	—	—	—	0.32	27	−223	−187	—	—
Forsterite refractory	R	Solid	2451	0.25	16-649	—	—	—	—	1888	—	—	—
Fuel oil (see Tables 2.5 and 2.10, Vol. I)	F	Liquid	833-977	—	—	0.50	—	—	—	—	—	—	80.6-83.4
Fusel oil	—	Liquid	—	—	—	0.56	0-100	—	—	—	—	—	—
Galena (PbS)	E	Solid	7481	0.0466	0-100	—	—	—	—	1121	1699	—	—
Gallium (Ga)	E	Solid	5879	0.080	12-113	—	—	—	—	30.1	70-90	—	70.1-81.2
Gasoline (commercial)	F	Liquid	732	—	—	0.5135	—	—	—	958	—	—	—
Germanium (Ge)	E	Solid	5367	0.0737	0-100	0.123	—	—	—	1010	—	47.9	—
German silver (see also Table A.12)	A	Solid	—	0.0946 / 0.109	0-100 / 16-1010	—	—	—	—	1199±	—	—	—
Glass	—	Solid	2307-2996	0.15-0.23	0-100	—	—	—	—	—	—	—	—
Glass block, expanded, foamglas	—	Solid	170	0.132 / 0.157 / 0.179	−82.7 / −29 / 44	—	—	—	—	—	—	—	—
Glass wool	I	Solid	22-77	0.210 / 0.279	66 / 329	—	—	—	—	—	—	—	—
Glycerine $(C_3H_8O_3)$ (glycerol)	O	Liquid	1261	—	—	0.576	15-100	—	—	−56	290	42.5	—
Gneiss	—	Solid	—	0.196	17-99	—	—	—	—	—	—	—	—

[a] A = alloy, E = element, F = fuel or fuel component, I = insulation, M = metal, O = organic compound, R = refractory, S = solution.

Table A.17 Metric. Densities and thermal properties of various substances (continued)

Name – Description	Other[a] classification	Normal state	Density kg/m³	Specific heat in cal/g°C						Melting (fusion) point C	Boiling point, C (at std barometric pressure)	Latent heat of fusion cal/g	Latent heat of vaporization cal/g
				Solid state		Liquid state		Gaseous state					
				Specific heat	Temp range C	Specific heat	Temp range C	Specific heat	Temp range C				
Gold (Au) (see also Table A.12)	E, M	Solid	19304	0.0316	0-100	0.034	–	–	–	1063	2971	16	16
Granite	–	Solid	2595-2804	0.192	12-100	–	–	–	–	–	–	–	–
Graphite	–	Solid	2216	0.201 / 0.38	0-100 / 21-1204	–	–	–	–	3482	4827	–	–
Gypsum	I	Solid	2323	0.259	10-100	–	–	–	–	1360	–	–	–
Hairfelt	I	Solid	–	0.334	64	–	–	–	–	–	–	–	–
Helium (He)	E	Gas	0.16709	–	–	–	–	1.25	236	<-271	-267	–	–
Hematite (Fe₂O₃)	–	Solid	–	0.1645	15-99	–	–	–	–	–	–	–	–
Heptane (C₇H₁₆)	O, F	Liquid	679	–	–	0.55	50	–	–	-90	90	–	77.84
Hexane (C₆H₁₄)	O, F	Liquid	654	–	–	0.600	20-100	–	–	-94	70	–	79.29
High-alumina refractory	R	Solid	2051	0.23	16-649	–	–	–	–	1810	–	–	–
Hornblende	–	Solid	–	0.195	0-100	–	–	–	–	–	–	–	–
Humus (soil) (see also earth)	–	Solid	–	0.44	0-100	–	–	–	–	–	–	–	–
Hydrochloric acid (HCl) (45.2%) + H₂O	–	Liquid	1474	–	–	–	–	–	–	-15.3	–	–	–
Hydrofluoric acid (HF) (35.35%) + H₂O	–	Liquid	1153	–	–	–	–	–	–	-33	120	–	–
Hydrogen (H₂)	E, F	Gas	0.0849	–	–	1.75-2.33	-79.6 -77.8	3.41	21-100	-259	-252.6	15	108
Hydrogen chloride (HCl)	–	Gas	1.5491	–	–	–	–	0.190	13-100	-111.3	-83	–	–
Hydrogen fluoride (HF)	–	Gas	1.2079	–	–	–	–	–	–	-92.3	-36.6	–	–
Hydrogen sulfide (H₂S)	–	Gas	1.4530	–	–	–	–	0.241	20-206	-87	-63	–	–
Ice (H₂O)	–	Solid	894-920	0.463	-75	–	–	–	–	0	–	80.1	–
India rubber (Para) (see also rubber)	–	Solid	–	0.27-0.48	0-100	–	–	–	–	–	–	–	–
Indium (In)	E	Solid	7305	0.0570	0-100	–	–	–	–	156.4	1443	–	–
Insulation, high temp block type	I	Solid	224-384	0.203 / 0.269	65 / 377	–	–	–	–	–	–	–	–
Iodine (I)	E	Solid	4934	0.0541	9-98	–	–	–	–	113.5	184	15.8	23.5
Iridium (Ir)	E	Solid	22428	0.0323	12-100	–	–	–	–	2454	5316	–	–
Iron, (see also Figure A.14 and Table A.12)	E, M	Solid	7866	0.11	20-100	–	–	–	–	1539	2999	65.1	–
gray cast	M	Solid	7097	0.119	20-100	–	–	–	–	1277	–	23.0	–
white cast	M	Solid	7690	0.119	20-100	–	–	–	–	1093	–	33.1	–
wrought	M	Solid	7802-7898	0.115	15-100	–	–	–	–	–	–	–	–

[a] A = alloy, E = element, F = fuel or fuel component, I = insulation, M = metal, O = organic compound, R = refractory, S = solution.

Table A.17 Metric. Densities and thermal properties of various substances (continued)

Name – Description	Other[a] classification	Normal state	Density kg/m³	Solid state Specific heat	Solid state Temp range C	Liquid state Specific heat	Liquid state Temp range C	Gaseous state Specific heat	Gaseous state Temp range C	Melting (fusion) point C	Boiling point, C (at std barometric pressure)	Latent heat of fusion cal/g	Latent heat of vaporization cal/g
Kaolin	R	Solid	2099	0.224 0.22	20-98 16-649	–	–	–	–	1760	–	–	–
Kapok fiber	I	Solid	14	0.320	18	–	–	–	–	–	–	–	–
Kerosene	F	Liquid	769-785	–	–	0.50	0-100	–	–	–	–	–	58.4-61.2
Krypton (Kr)	E	Gas	3.444	–	–	–	–	–	–	-169	-152	–	–
Lanthanum (La)	E	Solid	6152	0.1069	0-100	–	–	–	–	810	1799	–	–
Lava	–	Solid	–	0.197	25-100	–	–	–	–	–	–	–	–
Lead (Pb) (see also Table A.12)	M, E	Solid	11342	0.0319	16-256	0.041	310-360	–	–	327 ± 2	1621	5.5	–
Lead-antimony alloy (62.9 Pb, 37.1 Sb)	A	Solid	–	0.0388	10-98	–	–	–	–	–	–	–	–
Lead-bismuth alloy (40 Pb, 60 Bi)	A	Solid	–	0.0317	0-100	–	–	–	–	–	–	–	–
Lead oxide (PbO)	–	Solid	9195-9500	0.049	16-100	–	–	–	–	954	–	–	–
Lead slag wool	I	Solid	–	0.178 0.235	66 381	–	–	–	–	–	–	–	–
Light oil	F	Liquid	801	–	–	0.50	–	–	–	–	–	–	80.6-83.4
Limestone	–	Solid	2691-2804	0.216	15-100	–	–	–	–	–	–	12	–
Linotype (see also Table A.12)	A	Solid	–	0.036	16-252	–	–	–	–	252	–	–	–
Linseed oil	–	Liquid	929	–	–	0.441	–	–	–	-21	316	9.56	–
Lipowitz's metal (Pb26, Sn 13, Cd 10, Bi 51)	A	Solid	–	0.041	16-60	–	–	–	–	60	–	–	–
Lithium (Li)	E	Solid	529	1.0407	100	–	–	–	–	186	1400	159	–
Lodestone (magnetite)	–	Solid	5158	0.156	0-100	–	–	–	–	–	–	–	–
Machine oil	–	Liquid	–	–	–	0.40	0-100	–	–	–	–	–	–
Magnesia, (85%)	I	Solid	176-208	0.276 0.283	66 137	–	–	–	–	–	–	–	–
Magnesite refractory	R	Solid	2739	0.27	16-649	–	–	–	–	1971+	–	–	–
Magnesite refractory (unburned)	R	Solid	2932	0.26	16-649	–	–	–	–	1971+	–	–	–
Magnesite refractory (fused)	R	Solid	2868	0.27	16-649	–	–	–	–	1971+	–	–	–
Magnesium (Mg) (see also Table A.12)	M, E	Solid	1740	0.2492	20-100	0.266	–	–	–	651	1120	89	–
Magnesium oxide	–	Solid	3653	0.234 0.295	30-38 30-982	–	–	–	–	2799	3638	–	–

[a] A = alloy, E = element, F = fuel or fuel component, I = insulation, M = metal, O = organic compound, R = refractory, S = solution.

Table A.17 Metric. Densities and thermal properties of various substances (continued)

Name – Description	Other[a] classification	Normal state	Density kg/m³	Solid state Specific heat	Solid state Temp range C	Liquid state Specific heat	Liquid state Temp range C	Gaseous state Specific heat	Gaseous state Temp range C	Melting (fusion) point C	Boiling point, C (at std barometric pressure)	Latent heat of fusion cal/g	Latent heat of vaporization cal/g
Magnetite (Lodestone)	–	Solid	5158	0.156	0-100	–	–	–	–	–	–	–	–
Manganese (Mn) (see also Table A.12)	E, M	Solid	7433	0.1211	20-100	–	–	–	–	1230	1900	63.9	–
Marble	–	Solid	2595-2804	0.210	0-100	–	–	–	–	–	–	–	–
Mercuric chloride ($HgCl_2$)	–	Solid	5431	–	–	–	–	–	–	282	305	–	–
Mercurous chloride (Hg_2Cl_2)	–	Solid	7145	0.05	20	–	–	–	–	302	383	–	–
Mercury (Hg)	E, M	Liquid	13569	0.032	-87-+20	0.033	0-100	–	–	-38.87	357	2.82	65.05
Methane (CH_4)	O, F	Gas	0.67972	–	–	0.992	-113	–	–	–	-162	–	–
Methanol (CH_3OH, Methyl Alcohol)	F, O	Liquid	793	–	–	0.6	–	0.5929	18-208	–	64	–	261
Methyl chloride (CH_3Cl)	O	Gas	2.16	–	–	0.385	16	–	–	-98	-24	–	–
Mica	I	Solid	–	0.10	20	–	–	–	–	–	–	–	–
Mica, expanded (vermiculite)	–	Solid	–	0.205, 0.236	65, 143	–	–	–	–	–	–	–	–
Milk	–	Liquid	1028	–	–	0.847	–	–	–	–	–	–	–
Mineral wool board with binder	I	Solid	229	0.095, 0.247	-84, 66	–	–	–	–	–	–	–	–
Molasses	–	Liquid	1399	–	–	0.60	–	–	–	–	–	–	–
Molybdenum (Mo)	E, M	Solid	10205	0.0647	20-100	–	–	–	–	2535	3693	70.1	–
Monel metal (see also Table A.12)	A	Solid	8811	0.129	16-1324	0.139	–	–	–	1324	–	65.27	–
Naphtha	–	Liquid	660	–	–	0.493	–	–	–	–	152	–	102
Naphthalene ($C_6H_4C_4H_4$)	O	Solid	1150	0.325	20-60	0.427	128	–	–	79.9	217.9	35.6	75.45
Neat's-foot oil	–	Liquid	–	0.457	20-30	–	–	–	–	0	–	–	–
Neodymium (Nd)	E	Solid	6905	–	–	–	–	–	–	840	–	–	–
Neon (Ne)	E	Gas	0.8234	–	–	–	–	0.443	any	-253?	-239	–	–
Nichrome	A	Solid	8282	0.109	18-100	0.111	–	–	–	–	–	–	–
Nickel (Ni) (see also Table A.12)	E, M	Solid	8907	0.1608	1000	–	–	–	–	1452	2732	73.9	1485
Nickel steel	A	Solid	–	0.109	0-100	–	–	–	–	–	–	–	–
Nitric acid (HNO_3)	–	Liquid	1540	–	–	0.445	–	–	–	-42	86	–	115.2
Nitric acid (10%) & water	S	Liquid	–	–	–	0.768	–	–	–	–	–	–	–
Nitric acid (2%) & water	S	Liquid	–	–	–	0.930	–	–	–	–	–	–	–
Nitric acid (1%) & water	S	Liquid	–	–	–	0.963	–	–	–	–	–	–	–
Nitric oxide (NO)	–	Gas	1.266	–	–	0.580	-156- -158	0.2317	13-172	–	-151	–	–
Nitrobenzene ($C_6H_5O_2N$)	O	Liquid	1197	–	–	0.38	50	–	–	5	211	–	–

[a] A = alloy, E = element, F = fuel or fuel component, I = insulation, M = metal, O = organic compound, R = refractory, S = solution.

Table A.16 US. Densities and thermal properties of various substances (continued)

Specific heat in cal/g °C

Name – Description	Other[a] classification	Normal state	Density kg/m³	Solid state — Specific heat	Solid state — Temp range C	Liquid state — Specific heat	Liquid state — Temp range C	Gaseous state — Specific heat	Gaseous state — Temp range C	Melting (fusion) point C	Boiling point, C (at std barometric pressure)	Latent heat of fusion cal/g	Latent heat of vaporization cal/g
Nitrobenzole	–	Liquid	–	–	–	0.35	14	–	–	–	–	–	–
Nitrogen (N_2)	–	Gas	1.1871	–	–	0.475	-197- -209	0.2419	20-440	-211	-195	6.17	47.6
Nitrous oxide (N_2O)	–	Gas	1.874	–	–	–	–	0.2262	16-207	–	-89	–	–
Octane (C_8H_{18})	O, F	Liquid	698	–	–	0.52	16	–	–	–	130	–	70.06
Oil (see castor oil, coal tar oil, cottonseed oil, fuel oil, fusel oil, light oil, linseed oil, machine oil, oil of citron, oil of juniper, oil of orange, oil of turpentine, olive oil, paraffin oil, petroleum, and Part 2, Vol. I.)													
of citron	–	Liquid	852	–	–	0.438	6	–	–	–	–	–	–
of juniper	–	Liquid	–	–	–	0.477	–	–	–	–	–	–	–
of orange	–	Liquid	–	–	–	0.489	0	–	–	–	–	–	–
of turpentine	–	Liquid	860	–	–	0.411	–	–	–	-10	160	–	102
Olive oil	–	Liquid	920	–	–	0.471	7	–	–	20±	300±	–	–
Osmium (Os)	E	Solid	22460	0.0311	20-98	–	–	–	–	2700?	5482	–	–
Oxalic acid ($C_2H_2O_4 \cdot 2H_2O$)	O	Solid	1663	0.338 / 0.416	-18 / 38	–	–	–	–	189	150[h]	–	–
Oxygen (O_2)	E	Gas	1.3569	–	–	0.398	-66.7- -61.8	0.2175	13-207	-218	-182.7	3.32	50.9
Palladium (Pd)	E	Solid	11999	0.0714	0-1265	0.0714	–	–	–	1549 ± 5	2200	35.92	–
Paper, expanding blanket ("Kimsul")	I	Solid	–	0.349	64	–	–	–	–	–	–	–	–
Parafin	–	Solid	865-913	0.622	35-40	0.712	60-63	–	–	38-56	350-430	35-39	–
Parafin oil	O, F	Liquid	–	–	–	0.52	0-100	–	–	–	–	–	–
Pentane $n\text{-}C_5H_{12}$	F	Liquid	622	–	–	–	–	–	–	–	36	–	85.85
Petroleum	–	Liquid	753-881	–	–	0.511	21-58	–	–	–	–	–	–
Phenol (C_6H_6O) (see aldo resin)	–	Solid	1070	–	–	0.35-0.45	–	–	–	40.9	182	25.0	–
Phosphorus (P)	E	Solid	1823	0.1829	0-51	–	–	–	–	44.2	288	5.03	–
Pitch (coal tar)	F	Solid	993-1298	0.45	16-100	–	–	–	–	30-150	163	–	–
Plaster	–	Solid	1442	0.20	–	–	–	–	–	–	–	–	–
Platinum (Pt)	E, M	Solid	21387	0.0359	20-1300	–	–	–	–	1755 ± 5	4288	27.22	–
Porcelain	–	Solid	2291-2499	0.26	15-950	–	–	–	–	–	–	–	–
Porcelain, refractory	R	Solid	–	0.23	16-649	–	–	–	–	1171-1649	–	–	–
Potassium (K)	E	Solid	859	0.170	-185-+20	–	–	–	–	62.3	760	14.6	–
Potassium chlorate ($KClO_3$)	–	Solid	2323	0.205	50	–	–	–	–	372	–	–	–

[a] A = alloy, E = element, F = fuel or fuel component, I = insulation, M = metal, O = organic compound, R = refractory, S = solution.
[h] Sublimes.

Table A.17 Metric. Densities and thermal properties of various substances (continued)

Name – Description	Other[a] classification	Normal state	Density kg/m³	Specific heat in cal/g °C						Melting (fusion) point C	Boiling point, C (at std barometric pressure)	Latent heat of fusion cal/g	Latent heat of vaporization cal/g
				Solid state		Liquid state		Gaseous state					
				Specific heat	Temp range C	Specific heat	Temp range C	Specific heat	Temp range C				
Potassium hydroxide (KOH + ½ H₂O)	-	Liquid	-	-	-	0.876	18	-	-	-	-	-	-
Potassium nitrate (KNO₃)	-	Solid	2070	0.19	15-100	-	-	-	-	341	-	49	-
Praseodymium (Pr)	E	Solid	-	-	-	-	-	-	-	940	-	-	-
Propane, commercial	F, O	Gas	1.87	-	-	0.588	16	0.-04	16	-	-46	-	103
Propane, pure C₃H₈	F, O	Gas	1.86	-	-	0.592	16	0.=88	16	-	-42.1	-	102
Pyrex	-	Solid	-	0.196	20-100	-	-	-	-	-	-	-	-
Pyrites, copper	-	Solid	-	0.1291	15-99	-	-	-	-	-	-	-	-
Quartz	-	Solid	2643	0.17-0.28	0-100	-	-	-	-	-	-	-	-
Quicklime	-	Solid	-	0.217	0-100	-	-	-	-	-	-	-	-
Radium (Ra)	E	Solid	4998	-	-	-	-	-	-	950	1138	-	-
Redwood bark, shredded ("Palco Bark")	I	Solid	64	0.172 0.246	-88 43	-	-	-	-	-	-	-	-
Resin, phenol, pure[d]	-	Solid	1202-1298	0.33-0.37	-	-	-	-	-	75-100[f]	-	-	-
Resin, phenol, wood flour filler[d]	-	Solid	1298-1394	0.30-0.36	-	-	-	-	-	125-130[f]	-	-	-
Resin, phenol, asbestos filler[d]	-	Solid	1794-2003	0.38-0.40	-	-	-	-	-	130-150[f]	-	-	-
Resin, copale[e]	-	Solid	1041-1137	0.38-0.40	-	-	-	-	-	149-360	-	-	-
Rhodium (Rh)	E	Solid	12448	0.058	10-97	-	-	-	-	1950	>2527	-	-
Rock wool	-	Solid	112-192	0.201 0.250	65 344	-	-	-	-	-	-	-	-
Rose's metal (25 Pb, 25 Sn, 50 Bi)	A	Solid	-	0.043	16-110	0.041	-	-	-	110	-	10.2	-
Rosin	-	Solid	1089	0.525	20-232	-	-	-	-	77-100	-	-	-
Rubber	-	Solid	993-2003	0.481	16-100	-	-	-	-	120	-	-	-
Rubber board, expanded ("Rubatex")	-	Solid	78.5	0.152 0.273	-87 44	-	-	-	-	-	-	-	-
Rubidium (Rb)	E	Solid	1530	0.0802	0	-	-	-	-	38	696	-	-
Ruthenium (Ru), black	E	Solid	8603	0.0611	0-100	-	-	-	-	1943	-	-	-
Ruthenium (Ru), gray	-	Solid	12175	-	-	-	-	-	-	2450	-	-	-
Salt, rock	-	Solid	2163	0.219	13-45	-	-	-	-	813	-	-	-
Samarium (Sm)	E	Solid	7706	-	-	-	-	-	-	1300-1400	-	-	-
Sand	-	Solid	2595	0.195	15-100	-	-	-	-	-	-	-	-
Sea water	S	Liquid	1025	-	-	0.938	17.5	-	-	-	-	-	-

[a] A = alloy, E = element, F = fuel or fuel component, I = insulation, M = metal, O = organic compound, R = refractory, S = solution.
[d] Manufactured under trade names: Bakelite, Redmanol, Condensite, etc.
[e] Resins used in varnish making: Kauric, Congo, Zanzibar, and Manila copals.
[f] Softening point under load.

Table A.17 Metric. Densities and thermal properties of various substances *(continued)*

Name – Description	Other[a] classification	Normal state	Density kg/m³	Specific heat in cal/g °C						Melting (fusion) point °C	Boiling point, C (at std barometric pressure)	Latent heat of fusion cal/g	Latent heat of vaporization cal/g
				Solid state		Liquid state		Gaseous state					
				Specific heat	Temp range C	Specific heat	Temp range C	Specific heat	Temp range C				
Selenium (Se)	E	Solid	4806	0.068	-188-+18	–	–	–	–	217-220	690	–	–
Serpentine	–	Solid	–	0.25	0-100	–	–	–	–	–	–	–	–
Shellac (Lac)	–	Solid	1202-1218	0.40	16-100	–	–	–	–	77-82	–	–	–
Silica (SiO₂)	–	Solid	2884	0.1910	0-100	–	–	–	–	1750	–	–	–
Silica aerogel ("Santocel")	I	Solid	85	0.205 / 0.274	64 / 332	–	–	–	–	–	–	–	–
Silica refractory	R	Solid	1778	0.23	16-649	–	–	–	–	1682-1699	–	–	–
Silicon (Si)	E	Solid	2323	0.1833 / 0.2029	57 / 232	–	–	–	–	1420	2287	337	–
Silicon carbide (SiC)	–	Solid	3188	0.23	16-510	–	–	–	–	2250	–	–	–
Silicon carbide (clay-bonded) refractory	R	Solid	2179-2547	0.20	16-649	–	–	–	–	1866	–	–	–
Silk, raw	–	Solid	1298-1394	0.33	16-100	–	–	–	–	–	–	–	–
Sillimanite (mullite) refr.	R	Solid	2323-3236	0.23	16-649	–	–	–	–	1821-1838	–	–	–
Silver (Ag) (see also Table A.12)	E, M	Solid	10493	0.05987	17-507	–	–	–	–	960.5	1955	25.1	–
Slag, blast furnace (powdered)	–	Solid	360	–	–	–	–	–	–	–	–	50.0	–
Slag wool	I	Solid	151-300	0.17	25	–	–	–	–	–	–	–	–
Soda, baking	–	Solid	2195	0.231	0-100	–	–	–	–	–	–	–	–
Sodium (Na)	E	Solid	971	0.253	-185-+20	–	–	–	–	97.5	879	27.4	–
Sodium carbonate (Na₂CO₃)	–	Solid	2427	0.306	–	–	–	–	–	852	–	–	–
Sodium carbonate (2%) and water	S	Liquid	–	–	–	0.896	–	–	–	–	–	–	–
Sodium chloride (NaCl) (see also rock salt)	–	Solid	2163	–	–	–	–	–	–	800	–	–	–
Sodium chloride (10%) and water (see also sea water)	S	Liquid	1073	–	–	0.791	18	–	–	–	–	–	–
Sodium hydroxide (2% NaOH) and water	S	Liquid	1022	–	–	0.942	18	–	–	–	–	–	–
Sodium nitrate (NaNO₃)	–	Solid	2251	0.231	–	–	–	–	–	314	380	64.94	–
Sodium sulfate (Na₂SO₄)	–	Solid	–	–	–	–	–	–	–	884	–	–	–
Solder (Pb and Sn)	A	Solid	9292	0.040-0.051	–	–	–	–	–	183-312	–	6.45-17.0	–
Spermaceti (whale oil)	–	Solid	–	–	–	–	–	–	–	45±	–	37.01	–

[a] A = alloy, E = element, F = fuel or fuel component, I = insulation, M = metal, O = organic compound, R = refractory, S = solution.

Table A.17 Metric. Densities and thermal properties of various substances (continued)

Name – Description	Other[a] classification	Normal state	Density kg/m³	Specific heat in cal/°C						Melting (fusion) point C	Boiling point, C (at std barometric pressure)	Latent heat of fusion cal/g	Latent heat of vaporization cal/g
				Solid state		Liquid state		Gaseous state					
				Specific heat	Temp range C	Specific heat	Temp range C	Specific heat	Temp range C				
Steel (see also Figure A.14, iron, nickel steel)	A	Solid	7850	0.165[k]	16-1593	–	–	–	–	–	–	–	–
Sterotype (see Figure A.15)	A	Solid	10733	0.036	–	0.036	–	–	–	260	–	14.6	–
Stones, all kinds (see also marble, granite, limestone, sandstone)	–	Solid	2691	0.18-0.23	12-100	–	–	–	–	–	–	–	–
Sugar, cane; amorphous	–	Solid	–	0.342	20	–	–	–	–	–	–	–	–
Sugar, cane; crystaline	–	Solid	1634	0.301	20	–	–	–	–	160	–	–	–
Sugar, cane; (4%) and water	S	Liquid	–	0.7558	–	–	–	–	–	–	–	–	–
Sulfur (S)	E	Solid	1906-2083	0.190	15-54	0.2337	113-449	–	–	115	444.7	9.38[g]	362.3
Sulfur dioxide (SO₂)	E	Gas	2.78	–	–	0.36	50	0.144	16-202	-76	-10	–	–
Sulfuric acid (H₂SO₄)	–	Liquid	1855	–	–	0.336	0-100	–	–	10.5	338[b]	–	–
Talc	–	Solid	–	–	–	0.2092	20-98	–	–	–	–	–	–
Tallow, beef	–	Solid	910-969	–	–	0.79	26-42	–	–	27-38	–	–	–
						0.54	66-102						
Tantalum (Ta)	E	Solid	16581	0.043	1400	–	–	–	–	2900	–	–	–
Tartaric acid (C₄H₆O₆)	–	Solid	1666	0.287	36	–	–	–	–	170	–	–	–
Tellurium (Te)	E	Solid	6232	0.0483	15-100	–	–	–	–	452	1390	7.28	–
Thallium (Tl)	E	Solid	11855	0.0326	20-100	–	–	–	–	302	1649	–	–
Thorium (Th)	E	Solid	11198	0.0276	0-100	–	–	–	–	1843	>3000	–	–
Tile, hollow	–	Solid	1202	0.15	–	–	–	–	–	–	–	–	–
Tin (Sn) (see also Table A.12)	E, M	Solid	7289	0.0551	21-109	0.05799	250	–	–	232 ± 0.2	2270	–	–
						0.0758	1100						
Titanium (Ti)	E	Solid	4534	0.1125	0-100	–	–	–	–	1795	–	–	–
Toluene (C₆H₅)	O	Liquid	859	–	–	0.40	0-100	–	–	-92	110.3	–	83.57
Toluol (C₆H₆)	O	Liquid	–	–	–	0.490	65	–	–	–	110.3	–	86.07
Tufa	–	Solid	–	0.33	0-100	–	–	–	–	–	–	–	–
Tungsten (W)	E, M	Solid	19256	0.0336	0-100	–	–	–	–	3400	5830	44	–
				0.0337	1000								
Turpentine	–	Liquid	859	0.42	0-100	–	–	–	–	–	159.3	–	74.11
Type metal	A	Solid	–	0.0388	0-100	–	–	–	–	–	–	–	–
Uranium (U)	E	Solid	18695	0.028	0-98	–	–	–	–	<1840	3499	–	–
Vanadium (V)	E	Solid	6008	0.1153	0-100	–	–	–	–	1720	2999	–	–

[a] A = alloy, E = element, F = fuel or fuel component, I = insulation, M = metal, O = organic compound, R = refractory, S = solution.
[b] Decomposes.
[g] Transformation from rhombic to monoclinic absorbs 2.81 cal/g.
[k] Specific heat; steel is about 0.11 in the 16-316 C range.

Table A.17 Metric. Densities and thermal properties of various substances *(concluded)*

Name – Description	Other[a] classification	Normal state	Density kg/m³	Specific heat in cal/g °C — Solid state Specific heat	Solid state Temp range C	Liquid state Specific heat	Liquid state Temp range C	Gaseous state Specific heat	Gaseous state Temp range C	Melting (fusion) point C	Boiling point, C (at std barometric pressure)	Latent heat of fusion cal/g	Latent heat of vaporization cal/g
Varnish (see resins)	I	–	–	–	–	–	–	–	–	–	–	–	–
Vegetable fiberboard ("Celotex")	I	Solid	231	0.171, 0.279	-82, 43	–	–	–	–	–	–	–	–
Vermiculite (see mica)	I	–	–	–	–	–	–	–	–	–	–	–	–
Vulcanite	–	Solid	–	0.3312	20-100	–	–	–	–	–	–	–	–
Water (H₂O) (see also sea water) (see Tables A.8 thru A.11)	–	Liquid	999	0.480	<0	1.00	16	–	–	0	100	80.1	539.4
Wood (see redwood bark)	–	Solid	304-897	0.33-0.67	–	–	–	–	–	–	–	–	–
Wood fiber blanket ("Balsam Wool")	I	Solid	42	0.330	66	–	–	–	–	–	–	–	–
Wood fiberboard	I	Solid	192-304	0.341	64	–	–	–	–	–	–	–	–
Wood, oak	–	Solid	769	0.57	0-100	–	–	–	–	–	–	–	–
Wood, pine	–	Solid	481	0.67	0-100	–	–	–	–	–	–	–	–
Wood's metal (26Pb, 13Sn, 12Cd, 49Bi)	A	Solid	–	0.041	16-70	0.042	–	–	–	70	–	9.56	–
Wool (see also glass wool, mineral wool, rock wool, lead slag wool, slag wool, etc.)	–	Solid	1282-1330	0.325	–	–	–	–	–	–	–	–	–
Xenon (Xe)	E	Gas	5.542	–	–	–	–	–	–	-140	-109.1	–	–
Xylene	–	Liquid	870	0.42	50	–	–	–	–	-28	142	–	81.7
Yttrium (Y)	E	Solid	5495	–	–	–	–	–	–	1490	–	–	–
Zinc (Zn) (see Table A.12)	E, M	Solid	7129	0.0931, 0.1040	20-100, 300	–	–	–	–	419.4	906	26.0	421
Zinc chloride (ZnCl₂)	–	Solid	2908	–	–	–	–	–	–	365	732	–	–
Zinc oxide (ZnO)	–	Solid	5607	0.125	0-100	–	–	–	–	>1782	–	–	–
Zinc sulfate (ZnSO₄)	S	Liquid	3749	0.132	–	0.174	41	–	–	721[b]	–	–	–
Zircon	–	Solid	4694	–	–	–	–	–	–	2550	–	–	–
Zirconium (Zr)	E	Solid	6488	0.0660	0-100	–	–	–	–	1700	5050	–	–

[a] A = alloy, E = element, F = fuel or fuel component, I = insulation, M = metal, O = organic compound, R = refractory, S = solution.
[b] Decomposes.

Figure A.18. Absolute viscosities of gases and vapors. See Tables 2.6 (Vol. I) and C.1 for conversion of units. Based on data abstracted from the following references:

Jorgensen, Robert (ed.): Fan Engineering, 7th edition, pp. 88-89. Buffalo Forge Co., Buffalo, NY, 1970.
McPhee, C. W. (ed.): "ASHRAE Handbook of Fundamentals", pg. 85, American Society of Heating, Refrigerating & Air Conditioning Engineers, Inc., New York, NY, 1967.
Shnidman, L., Farmer, & Gustafson: Properties of Elements, Common Substances, and Materials, in **Segeler, C. G.** (ed.), "Gas Engineers' Handbook", pg. 1/30, The Industrial Press, New York, NY, 1965.

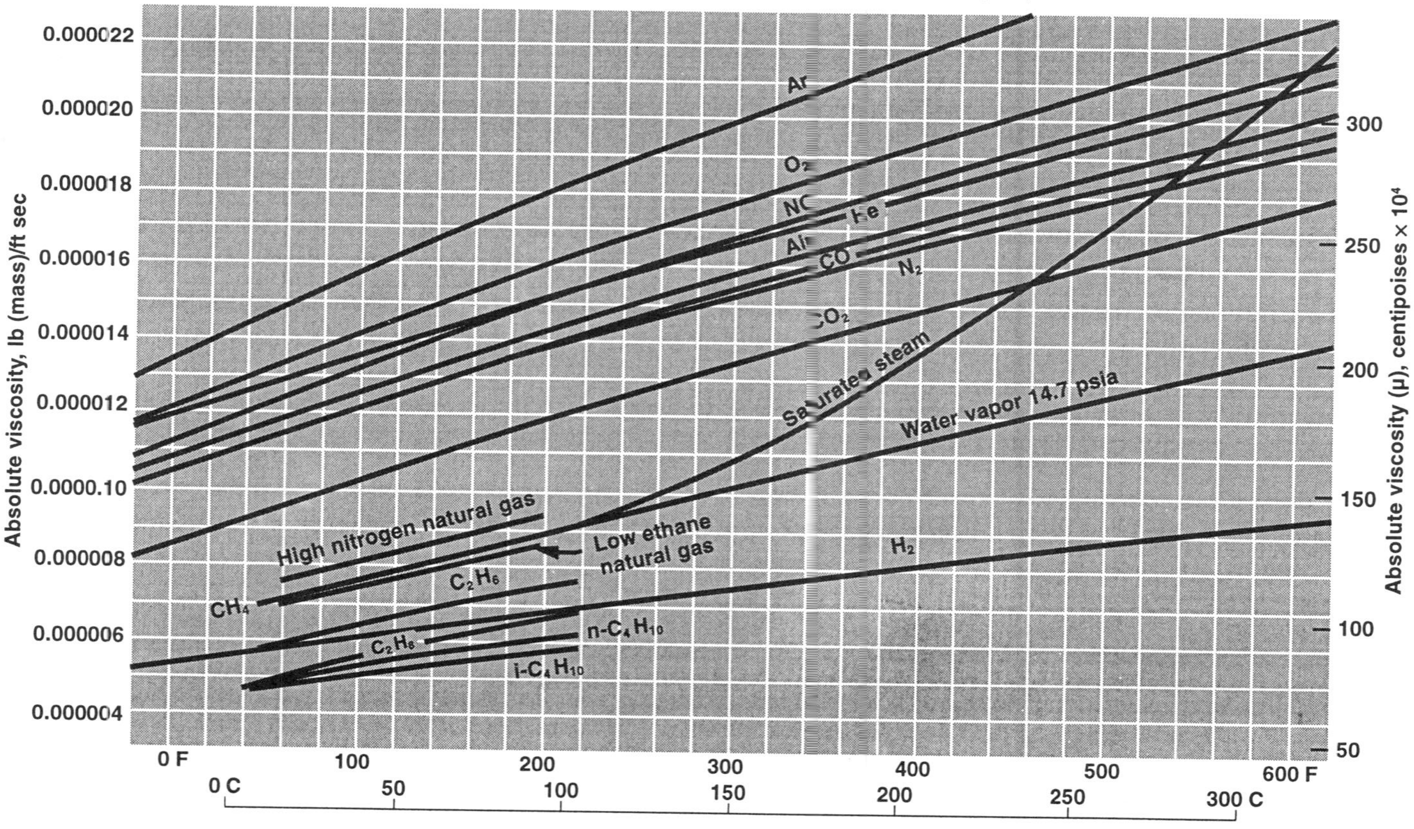

Table A.19 US. Properties of some heat transfer fluids

	Max temp, F	Pour point, F	Melt point, F	Specific Gravity	Sp. Ht. Btu/lb °F at 70 F	Sp. Ht. at ( ) F	Thermal conductivity, Btu ft/ft² hr °F at 70 F	Thermal conductivity at ( ) F	Viscosity, centipoise at 70 F	Viscosity at ( ) F
Ethylene glycol	325-400	–	–60	1.12	0.625	0.710 (300)	0.167	–	20.0	1.0 (300)
Polyalkylene glycols	400-565	0 to –45	–	0.99-1.08	0.44	0.64 (500)	0.099 to 0.121	0.095 to 0.099 (300)	90 to 320	1.5 to 1.85 (500)
Water	450	–	32	0.8†	1.00	0.66 (600)	0.349	0.78 (600)	0.98	0.81 (600)
Aromatic base hydrocarbons	425-850	–18 to –80	*	0.89-1.13	0.37 to 0.46	0.63 to 0.72 (600)	0.071 to 0.083	0.063 (600)	6.2 to 387	0.29 to 0.42 (600)
Triaryldimethane	500	–31	–	1.03	0.36	0.60 (572)	0.17	0.14 (600)	32	0.37 (600)
Mineral oils	525-680	15 to 25	–	0.85-0.94	0.38 to 0.48	0.62 to 0.72 (600)	0.070 to 0.078	0.059 to 0.065 (600)	190 to 10 000	0.25 to 1.0 (600)
Mercury	600	–	–38	12.86†	0.033	0.0325 (850)	4.8	8.1 (600)	–	1.0 (400)
Diphenyl/diphenyl oxides	700-805	–	12 to 22	1.06	0.379	0.591 (700)	0.081	0.057 (700)	4.5	0.30 (700)
Salts (HTS)	800	–	300	1.83‡	–	0.37 (?)	–	0.42 (600)	–	2.0 (600)
Sodium	1500	–	208	0.89‡	0.253	0.32 (1000)	82.08	37 (1000)	–	0.3 (600)
NaK (78 wt %K)	1500	–	10	0.81‡	–	0.20 (1000)	–	15 (1000)	–	0.2 (600)

* Except Therminol 88 melts at 293 F.
† At 600 F compared to 62.35 lb/ft³ water at 60 F.
‡ At 500 F compared to 62.35 lb/ft³ water at 60 F.

REFERENCES:
Boyen, John L.: "Practical Heat Recovery", pp. 102-105, John Wiley & Sons, Inc., New York, NY, 1975.
Geiringer, Paul L.: "Handbook of Heat Transfer Media", Reinhold Publishing Corp., New York, NY, 1962.

Table A.20. More properties of heat transfer fluids

Fluid Composition[a]	Max. use temperature[c]	Min. pumping temperature	Film coefficient Btu/(h·ft²·°F) (w/h·cm²·°K)[b] @ 5 ft/s 1.5 m/s	@ 7 ft/s (2.1 m/s)	Pressure drop psi/100 ft (kPa/m)[b] @ 3 ft/s (0.91 m/s)	@ 7 ft/s (2.1 m/s)
Synthetic Organic Fluids						
Mixture of diphenyl and diphenyl oxide (A)	750 F (399 C)	53.6 F (12 C)	464 (263)	607 (3440)	1.3 (0.29)	6.9 (1.6)
Mixture of diphenyl and diphenyl oxide (B)	750 F (399 C)	53.6 F (12 C)	449 (255)	588 (3330)	1.3 (0.29)	7.0 (1.6)
Mixture of di-and triaryl ethers	700 F (371 C)	-4 F (-20 C)	387 (-19)	506 (2870)	1.5 (0.34)	7.6 (1.7)
Hydrogenated terphenyl (A)	650 F (343 C)	31 F (-0.6 C)	353 (300)	462 (2620)	1.4 (0.32)	7.2 (1.6)
Hydrogenated terphenyl (B)	650 F (343 C)	30 F (-1.1 C)	319 (181)	417 (2360)	1.4 (0.32)	7.3 (1.6)
Aromatic blend	650 F (343 C)	-50 F (-46 C)	384 (212)	503 (2850)	1.3 (0.29)	6.9 (1.6)
Alkylated aromatic	600 F (316 C)	<-100 F (<-73 C)	522 (296)	684 (3880)	0.9 (0.20)	4.9 (1.1)
Polyaromatic	600 F (316 C)	-50 F (-46 C)	355 (201)	465 (2640)	1.4 (0.32)	7.0 (1.6)
Synthetic hydrocarbon	600 F (316 C)	4 F (-16 C)	302 (171)	395 (2340)	1.2 (0.27)	6.2 (1.4)
Isomeric dibenzyl benzenes	662 F (350 C)	-18 F (-28 C)	366 (208)	479 (2720)	1.4 (0.32)	7.4 (1.7)
Isomeric dimethyl diphenyl oxide	626 F (330 C)	-46 F (-43 C)	359 (204)	469 (2660)	1.4 (0.32)	7.1 (1.6)
Alkyl diphenyl	707 F (375 C)	-22 F (-30 C)	350 (199)	458 (2600)	1.4 (0.32)	6.9 (1.6)
Alkyl benzene	590 F (310 C)	-49 F (-45 C)	330 (187)	433 (2460)	1.3 (0.29)	6.4 (1.4)
Siloxanes						
Dimethyl siloxane	750 F (399 C)	<-50 F (-46 C)	239 (130)	313 (1770)	1.2 (0.27)	6.2 (1.4)
Polydimethyl siloxane[d]	500 F (260 C)	<-148 F (-82 C)	214 (12 0)	280 (1590)	0.97 (0.22)	5.02 (1.16)
Inhibited Glycols						
50% Inhibited EG/H$_2$O[d]	350 F (177 C)	-60 F (-51 C)	1358 (7790)	1778 (10080)	1.60 (0.36)	8.47 (1.92)
50% Inhibited PG/H$_2$O[d]	325 F (163 C)	-50 F (-46 C)	1030 (580)	1348 (7640)	1.65 (0.37)	8.42 (1.90)
Paraffinic and Mineral Oils						
Paraffinic oil (A)	600 F (316 C)	<15 F[e] (<-9.4 C[e])	268 (152)	351 (1940)	1.3 (0.29)	6.5 (1.5)
Paraffinic oil (B)	600 F (316 C)	<40 F[e] (<4.4 C[e])	280 (159)	367 (2080)	1.3 (0.29)	6.4 (1.4)
Mineral oil (A)	600 F (316 C)	<35 F (<1.7 C)	319 (181)	417 (2360)	1.4 (0.32)	7.0 (1.6)
Mineral oil (B)	600 F (316 C)	<25 F[e] (<-3.9 C[e])	310 (176)	406 (2300)	1.3 (0.29)	6.3 (1.4)
Mineral oil (C)	600 F (316 C)	20 F (-6.7 C)	298 (169)	390 (2200)	1.3 (0.29)	6.3 (1.4)

[a] (A), (B), and (C) denote different product formulations.   [b] at 600 F (316 C).   [c] Maximum use temperature recommended by fluid manufacturer.
[d] Properties are evaluated at the maximum use temperature.   [e] Reported pour-point temperature.
Courtesy of John Cuthbert, Dow Chemical Co. Reference: Chemical Engineering Progress, July 1994, pp 29-37.

Table A.21. Effect of humidity on oxygen content of air

% (volume) oxygen in air		% Relative humidity					
		0	20	40	60	80	100
54.4 C	130 F	20.96	20.34	19.72	19.09	18.47	17.85*
48.9 C	120 F	20.96	20.48	20.01	19.53	19.05	18.57*
43.3 C	110 F	20.96	20.61	20.25	19.89	19.53	19.18
37.8 C	100 F	20.97	20.70	20.43	20.17	19.90	19.64
32.2 C	90 F	20.97	20.77	20.58	20.38	20.19	19.99
26.7 C	80 F	20.97	20.83	20.68	20.54	20.39	20.25
21.1 C	70 F	20.97	20.87	20.77	20.67	20.57	20.47
15.6 C	60 F	20.98	20.91	20.84	20.77	20.70	20.63
4.4 C	40 F	20.98	20.95	20.92	20.88	20.85	20.82
–6.7 C	20 F	20.98	20.97	20.96	20.94	20.93	20.92
–17.8 C	0 F	20.99	20.98	20.98	20.97	20.97	20.96

(Air temperature)

Interpolated from data by GTE.
* Extrapolated.

Table A.22.   Calorific (heating) values. (Adapted from "Incineration of Hazardous, Toxic, and Mixed Wastes" by Gill and Quiel, published by North American Mfg. Co., Cleveland, OH 44105-5600.) For organic compounds, see pages 206-231 of that reference.

Substance[a]	Formula	(b)	(c)	kcal/kg[c]	Btu/lb[c]
Acetaldehyde	CH_3CHO	daf	n	6340	11410
Acetic Acid	CH_3CO_2H	daf	n	3490	6280
Acetone	$(CH_3)_2CO$	daf	n	7360	13250
Acetylene	CHCH	daf	n	12000	21600
Aluminum	(to Al_2O_3)	daf	g	7420	13350
Aniline	$C_6H_5NH_2$	daf	n	8730	15710
Asphalt		ar	n	9530	17190
Bagasse, 12m		ar	n	4050	7300
Bamboo, 10m		ar	n	4110	7410
Benzaldehyde	C_6H_5CHO	daf	n	7940	14290
Benzene	C_6H_6	daf	n	10030	18050
Books, 24a		daf	g	4200	7560
Brown Paper, 6m, 1a		ar	g	4030	7260
Brown Paper, 6m, 1a		d	g	4280	7710
Brown Paper, 6m, 1a		daf	g	4330	7800
Brown Peanut Skins		d	g	5800	10430
Brush, 40m, 5a		ar	g	2640	4740
Brush, 40m, 5a		d	g	4390	7900
Brush, 40m, 5a		daf	g	4780	8600
iso-Butyl Alcohol	$(CH_3)_2CH_2CH_2OH$	daf	n	8510	15320
n-Butyl Alcohol	C_4H_9OH	daf	n	8630	15530
Butyl sole composition, 1m, 30a		ar	g	6060	10900
Carbon		ar	g	7830	14090
Carbon Disulfide	CS_2	daf	n	3240	5840
Carbon Tetrachloride	CCl_4	daf	n	240	440
Cardboard		daf	g	4650	8370
Castor Oil		d	g	8860	15950
Castor Oil		daf	g	8860	15950
Charcoal, 4m		ar	n	7260	13090
Chloroform	$CHCl_3$	daf	n	750	1340
Citrus Rinds & Seeds, 79m, 1a		ar	g	950	1710
Citrus Rinds & Seeds, 79m, 1a		d	g	4450	8020
Citrus Rinds & Seeds, 79m, 1a		daf	g	4610	8300
Coal, Sub-bituminous B, 15m, 7a		ar	g	5690	10240
Coal, Sub-bituminous B, 15m, 7a		d	g	6720	12100
Coal, Sub-bituminous B, 15m, 7a		daf	g	7200	12960
Coal, Bituminous-high volatile B, 9m, 8a		ar	g	6800	12240
Coal, Bituminous-high volatile B, 9m, 8a		d	g	7440	13390
Coal, Bituminous-high volatile B, 9m, 8a		daf	g	8190	14740
Coal, Bituminous-volatile, 4m, 5a		ar	g	8030	14450
Coal, Bituminous-volatile, 4m, 5a		d	g	8330	14990
Coal, Bituminous-volatile, 4m, 5a		daf	g	8780	15800

(continued)

[a] a = % ash by weight
    m = % moisture by weight

[b] ar = as received
    d = dry
    daf = dry & ash-free

[c] g = gross or higher HV
    n = net or lower HV

Table A.22. Calorific (heating) values. *(continued)*

Substance[a]	Formula	[b]	[c]	kcal/kg[c]	Btu/lb[c]
Coal, Anthracite, 5m, 14a		ar	g	6170	11110
Coal, Anthracite, 5m, 14a		d	g	6460	11630
Coal, Anthracite, 5m, 14a		daf	g	7600	13680
Coffee Grounds, 20m, 2a		d	g	5590	10060
Copper	(to CuO)	daf	g	600	1090
Corn Cobs, 5m, 3a		ar	g	4440	8000
Corrugated Boxes, 5.2m		ar	g	3910	7040
Corrugated Boxes, 5.2m		d	g	4130	7430
Corrugated Boxes, 5.2m		daf	g	4360	7850
Cotton Seed Hulls, 10m, 2a		ar	g	4440	8000
Cresol (av. o, m, p)	$CH_3C_6H_4OH$	daf	n	8170	14700
Ethane	C_2H_6	daf	n	12280	22100
Ethyl Acetate	$CH_3CO_2C_2H_5$	daf	n	6100	10980
Ethyl Alcohol	C_2H_5OH	daf	n	7120	12820
Ethylene (ethene)	CH_2CH_2	daf	n	11840	21320
Evergreen Shrubs, 69m, 1a		ar	g	1500	2710
Evergreen Shrubs, 69m, 1a		d	g	4850	8740
Evergreen Shrubs, 69m, 1a		daf	g	4980	8960
Excelsior, 0.77a		daf	g	4790	8620
Fats (Animal)		ar	n	9500	17130
Fats, Cooking		daf	g	9000	16200
Fats, Fried		ar	g	9150	16470
Fats, Fried		d	g	9150	16470
Fats, Fried		daf	g	9150	16470
Flowering Plants, 54m, 2a		ar	g	2050	3700
Flowering Plants, 54m, 2a		d	g	4460	8030
Flowering Plants, 54m, 2a		daf	g	4700	8460
Fuel Oil, #2		ar	g	10870	19570
Fuel Oil, #2		d	g	10870	19570
Fuel Oil, #2		daf	g	10870	19570
Fuel Oil, #6, 2m, 0.1a		ar	g	10150	18260
Fuel Oil, #6, 2m, 0.1a		d	g	10300	18540
Fuel Oil, #6, 2m, 0.1a		daf	g	10380	18680
Grass, Dirt, Leaves, 21-62m		d	g	3490	6280
Grass, Dirt, Leaves, 21-62m		daf	g	4990	8990
Glycerol (glycerin)	$(CH_2OH)_2$	daf	n	6400	11530
n-Heptane	C_7H_{16}	daf	n	11500	20700
n-Hexane	C_6H_{14}	daf	n	11510	20720
Household Dirt, 3m, 70a		ar	g	2040	3670
Household Dirt, 3m, 70a		d	g	2110	3790
Household Dirt, 3m, 70a		daf	g	7580	13650
Hydrogen		ar	g,n	33890	61000

(continued)

---

[a] a = % ash by weight
m= % moisture by weight

[b] ar = as received
d   = dry
daf = dry & ash-free

[c] g = gross or higher HV
n = net or lower HV

Table A.22.   Calorific (heating) values. *(continued)*

Substance[a]	Formula	(b)	(c)	kcal/kg[c]	Btu/lb[c]
Iron	(to Fe_2O_3)	daf	g	1760	3160
Iron	(to $FeO_{.947}$)	daf	g	1200	2160
Iron	(to Fe_3O_4)	daf	g	1590	2870
Kerosene, 0.5a		ar	g	10500	18900
Latex		ar	g	5560	10000
Lead	(to PbO)	daf	g	250	450
Leather, 10m, 9a		ar	g	4420	7960
Leather, 10m, 9a		d	g	4920	8850
Leather, 10m, 9a		daf	g	5470	9850
Lignin		daf	g	5850	10530
Lignite		ar	g	3930	7070
Lignite		d	g	6130	11030
Lignite		daf	g	7560	13610
Linoleum, 2m, 27a		ar	g	4520	8150
Linoleum, 2m, 27a		d	g	4620	8310
Linoleum, 2m, 27a		daf	g	6060	11450
Logs, Green, 50m, 0.5a		ar	g	1170	2100
Logs, Green, 50m, 0.5a		d	g	2340	4200
Logs, Green, 50m, 0.5a		daf	g	2360	4250
Magazines, 4m, 22a		ar	g	2920	5250
Magazines, 4m, 22a		d	g	3040	5480
Magazines, 4m, 22a		daf	g	3970	7150
Magnesium	(to MgO)	ar	n	4730	8530
Magnesium	(to MgO)	daf	g	5910	10640
Meat Scraps, 39m, 3a		ar	g	4240	7620
Meat Scraps, 39m, 3a		d	g	6910	12440
Meat Scraps, 39m, 3a		daf	g	7280	13110
Methane	CH_4	daf	n	13180	23720
Methyl Alcohol	CH_3OH	daf	n	5340	9610
Methyl Ethyl Ketone	$CH_3COC_2H_5$	daf	n	8090	14560
Methylene Chloride	CH_2Cl_2	daf	n	1260	2260
Naphtha		ar	g	8330	15000
Newsprint, 6m, 1.4a		ar	g	4430	7970
Newsprint, 6m, 1.4a		d	g	4710	8480
Newsprint, 6m, 1.4a		daf	g	4780	8600
Oats		d		4440	8000
n-Octane	C_8H_{18}	daf	n	11430	20570
Oil, cotton seed		ar	n	9500	17130
Oils, Paints, 16a		ar	g	7440	13400
Oils, Paints, 16a		d	g	7440	13400
Oils, Paints, 16a		daf	g	8890	16000

(continued)

[a] a = % ash by weight
   m = % moisture by weight

[b] ar = as received
   d = dry
   daf = dry & ash-free

[c] g = gross or higher HV
   n = net or lower HV

Table A.22. Calorific (heating) values. *(continued)*

Substance[a]	Formula	(b)	(c)	kcal/kg[c]	Btu/lb[c]
Paper Food Cartons		ar	g	4030	7260
Paper Food Cartons		d	g	4290	7730
Paper Food Cartons		daf	g	4580	8250
Paraffin		ar	n	10340	18650
Peat		ar	g	1800	3240
Peat		d	g	5030	9050
Peat		daf	g	7000	12590
Phenol	C_6H_5OH	daf	n	7790	14020
Phenol formaldehyde		daf	g	6220	11190
Phthalic Acid	$C_8H_4(CO_2H)_2$	daf	n	4640	8360
Pitch		ar	n	8400	15150
Plastic Coated Paper, 4.71m		ar	g	4080	7340
Plastic Coated Paper, 4.71m		d	g	4280	7700
Plastic Coated Paper, 4.71m		daf	g	4410	7940
Plastic Film, 3-20m		d	g	7690	13850
Plastic Film, 3-20m		daf	g	8260	14870
Plastics, Mixed, 2m, 10a		ar	g	7830	14100
Plastics, Mixed, 2m, 10a		d	g	7980	14370
Plastics, Mixed, 2m, 10a		daf	g	8890	16000
Polyethylene, 0.2m, 1a		ar	g	10930	19680
Polyethylene, 0.2m, 1a		d	g	10960	19730
Polyethylene, 0.2m, 1a		daf	g	11110	20000
Polypropylene		daf	g	11080	19950
Polystyrene, 0.2m, 0.5a		ar	g	9120	16420
Polystyrene, 0.2m, 0.5a		d	g	9140	16450
Polystyrene, 0.2m, 0.5a		daf	g	9170	16510
Polyurethane, 0.2m, 4a		ar	g	6220	11200
Polyurethane, 0.2m, 4a		d	g	6240	11220
Polyurethane, 0.2m, 4a		daf	g	6520	11730
Polyurethane (foamed)		ar	g	7220	13000
Polyurethane (foam)		daf	g	5700	10260
Polyvinyl Chloride, 0.2m, 2a		ar	g	5420	9750
Polyvinyl Chloride, 0.2m, 2a		d	g	5430	9780
Polyvinyl Chloride, 0.2m, 2a		daf	g	5560	10000
Propane	C_3H_8	daf	n	11960	21530
iso-Propyl Alcohol	$(CH_3)_2CHOH$	daf	n	7910	14240
n-Propyl Alcohol	C_3H_7OH	daf	n	8010	14420
Propylene	CH_3CHCH_2	daf	n	11670	21010
Rags, 10m, 2a		ar	g	3830	6900
Rags, 10m, 2a		d	g	4250	7650
Rags, 10m, 2a		daf	g	4360	7840

(continued)

(a) a = % ash by weight
   m= % moisture by weight

(b) ar = as received
   d = dry
   daf = dry & ash-free

(c) g = gross or higher HV
   n = net or lower HV

Table A.22.   Calorific (heating) values. *(continued)*

Substance[a]	Formula	(b)	(c)	kcal/kg[c]	Btu/lb[c]
Rags, cellulose acetate, 2.19a		daf	g	4440	8000
Rags, cotton		daf	g	4000	7200
Rags, linen		d	g	3960	7130
Rags, mixed, 2.19a		daf	g	4191	7540
Rags, nylon		daf	g	7330	13190
Rags, rayon		daf	g	420	750
Rags, silk		d	g	4660	8390
Rags, wool		daf	g	5440	9800
Rubber, 1m, 10a		ar	g	6220	11200
Rubber, 1m, 10a		d	g	6290	11330
Rubber, 1m, 10a		daf	g	7000	12600
Rubber Waste		ar	g	5560	10000
Sawdust (pine)		d	g	5380	9680
Sawdust (fir)		d	g	4580	8250
Shellac		daf	g	7540	13580
Shoe, Heel & Sole, 1m, 30a		ar	g	6060	10900
Shoe, Heel & Sole, 1m, 00a		d	g	6130	11030
Shoe, Heel & Sole, 1m, 30a		daf	g	8770	15790
Starch		d	g	4180	7520
Starch		daf	g	4180	7520
Street Sweepings, 20m, 20a		ar	g	2670	4800
Street Sweepings, 20m, 20a		d	g	3330	6000
Street Sweepings, 20m, 20a		daf	g	4440	8000
Styrene-butadiene copolymer		daf	g	9830	17700
Sugar (sucrose)		d	g	3940	7100
Sugar (sucrose)		daf	g	3940	7100
Sulfur (rhombic)		ar	n	2200	3970
Tar or asphalt, 1m		ar	g	9440	17000
Tar 1/3, Paper 2/3, 1m, 2a		ar	g	6110	11000
Textiles, 15-31m		d	g	4460	8040
Textiles, 15-31m		daf	g	4610	8300
Tin	(to SnO_2)	daf	g	1170	2100
Tires, 1m, 7a		ar	g	7670	13800
Tires, 1m, 7a		d	g	7730	13910
Tires, 1m, 7a		daf	g	8280	14900
Toluene	$CH_3C_6H_5$	daf	n	10150	18280
Turpentine		ar	g	9440	17000
Upholstery, 7m, 3a		ar	g	3870	6960
Upholstery, 7m, 3a		d	g	4160	7480
Upholstery, 7m, 3a		daf	g	4270	7690
Urea	$(NH_2)_2CO$	daf	n	2530	4550

(continued)

(a) a = % ash by weight
    m= % moisture by weight

(b) ar = as received
    d   = dry
    daf = dry & ash-free

(c) g = gross or higher HV
    n = net or lower HV

Table A.22.   Calorific (heating) values. *(concluded)*

Substance[a]	Formula	(b)	(c)	kcal/kg[c]	Btu/lb[c]
Vegetable Food Waste, 78m		ar	g	1000	1790
Vegetable Food Waste, 78m		d	g	4590	8270
Vegetable Food Waste, 78m		daf	g	4830	8700
Vinyl chloride/acetate copolymer		daf	g	4910	8830
Waste, Type 0, 5a, 10m		ar	g	4720	8500
Waste, Type 1, 10a, 25m		ar	g	3610	6500
Waste, Type 2, 7a, 50m		ar	g	2390	4300
Waste, Type 3, 5a, 70m		ar	g	1390	2500
Waste, Type 4, 5a, 85m		ar	g	560	1000
Waxed Cartons		daf	g	6670	12000
Wax paraffin		ar	g	10350	18620
Wheat		d	g	4180	7530
Wood and Bark, 20m, 0.8a		ar	g	3830	6900
Wood and Bark, 20m, 0.8a		d	g	4790	8610
Wood and Bark, 20m, 0.8a		daf	g	4830	8700
Wood, Balsam, Spruce, 74m, 1a		ar	g	1360	2450
Wood, Balsam, Spruce, 74m, 1a		d	g	5300	9540
Wood, Balsam, Spruce, 74m, 1a		daf	g	5470	9850
Wood, beech, 13m		d	g	3640	6540
Wood, birch, 11.8m		d	g	3720	6690
Wood, Demolition Softwood, 7.7m, .8a		ar	g	4060	7300
Wood, Demolition Softwood, 7.7m, .8a		d	g	4400	7920
Wood, Demolition Softwood, 7.7m, .8a		daf	g	4440	8000
Wood, Furniture, 6m, 1a		ar	g	4080	7350
Wood, Furniture, 6m, 1a		d	g	4340	7810
Wood, Furniture, 6m, 1a		daf	g	4410	7940
Wood, oak, 13m		ar	n	3990	7200
Wood, pine, 12m		ar	n	4420	7970
Wood, Rotten Timbers, 27m, 2a		ar	g	2620	4710
Wood, Rotten Timbers, 27m, 2a		d	g	3540	6370
Wood, Rotten Timbers, 27m, 2a		daf	g	3640	6560
Wood, Waste Hardwood, 12m, 0.5a		ar	g	3570	6430
Wood, Waste Hardwood, 12m, 0.5a		d	g	4060	7300
Wood, Waste Hardwood, 12m, 0.5a		daf	g	4080	7340
Xylene (av. o, m, p)	$(CH_3)_2C_6H_4$	daf	n	10280	18500
Zinc	(to NzO)	daf	g	1280	2300

---

[a] a = % ash by weight
  m= % moisture by weight

[b] ar = as received
  d = dry
  daf = dry & ash-free

[c] g = gross or higher HV
  n = net or lower HV

Table A.23.   Periodic table of the elements

Element	Symbol	Atomic Number	Column, Row	Atomic Weight
Actinium	Ac	89	c,7	(227)
Aluminum	Al	13	m,3	26.982
Americium	Am	95	j,9A	(243)
Antimony	Sb	51	o,5	121.750
Argon	Ar	18	r,3	39.948
Arsenic	As	33	o,4	74.922
Astatine	At	85	q,6	(210)
Barium	Ba	56	b,6	137.340
Berkelium	Bk	97	l,9A	(247)
Berylium	Be	4	b,2	9.012
Bismuth	Bi	83	o,6	208.980
Boron	B	5	m,2	10.811
Bromine	Br	35	q,4	70.000
Cadmium	Cd	48	l,5	112.400
Calcium	Ca	20	b,4	40.080
Californium	Cf	98	m,9A	(251)
Carbon	C	6	n,2	12.011
Cerium	Ce	58	e,8L	140.120
Cesium	Cs	55	a,6	132.905
Chlorine	Cl	17	q,3	35.453
Chromium	Cr	24	f,4	51.996
Cobalt	Co	27	i,4	58.933
Copper	Cu	29	k,4	63.540
Curium	Cm	96	k,9A	(247)
Dysprosium	Dy	66	m,8L	162.500
Einsteinium	Es	99	n,9A	(254)
Erbium	Er	68	o,8L	167.260
Europium	Eu	63	j,8L	151.960
Fermium	Fm	100	o,9A	(253)
Fluorine	F	9	q,2	18.998
Francium	Fr	87	a,7	(223)
Gadolinium	Gd	64	k,8L	157.250
Gallium	Ga	31	m,4	69.720
Germanium	Ge	32	n,4	72.590
Gold	Au	79	k,6	196.967

(continued)

Table A.23. *(continued)*

Element	Symbol	Atomic Number	Column, Row	Atomic Weight
Hafnium	Hf	72	d,6	178.490
(Hahnium)	(Ha)	105	e,7	(260)
Helium	He	2	r,1	4.003
Holmium	Ho	67	n,8L	164.930
Hydrogen	H	1	a,1	1.008
Indium	In	49	m,5	114.820
Iodine	I	53	q,5	126.904
Iridium	Ir	77	i,6	192.200
Iron	Fe	26	h,4	55.847
Krypton	Kr	36	r,4	83.800
Lanthanum	La	57	c,6	138.910
(Lawrencium)	(Lw)	103	r,9A	(257)
Lead	Pb	82	n,6	207.190
Lithium	Li	3	a,2	6.939
Lutetium	Lu	71	r,8L	174.970
Magnesium	Mg	12	b,3	24.312
Manganese	Mn	25	g,4	54.938
Mendelevium	Md	101	p,9A	(256)
Mercury	Hg	80	1,6	200.590
Molybdenum	Mo	42	f,5	95.940
Neodymium	Nd	60	g,8L	144.240
Neon	Ne	10	r,2	20.183
Neptunium	Np	93	h,9A	(237)
Nickel	Ni	28	j,4	58.710
Niobium	Nb	41	e,5	92.906
Nitrogen	N	7	o,2	14.007
Nobelium	No	102	q,9A	(254)
Osmium	Os	76	h,6	190.200
Oxygen	O	8	p,2	15.999
Palladium	Pd	46	j,5	106.400
Phosphorous	P	15	o,3	30.974
Platinum	Pt	78	j,6	195.090
Plutonium	Pu	94	i,9A	(242)
Polonium	Po	84	p,6	(210)

(continued)

Table A.23. *(concluded)*

Element	Symbol	Atomic Number	Column, Row	Atomic Weight
Potassium	K	19	a,4	39.102
Praseodymium	Pr	59	f,8L	140.907
Promethium	Pm	61	h,8L	(147)
Protactinium	Pa	91	f,9A	231.036
Radium	Ra	88	b,7	226.025
Radon	Rn	86	r,6	(222)
Rhenium	Re	75	g,6	186.200
Rhodium	Rh	45	i,5	102.905
Rubidium	Rb	37	a,5	85.470
Ruthenium	Ru	44	h,5	101.070
(Rutherfordium)	(Rf)	104	d,7	(261)
Samarium	Sm	62	l,8L	150.350
Scandium	Sc	21	c,4	44.956
Selenium	Se	34	p,4	78.960
Silicon	Si	14	n,3	28.086
Silver	Ag	47	k,5	107.870
Sodium	Na	11	a,3	22.990
Strontium	Sr	38	b,5	87.620
Sulfur	S	16	p,3	32.064
Tantalum	Ta	73	e,6	180.948
Technetium	Tc	43	g,5	(99)
Tellurium	Te	52	p,5	127.600
Terbium	Tb	65	l,8L	158.925
Thallium	Tl	81	m,6	204.370
Thorium	Th	90	e,9A	232.038
Thulium	Tm	69	p,8L	168.934
Tin	Sn	50	n,5	118.690
Titanium	Ti	22	d,4	47.900
Tungsten (see Wolfram)				
Uranium	U	92	g,9A	238.030
Vanadium	V	23	e,4	50.94
Wolfram (Tungsten)	W	74	f,6	183.850
Xenon	Xe	54	r,5	131.300
Ytterbium	Yb	70	q,8L	173.040
Yttrium	Y	39	c,5	88.905
Zinc	Zi	30	l,4	65.370
Zirconium	Zr	40	d,5	91.220

Table B.1. Approximate temperature ranges of industrial heating processes

Material	Operation	Temperature, F/K
Aluminum	Melting	1200-1400 / 920-1030
Aluminum alloy	Ageing	250-460 / 395-510
Aluminum alloy	Annealing	450-775 / 505-685
Aluminum alloy	Forging	650-970 / 616-794
Aluminum alloy	Heating for rolling	850 / 728
Aluminum alloy	Homogenizing	850-1175 / 720-900
Aluminum alloy	Solution h.t.	820-1080 / 708-800
Aluminum alloy	Stress Relieving	650-1200 / 615-920
Antimony	Melting point	1166 / 903
Asphalt	Melting	350-450 / 450-505
Babbitt	Melting (1)	600-800 / 590-700
Brass	Annealing	600-1000 / 590-811
Brass	Extruding	1400-1450 / 1030-1060
Brass	Forging	1050-1400 / 840-1030
Brass	Rolling	1450 / 1011
Brass	Sintering	1550-1600 / 1116-1144
Brass, red	Melting (1)	1830 / 1270
Brass, yellow	Melting	1705 / 1200
Bread	Baking	300-500 / 420-530
Brick	Burning	1800-2600 / 1255-1700
Brick, refractory	Burning	2400-3000 / 1589-1920
Bronze	Sintering	1400-1600 / 1033-1144
Bronze, 5% aluminum	Melting (1)	1940 / 1330
Bronze, manganese	Melting	1645 / 1170
Bronze, phosphor	Melting	1920 / 1320
Bronze, Tobin	Melting	1625 / 1160
Cadmium	Melting point	610 / 595
Cake (food)	Baking	300-350 / 420-450
Calcium	Melting point	1562 / 1123
Calender rolls	Heating	300 / 420
Candy	Cooking	225-300 / 380-420
Cement	Calcining kiln firing	2600-3000 / 1700-1922
China, porcelain	Bisque firing	2250 / 1505
Chine, porcelain	Decorating	1400 / 1033
China, porcelain	Glazing, glost firing	1500-2050 / 1088-1394
Clay, refractory	Burning	2200-2600 / 1480-1700
Cobalt	Melting point	2714 / 1763
Coffee	Roasting	600-800 / 590-700
Coke	By-product oven	1830-2730 / 1270-1770
Cookies	Baking	375-450 / 460-505
Copper	Annealing	800-1200 / 700-920
Copper	Forging	1800 / 1255
Copper	Melting (1)	2100-2300 / 1420-1530
Copper	Refining	2100-2600 / 1420-1700
Copper	Rolling	1600 / 1144
Copper	Sintering	1550-1650 / 1116-1172
Copper	Smelting	2100-2600 / 1420-1700

(continued)

Table B.1.   *(continued)*

Material	Operation	Temperature, F/K
Cores, sand	Baking	250-550 / 395-560
Cupronickel, 15%	Melting	2150 / 1450
Cupronickel, 30%	Melting	2240 / 1500
Electrotype	Melting	740 / 665
Enamel, organic	Baking	250-450 / 395-505
Enamel, vitreous	Enameling	1200-1800 / 922-1255
Everdur 1010	Melting	1865 / 1290
Ferrites		2200-2700 / 1478-1755
Frit	Smelting	2000-2400 / 1365-1590
German silver	Annealing	1200 / 922
Glass	Annealing	800-1200 / 700-920
Glass	Melting, pot furn	2300-2500 / 1530-1645
Glass, bottle	Melting, tank furn	2500-2900 / 1645-1865
Glass, flat	Melting, tank furn	2500-3000 / 1645-1920
Gold	Melting	1950-2150 / 1340-1450
Iron	Melting, blast furnace tap	2500-2800 / 1645-1810
Iron	Melting, cupola (1)	2600-2800 / 1700-1810
Iron, cast (2)	Annealing	1300-1750 / 978-1228
Iron, cast	Austenitizing	1450-1700 / 1060-1200
Iron, cast	Malleablizing	1650-1800 / 1170-1255
Iron, cast	Melting, cupola (2)	2600-2800 / 1700-1800
Iron, cast	Normalizing	1600-1725 / 1145-1210
Iron, cast	Stress relieving	800-1250 / 700-945
Iron, cast	Tempering	300-1300 / 420-975
Iron, cast	Vitreous enameling	1200-1300 / 920-975
Iron, malleable	Melting (1)	2400-3100 / 1590-1980
Iron, malleable	Annealing, long cycle	1500-1700 / 1090-1200
Iron, malleable	Annealing, short cycle	1800 / 1255
Iron	Sintering	1283-1422 / 1850-2100
Japan	Baking	180-450 / 355-505
Lacquer	Drying	150-300 / 340-422
Lead	Melting (1)	620-750 / 600-670
Lead	Blast furnace	1650-2200 / 1170-1480
Lead	Refining	1800-2000 / 1255-1365
Lead	Smelting	2200 / 1477
Lime	Burning, roasting	2100 / 1477
Limestone	Calcining	2500 / 1644
Magnesium	Ageing	350-400 / 450-480
Magnesium	Annealing	550-850 / 156-728
Magnesium	Homogenizing	700-800 / 644-700
Magnesium	Solution h.t	665-1050 / 625-839
Magnesium	Stress relieving	300-1200 / 422-922
Magnesium	Superheating	1450-1650 / 1060-1170
Meat	Smoking	100-150 / 310-340
Mercury	Melting point	38 / 234
Molybdenum	Melting point	2898 / 4757

(continued)

Table B.1. *(continued)*

Material	Operation	Temperature, F/K
Monel metal	Annealing	865-1075 / 1100-1480
Monel metal	Melting (1)	2800 / 1810
Moulds, foundry	Drying	400-750 / 475-670
Muntz metal	Melting	1660 / 1175
Nickel	Annealing	1100-1480 / 865-1075
Nickel	Melting (1)	2650 / 1725
Nickel	Sintering	1850-2100 / 1283-1422
Palladium	Melting point	2829 / 1827
Petroleum	Cracking	750 / 670
Phosphorus, yellow	Melting point	111 / 317
Pie	Baking	500 / 530
Pigment	Calcining	1600 / 1150
Platinum	Melting	3224 / 2046
Porcelain	Burning	2600 / 1700
Potassium	Melting point	145 / 336
Potato chips	Frying	350-400 / 450-480
Primer	Baking	300-400 / 420-480
Sand, cove	Baking	450 / 505
Silicon	Melting point	2606 / 1703
Silver	Melting	1750-1900 / 1225-1310
Sodium	Melting point	208 / 371
Solder	Melting (1)	400-600 / 480-590
Steel	Annealing	1250-1650 / 950-1172
Steel	Austenitizing	1400-1700 / 1033-1200
Steel	Bessemer converter	2800-3000 / 1810-1920
Steel	Calorizing (baking in aluminum powder)	1700 / 1200
Steel	Carbonitriding	1300-1650 / 778-1172
Steel	Carburizing	1500 / 1750
Steel	Case hardening	1600-1700 / 1140-1200
Steel	Cyaniding	1400-1800 / 1030-1250
Steel	Drawing forgings	850 / 725
Steel	Drop-forging	2200-2400 / 1475-1590
Steel	Forging	1700-2150 / 1200-1450
Steel	Form-bending	1600-1800 / 1140-1250
Steel	Galvanizing	800-900 / 700-760
Steel	Heat treating	700-1800 / 650-1250
Steel	Lead hardening	1400-1800 / 1030-1250
Steel	Melting, open hearth (1)	2800-3100 / 1810-1975
Steel	Melting, electric furnace (1)	2400-3200 / 1590-2030
Steel	Nitriding	950-1051 / 783-838
Steel	Normalizing	1650-1900 / 1170-1310
Steel	Open Hearth	2800-2900 / 1810-1866
Steel	Pressing, die	2200-2370 / 1478-1572
Steel	Rolling	2200-2300 / 1478-1533
Steel	Sintering	2000-2350 / 1366-1561
Steel	Soaking pit, heating for rolling	1900-2100 / 1310-1420

(continued)

Table B.1. *(continued)*

Material	Operation	Temperature, F/K
Steel	Spheroidizing	1250-1330 / 950-994
Steel	Stress relieving	450-1200 / 505-922
Steel	Tempering (drawing)	300-1400 / 422-1033
Steel	Upsetting	2000-2300 / 1365-1530
Steel	Welding	2400-2800 / 1590-1810
Steel bars	Heating	1900-2200 / 1310-1480
Steel billets	Rolling	1750-2275 / 1228-1519
Steel blooms	Rolling	1750-2275 / 1228-1519
Steel bolts	Heading	2200-2300 / 1480-1530
Steel castings	Annealing	1300-1650 / 978-1172
Steel flanges	Heating	1800-2100 / 1250-1420
Steel ingots	Heating	2000-2200 / 1365-1480
Steel nails	Blueing	650 / 615
Steel pipes	Butt welding	2400-2600 / 1590-1700
Steel pipes	Normalizing	1650 / 1172
Steel rails	Hot bloom reheating	1900-2050 / 1310-1400
Steel rivets	Heating	1750-2275 / 1228-1519
Steel rods	Mill heating	1900-2100 / 1310-1420
Steel shapes	Heating	1900-2200 / 1310-1480
Steel, sheet	Blue annealing	1400-1600 / 1030-1140
Steel, sheet	Box annealing	1500-1700 / 1090-1200
Steel, sheet	Bright annealing	1250-1350 / 950-1000
Steel, sheet	Job mill heating	2000-2100 / 1365-1420
Steel, sheet	Mill heating	1800-2100 / 1250-1420
Steel, sheet	Normalizing	1750 / 1228
Steel, sheet	Open annealing	1500-1700 / 1090-1200
Steel, sheet	Pack heating	1750 / 1228
Steel, sheet	Pressing	1920 / 1322
Steel, sheet	Tin plating	650 / 615
Steel, sheet	Vitreous enameling	1400-1650 / 1030-1170
Steel skelp	Welding	2550-2700 / 1673-1755
Steel slabs	Rolling	1750-2275 / 1228-1519
Steel spikes	Heating	2000-2200 / 1365-1480
Steel springs	Annealing	1500-1650 / 1090-1170
Steel strip, cold rolled	Annealing	1250-1400 / 950-1033
Steel, tinplate sheet	Box annealing	1200-1650 / 920-1170
Steel, tinplate sheet	Hot mill heating	1800-2000 / 1250-1365
Steel, tinplate sheet	Lithographing	300 / 420
Steel tubing (see Steel skelp)		
Steel wire	Annealing	1200-1400 / 920-1030
Steel wire	Baking	300-350 / 420-450
Steel wire	Drying	300 / 422
Steel wire	Patenting	1600 / 1144
Steel wire	Pot annealing	1650 / 1170
Steel, alloy, tool	Hardening	1425-2150 / 1050-1450
Steel, alloy, tool	Preheating	1200-1500 / 920-1900

(continued)

Table B.1. *(continued)*

Material	Operation	Temperature, F/K
Steel, alloy, tool	Tempering	325-1250 / 435-950
Steel, carbon	Hardening	1360-1550 / 1010-1120
Steel, carbon	Tempering	300-1100 / 420-870
Steel, carbon, tool	Hardening	1450-1500 / 1060-1090
Steel, carbon, tool	Tempering	300-550 / 420-560
Steel, chromium	Melting	2900-3050 / 1867-1950
Steel, high–carbon	Annealing	1400-1500 / 1030-1090
Steel, high–speed	Hardening	2200-2375 / 1478-1575
Steel, high–speed	Preheating	1450-1600 / 1060-1150
Steel, high–speed	Tempering	630-1150 / 605-894
Steel, manganese, castings	Annealing	1900 / 1311
Steel, medium carbon	Heat treating	1550 / 1117
Steel, spring	Rolling	2000 / 1367
Steel, S.A.E.	Annealing	1400-1650 / 1030-1170
Steel, stainless	Annealing (3)	1750-2050 (3) / 1228-1505
Steel, stainless	Annealing (4)	1200-1525 (4) / 922-1103
Steel, stainless	Annealing (5)	1525-1650 (5) / 1103-1172
Steel, stainless	Austenitizing (5)	1700-1950 (5) / 1200-1339
Steel, stainless	Bar and pack heating	1900 / 1311
Steel, stainless	Forging	1650-2300 / 1172-1533
Steel, stainless	Nitriding	975-1025 / 797-825
Steel, stainless	Normalizing	1700-2000 / 1200-1367
Steel, stainless	Rolling	1750-2300 / 1228-1533
Steel, stainless	Sintering	2000-2350 / 1366-1561
Steel, stainless	Stress relieving (6)	400-1700 / 478-1200
Steel, stainless	Tempering (drawing)	300-1200 / 422-922
Steel, tool	Rolling	1900 / 1311
Tin	Melting	500-650 / 530-615
Titanium	Forging	1400-2160 / 1033-1450
Tungsten, Ni-Cu, 90-6-4	Sintering	2450-2900 / 1616-1866
Tungston carbide	Sintering	2600-2700 / 1700-1755
Type metal	Stereotyping	525-650 / 530-615
Type metal	Linotyping	550-650 / 545-615
Type metal	Electrotyping	650-750 / 615-670
Varnish	Cooking	520-600 / 545-590
Zinc	Melting (1)	800-900 / 700-760
Zinc alloy	Die-casting	850 / 730

(1) Refer to Appendices A.12-A.15 for typical pouring temperatures.
(2) Includes gray and ductile iron.
(3) Austenitic stainless steels only. (AISI 200 and 300 series.)
(4) Ferritic stainless steels only. (AISI 400 series.)
(5) Martensitic stainless steels only. (AISI 400 series.)
(6) Austenitic and martensitic stainless steels only.

(continued)

Table B.1. *(concluded)*

SOURCES

Lukasiewicz, M. A.: "Natural Gas-Fired Radiant Tubes", Gas Research Institute, Chicago, IL 60631, July 16, 1991.

Schneidewind, A.: private communication, "Lindberg unit of General Signal", Watertown, WI, 1990.

Pritchard, R., J. T. Guy, and N. E. Connor: "Industrial Gas Utilization", Bowker Publishing Company Limited, Epping, Essex, UK (1977).

Dryden. I. G. C. (ed.): "The Efficiency Use of Energy", IPC Science and Technology Press, Guildford, England, 1975.

Pohl, J. H.: "The Potential for Conserving Oil and Natural Gas Used in Industry", Sandia National Laboratories, SAND 79-8290, September 1980.

Pohl, J. H., J. Lee, J. Clough, and J. Dan: "Technology Research Needed in Industrial Combustion Processes", Prepared for EG & G Idaho Inc. under Contract C84-1303490–GHL–44–84, 1984.

Table B.2.   Boiler load terminology and data

1 boiler horsepower = 33 475 Btu/hr *(8442 kcal/h)* heat to steam
              = 34.5 pounds/hr *(15.65 kg)* steam evaporated from and at 212 F *(100 C)*
              = 44 633 Btu/hr *(11 256 kcal/h)* fuel input for 75% overall efficiency
              = 41 850 Btu/hr *(10 555 kcal/h)* fuel input for 80% overall efficiency
              = 39 400 Btu/hr *(9937 kcal/h)* fuel input for 85% overall efficiency
              = approximately 10 ft² *(0.929 m²)* of boiler heating surface *(basis for boiler ratings)*
              = 139.4 ft² *(12.95 m²)* of equivalent direct radiation

OVERRATE firing (common practice):

Steel firebox boilers (gas or oil firing) .... 200% of rating
Steel firebox boilers (coal fired) .... 150% of rating
Steel firebox boilers (simultaneous coal and oil or gas) .... 200% of rating
Scotch Marine boilers (conventional type) .... 250% of rating
Water tube boilers (small) .... 300% of rating
Water tube boilers (large, power type) .... 600% of rating

FURNACE HEAT RELEASE, Btu/ft³ *(kcal/m³)*:

Steel firebox boilers (gas or oil firing) .... 65 000 Btu/ft³ *(578 400 kcal/m³)*
Steel firebox boilers (coal fired) .... 60 000 Btu/ft³ *(533 900 kcal/m³)*
Scotch Marine boilers (gas or oil fired) .... 100 000 Btu/ft³ *(889 900 kcal/m³)* max.

EQUIVALENT DIRECT RADIATION (EDR):

1 ft² EDR = 240.0 Btu/hr for steam heating
          = 150.0 Btu/hr for hot water heating (open system)
          = 180.0 Btu/hr for hot water heating (closed system)

1 m² EDR = 651.28 kcal/h for steam heating
            = 407.05 kcal/h for hot water heating (open system)
            = 438.46 kcal/h for hot water heating (closed system)

OVERALL BOILER EFFICIENCY, %:

$$= \frac{(\text{steam}^a) \times [(\text{heat content, } h_g, \text{ of delivered steam}^b) - (\text{heat content, } h_f, \text{ of feedwater}^c)]}{(\text{fuel input rate}^d) \times (\text{gross heating value of the fuel}^e)} \times 100$$

	Gas fired American units	Gas fired Metric units	Oil fired American units	Oil fired Metric units	Coal fired American units	Coal fired Metric units
[a] Steam	lb/hr	kg/h	lb/hr	kg/h	lb/hr	kg/h
[b] Heat content of delivered steam	Btu/lb	MJ/kg	Btu/lb	MJ/kg	Btu/lb	MJ/kg
[c] Heat content of feedwater	Btu/lb	MJ/kg	Btu/lb	MJ/kg	Btu/lb	MJ/kg
[d] Fuel input rate	ft³/hr	m^3/h	gal/hr	dm^3/h	lb/hr	kg/h
[e] Gross heating value of the fuel	Btu/ft³	MJ/m^3	Btu/hr	MJ/dm^3	Btu/lb	MJ/kg

1 Btu/lb *= 0.002 326 MJ/kg*
1 cal/g = 0.004 183 MJ/kg
1 US gal = 3.785 L = 3.785 dm³
$h_f = h_g - h_{f_g}$. Values of h_g and h_{f_g} can be determined from Tables A.8-A.11.

Table B.3 US.   Heat requirements for drying

The pounds of water to be evaporated per ton of dried material is $\dfrac{2000 \times M}{100 - M}$ where M is the percent weight of water to be removed from the wet material.

The (Btu) heat required to produce one ton of dried material (in driers with no heat recovery equipment) is approximately

$$\frac{\text{available Btu required}}{\text{ton of dried material}} = \frac{2000 \times M}{100 - M} (h_2 - hf_1) + 2000\ c\ (T_{m_2} - T_{m_1})$$

where: M   is the percent weight of water to be removed
 h_2 is the heat content of the water vapor as it leaves the drier in Btu/lb (from Table A.10 US at the vapor exit temperature)
 hf_1 is the heat content of the liquid water in the material entering the drier in Btu/lb
       $= T_{m_1} - 32$
 c   is the specific heat of the dry material (from Table A.16 US)
 T_{m_2} is the temperature at which the dried material leaves the drier, F
 T_{m_1} is the temperature at which the moist material enters the drier, F

Excess air is usually used, so the appropriate excess air curve from Figure 3.10 (Volume I) must be used in determining % available.

For materials with a specific heat of 0.20 (most sands and slags) and for material entering the drier at 60 F, the above formula may be simplified to

$$\frac{\text{available Btu required}}{\text{ton of dried material}} = \frac{(2000 \times M \times h_2) - (56\ 000 \times M)}{(100 - M)} + 400\ T_{m_2} - 24\ 000.$$

The table below lists results from the formula immediately above for selected percents of moisture removed, material exit temperatures, and vapor (flue gas) exit temperatures. Both formula and table assume thorough exposure of the material being heated to the products of combustion to allow sufficient heat and mass transfer.

Approximate available heat requirement, 1000s of Btu per ton of dried material	Exit temperature of dried material, T_{m_2}, F														
	150			200			250			300			350		
	Exit temperature of vapor, F														
	150	200	250	200	250	300	250	300	350	300	350	400	350	400	450
2	81	82	83	102	103	104	123	124	125	144	145	146	165	166	167
4	127	129	131	149	151	153	171	173	175	193	195	197	215	217	219
6	175	179	182	199	202	205	222	225	228	245	248	251	268	271	274
8	226	230	234	250	254	258	274	279	283	299	303	307	323	327	331
10	279	284	290	304	310	319	330	335	340	355	360	365	380	385	391
12	334	340	347	360	367	374	387	394	400	414	420	427	440	447	453
14	392	400	407	420	427	435	447	455	463	475	483	491	503	511	578
16	452	462	471	482	491	500	511	520	529	540	549	558	569	578	587
18	515	526	537	546	557	567	577	587	598	607	618	628	638	648	659
20	582	595	606	615	626	639	646	659	670	679	690	702	710	722	734

(leftmost column label, rotated: % weight of moisture removed)

Table B.4 Metric.   Heat requirements for drying

The kg of water to be evaporated per metric ton of dried material is $\dfrac{1000 \times M}{100 - M}$ where M is the percent weight of water to be removed from the wet material.

The heat required to produce one metric ton of dried material (in driers with no heat recovery equipment) is approximately

$$\frac{\text{available kcal required}}{\text{metric ton of dried material}} = \frac{1000 \times M}{100 - M}(h_2 - hf_1) + 1000\, c\, (Tm_2 - Tm_1)$$

where: M   is the percent weight of water to be removed
     h_2 is the heat content of the water vapor as it leaves the drier in kcal/kg (from Table A.11 SI at the vapor exit temperature)
     hf_1 is the heat content of the liquid water in the material entering the drier in kcal/kg
        $= Tm_1$
     c   is the specific heat of the dry material (from Table A.17 SI)
     Tm_2 is the temperature at which the dried material leaves the drier, C
     Tm_1 is the temperature at which the moist material enters the drier, C

Excess air is usually used, so the appropriate excess air curve from Figure 3.10 (Volume I) must be used in determining % available.

For materials with a specific heat of 0.20 (most sands and slags) and for material entering the drier at 16 C, the above formula may be simplified to

$$\frac{\text{available kcal required}}{\text{metric ton of dried material}} = \frac{(1000 \times M \times h_2) - (16\,000 \times M)}{(100 - M)} + 200\, Tm_2 - 32\,000.$$

The table below lists results from the formula immediately above for selected percents of moisture removed, material exit temperatures, and vapor (flue gas) exit temperatures. Both formula and table assume thorough exposure of the material being heated to the products of combustion to allow sufficient heat and mass transfer.

Approximate available heat requirement, 1000s of kcal per metric ton of dried material	Exit temperature of dried material, Tm_2, C									
	60		90		120		150		180	
	Exit temperature of vapor, C									
	100	120	120	150	150	180	180	210	210	240
2	21.5	21.7	27.7	28.0	34.0	34.3	40.3	40.6	46.6	46.9
4	34.3	34.7	41.0	41.2	47.2	47.8	53.8	54.4	60.4	61.0
6	47.0	47.6	53.6	54.5	60.5	61.3	67.3	68.2	74.2	75.1
8	59.8	60.6	66.6	67.7	73.7	74.9	80.9	82.0	88.0	89.1
10	72.5	73.5	79.5	80.9	86.9	88.4	94.4	95.8	102	103
12	85.3	86.5	92.5	94.1	100	102	108	110	116	117
14	98.0	99.4	105	107	114	115	121	123	129	131
16	111	112	118	121	127	129	135	137	143	145
18	124	125	131	134	140	142	148	151	157	160
20	136	138	144	147	153	156	162	165	171	174

% weight of moisture removed

Table B.5 US. Heat requirements for direct-fired air heating

The table below lists the gross Btu/hr of fuel input required to heat one standard cubic foot of air from a given inlet temperature to a given outlet temperature. It is based on natural gas at 60 F, having 1000 gross Btu/ft³, 910 net Btu/ft³, and stoichiometric air/gas ratio of 9.4:1. The oxygen for combustion is supplied by the air that is being heated. The hot outlet "air" includes combustion products obtained from burning sufficient natural gas to raise the air to the indicated outlet temperature.

Gross Btu of fuel input per scf of air — Inlet air temperature, F	Outlet air temperature, F														
	100	200	300	400	500	600	700	800	900	1000	1100	1200	1300	1400	1500
−20	2.39	4.43	6.51	8.63	10.8	13.0	15.2	17.5	19.9	22.2	24.7	27.1	29.7	32.2	34.9
0	2.00	4.04	6.11	8.23	10.4	12.6	14.8	17.1	19.5	21.8	24.3	26.7	29.3	31.8	34.4
+20	1.60	3.64	5.71	7.83	9.99	12.2	14.4	16.7	19.0	21.4	23.8	26.3	28.8	31.4	34.0
40	1.20	3.24	5.31	7.43	9.58	11.8	14.0	16.3	18.6	21.0	23.4	25.9	28.4	31.0	33.6
60	0.802	2.84	4.91	7.02	9.18	11.4	13.6	15.9	18.2	20.6	23.0	25.5	28.0	30.6	33.2
80	0.402	2.43	4.51	6.62	8.77	11.0	13.2	15.5	17.8	20.2	22.6	25.1	27.6	30.1	32.7
100		2.03	4.10	6.21	8.36	10.6	12.8	15.1	17.4	19.8	22.2	24.6	27.2	29.7	32.3
200			2.06	4.17	6.31	8.50	10.7	13.0	15.5	17.7	20.1	22.5	25.0	27.6	30.2
300				2.10	4.23	6.41	8.63	10.9	13.2	15.5	17.9	20.4	22.9	25.4	28.0
400					2.13	4.30	6.51	8.76	11.1	13.4	15.8	18.2	20.7	23.2	25.8
500						2.16	4.36	6.61	8.90	11.2	13.6	16.0	18.5	21.0	23.6
600							2.19	4.43	6.71	9.03	11.4	13.8	16.3	18.8	21.3
700								2.23	4.50	6.81	9.16	11.6	14.0	16.5	19.0
800									2.25	4.56	6.91	9.30	11.7	14.2	16.7
900										2.29	4.63	7.01	9.43	11.9	14.4
1000											2.32	4.69	7.11	9.57	12.1

Example: Find the amount of natural gas required to heat 1000 scfm of air from 100 F to 1400 F.

Solution: From the table, read 23.2 gross Btu/scf air. Then

$$\left(\frac{23.2\text{ gross Btu}}{\text{scf air}} \times \frac{1000\text{ scf air}}{\text{min}} \times \frac{60\text{ min}}{1\text{ hr}}\right) \div \frac{1000\text{ gross Btu}}{\text{ft}^3\text{ gas}} = 1392\text{ cfh gas.}$$

The conventional formula derived from the specific heat equation is: $Q = wc\Delta T$; so Btu/hr = weight/hr × specific heat × temp rise

$$= \frac{\text{scf}}{\text{min}} \times \frac{60\text{ min}}{\text{hr}} \times \frac{0.076\text{ lb}}{\text{ft}^3} \times \frac{0.24\text{ Btu}}{\text{lb °F}} \times °\text{rise} = \text{scfm} \times 1.1 \times °\text{rise.}$$

The table above incorporates many refinements not considered in the conventional formulas: (a) % available heat which corrects for heat loss to dry flue gases and the heat loss due to heat of vaporization in the water formed by combustion, (b) the specific heats of the products of combustion (N_2, CO_2, and H_2O) are not the same as that of air, and (c) the specific heats of the combustion products change at higher temperatures.

For the example above, the rule of thumb would give 1000 scfm × 1.1 × (1400 − 400) = 1 100 000 gross Btu/hr; whereas the example finds 1392 × 1000 = 1 392 000 gross Btu/hr required. *Reminder:* The fuel being burned adds volume and weight to the stream being heated.

Table B.6 Metric. Heat requirements for direct-fired air heating

The table below lists the gross fuel input required to heat one cubic metre of air (at 16 C) from a given inlet temperature to a given outlet temperature. It is based on natural gas at 16 C, having 8899 kcal/m³ gross, 8098 kcal/m³ net (37.22 MJ/m³ gross, 33.63 MJ/m³ net), and stoichiometric air/gas ratio of 9.4:1. The oxygen for combustion is supplied by the air that is being heated. The hot outlet "air" includes combustion products obtained from burning sufficient natural gas to raise the air to the indicated outlet temperature.

Gross kcal/MJ of fuel input per m³ of air / Inlet air temp., C/K	Outlet air temperature, C/K							
	100C/373K	200C/473K	300C/573K	400C/673K	500C/773K	600C/873K	700C/973K	800C/1093K
−20/253	38.9/0.163	72.4/0.303	107/0.449	143/0.599	181/0.757	220/0.922	260/1.09	300/1.26
0/273	33.2/0.139	66.0/0.276	101/0.422	137/0.572	174/0.728	213/0.890	252/1.06	294/1.23
+20/293		58.8/0.246	94.5/0.396	130/0.545	167/0.701	206/0.862	246/1.03	287/1.20
50/323		39.6/0.166	84.7/0.354	120/0.504	158/0.660	195/0.817	236/0.988	277/1.16
100/373			68.0/0.285	104/0.434	141/0.589	179/0.750	218/0.915	260/1.09
200/473				69.9/0.292	107/0.447	144/0.602	184/0.770	225/0.941
300/573					71.8/0.300	109/0.458	149/0.623	189/0.791
400/673					36.2/0.152	73.7/0.308	112/0.470	152/0.637
500/773						37.2/0.156	75.6/0.316	115/0.483

Example: Find the amount of natural gas required to heat 464 000 m³/h of air from 200 C to 800 C.

Solution: From the table, read 225 gross kcal/m³ air. Then $\left(\dfrac{225 \text{ gross kcal}}{\text{m}^3 \text{ air}} \times \dfrac{464\,000 \text{ m}^3 \text{ air}}{\text{hr}} \right) \div \dfrac{8899 \text{ gross kcal}}{\text{m}^3 \text{ gas}} = 11\,732 \text{ m}^3/\text{h gas.}$

The conventional formula derived from the specific heat equation is: $Q = wc\Delta T$, where Q = required available heat input, w = weight flow rate of gases being heated, c = specific heat of gases, and ΔT = temperature rise of the gases.

The table above incorporates many refinements not considered in the conventional formulas: (a) % available heat which corrects for heat loss to dry flue gases and the heat loss due to heat of vaporization in the water formed by combustion, (b) the specific heats of the products of combustion (N_2, CO_2, and H_2O) are not the same as that of air, and (c) the specific heats of the combustion products change at higher temperatures.

Reminder: The fuel being burned adds volume and weight to the stream being heated.

Table B.7. Sizes of crucibles*

Crucible number	Approx. height outside in inches†	Approx. diam top outside in inches†	Approx. working capacity in lb red brass‡	Approx. working capacity in lb aluminum‡	Approx. working capacity in lb magnesium
1	$3^5/_8$	$3^1/_4$	3.0	0.93	0.60
$1^1/_2$	$4^1/_{16}$	$3^9/_{16}$	4.3	1.3	0.83
2	$4^1/_2$	$3^3/_4$	4.7	1.4	0.90
3	$5^3/_8$	$4^1/_4$	7.7	2.3	1.5
4	$5^3/_4$	$4^5/_8$	10	3.1	2.0
5	$6^1/_8$	$4^7/_8$	14	4.3	2.8
6	$6^1/_2$	$5^1/_4$	15	4.6	3.0
7	$6^3/_4$	$5^1/_2$	21	6.5	4.2
8	$7^1/_8$	$5^7/_8$	24	7.4	4.8
9	$7^5/_8$	$5^7/_8$	26	8.0	5.1
10	$8^1/_{16}$	$6^1/_{16}$	36	11	7.1
12	$8^1/_2$	$6^3/_8$	42	12	7.7
14	$8^7/_8$	$6^{11}/_{16}$	47	14	9.0
16	$9^1/_4$	$6^{15}/_{16}$	53	16	10
18	$9^{13}/_{16}$	$7^5/_{16}$	64	18	12
20	$10^5/_{16}$	$7^{11}/_{16}$	74	20	13
25	$10^{15}/_{16}$	$8^3/_{16}$	89	25	16
30	$11^1/_2$	$8^5/_8$	100	30	19
35	12	9	120	35	23
40	$12^1/_2$	$9^3/_8$	130	40	26
45	$13^3/_{16}$	$9^7/_8$	160	45	29
50	$13^3/_4$	$10^1/_4$	180	50	32
60	$14^7/_{16}$	$10^{13}/_{16}$	210	60	39
70	$15^1/_{16}$	$11^1/_4$	240	70	45
80	$15^5/_8$	$11^{11}/_{16}$	270	80	51
90	$16^3/_{16}$	$12^1/_8$	300	90	58
100	$16^{11}/_{16}$	$12^1/_2$	330	100	64
125	$17^3/_8$	13	370	125	80
150	$18^3/_8$	$13^3/_4$	470	150	96
175	$19^1/_4$	$14^3/_8$	520	175	110
200	20	15	600	200	130
225	$20^3/_4$	$15^1/_2$	670	225	140
250	$21^3/_8$	16	750	250	160
275	22	$16^7/_{16}$	820	275	180
300	$22^1/_2$	$16^7/_8$	900	300	190
400	$24^5/_{16}$	$18^3/_{16}$	1200	400	260

* Crucibles above size 400 are rarely used for brass, since the weight of so much brass leads to structural difficulties in the crucible. Large crucibles up to size 1000 are frequently used for light metals such as aluminum.

† For carbon-bonded (silicon carbide) crucibles, and for clay-bonded (graphite) crucibles; but consult the crucible manufacturer for detailed dimensions.

‡ Rough approximations of metal working capacity for bilge shape crucibles, through size number 800, are: pounds of aluminum = 1 × crucible number; pounds of brass = 3 × crucible number.

Table C.1. Thermophysical constants in US and *Metric* units

Acceleration of gravity = 32.17 ft/sec^2 or *9.806 m/s^2*

Gross heat release with 1 ft^3 air + stoichiometric amount of natural gas or oil = about 100 Btu

Gross heat release with 1 m^3 *air + stoichiometric amount of natural gas or oil = about 900 kcal*

Stefan-Boltzmann constant = 0.1713 × 10^{-8} Btu/ft^2 hr °R^4 or *4.88 × 10^{-8} kcal/m^2 h °K^4*

Universal gas constant, $\overline{\text{MR}}$, in pv = $\dfrac{\overline{\text{MRT}}}{M}$, (M is molar weight) = 1544 $\dfrac{\text{ft lb mass}}{\text{°R lb-mole}}$, or *84.7 $\dfrac{m\ kg\ mass}{°K\ kg\text{-}mole}$*, or *8.314 $\dfrac{joule}{°K\ g\text{-}mole}$*

Velocity of sound in air at 68 F (20 C) and 29.92" *(760 mm)* Hg = 1129 ft/sec = 769.5 mph = *344.1 m/s*

Volume of 1 lb mole at *15.6 C* (60 F) and 1 atmosphere = 379 ft^3

Volume of 1 kg mole at *15.6 C* (60 F) and 1 atmosphere = *23.7 m^3*

PROPERTIES of water, moist air, and a typical natural gas at 1 atmosphere *(760 mm Hg* = 14.696 psia) and *15.6 C* (60 F):

	Water		Air (moist)		Natl gas	
Specific weight, lb/ft^3; *kg/m^3*	62.35*	*1000*	0.0763	*1.225*	0.046	*0.738*
Specific heat, Btu/lb °F; *cal/g °C*	1.0	*1.0*	0.240	*0.240*	0.55	*0.55*
Thermal conductivity, Btu ft/hr ft^2 °F; *watt/m °K*	0.344	*0.595*	0.0148	*0.0256*	0.02	*0.03*
Absolute viscosity, lb mass/ft hr; *kg/m h*	2.72	*4.05*	0.0434	*0.0646*	0.0266	*0.0396*
Kinematic viscosity, ft^2/hr; *(m^2/s)* × 10^6	0.0436	*1.126*	0.567	*14.68*	0.578	*14.93*

* Water weighs 8.335 lb/gal.

Table C.2. Formulas, definitions, converting units

Force = Mass × Acceleration

1 lb force	= the force that accelerates 1 lb mass at a rate of 32.17 ft/sec²	
1 poundal	= " " " " 1 lb mass " " " "	1 ft/sec²
1 lb force	= " " " " 1 slug " " " "	1 ft/sec²
1 newton	= " " " " 1 kg " " " "	1 m/s² = 10⁵ dynes
1 dyne	= " " " " 1 g " " " "	1 cm/s²

1 lb force = the force that accelerates 1 lb mass at a rate of 32.17 ft/sec²
1 poundal = " " " " 1 lb mass " " " " 1 ft/sec²
1 lb force = " " " " 1 slug " " " " 1 ft/sec²
1 newton = " " " " 1 kg " " " " $1 \text{ m/s}^2 = 10^5$ dynes
1 dyne = " " " " 1 g " " " " 1 cm/s²

Pressure = Force per unit Area

1 newton/m² = 1 pascal (Pa) (See pressure conversions in Table C.5.)
1 pound/in.² = 1 psi = 144 psf = 144 lb/ft²

Work = Force × Distance (Work and heat are convertible forms of energy)

1 ft lb   = work done by 1 lb   force through a   1 ft   = 1.3558 joules = $\dfrac{\text{Btu}}{778.8}$

1 dyne cm   = "   "   " 1 newton   "   "   "   1 cm = 1 erg
1 newton metre = "   "   " 1 dyne   "   "   "   1 m = 10⁷ erg = 1 joule = 1 watt sec

Power = rate of doing Work   (Heat flow rate can be expressed in the same units)

1 newton metre = 1 watt = 1 joule = $\dfrac{1}{746}$ hp

$$1 \text{ kW} = 1000 \text{ joule} = 3413 \text{ Btu/hr} = 860.4 \text{ kcal/h}$$

Converting units

Multiply a known measurement by conversion fraction. Make conversion fractions from the equalities in Appendix Tables A.4, A.5 and A.6, putting numbers and units from each side of the = sign above and below the fraction line. You can multiply anything by 1/1 = 1 without changing its value. All you have done is change its units. If you can't cancel out units (as you would cancel fractions ½ × ⅔ × ¾ = ¼), turn the conversion fraction upside down. For example:

$$3.5\text{"wc} \times \frac{25.4 \text{ mm}}{1\text{"}} = 88.9 \text{ mm H}_2\text{O.}\dagger$$

$$10\,000 \text{ kW} \times \frac{3413 \text{ Btu/hr}}{1 \text{ kW}} \times \frac{1 \text{ kcal}}{3.968 \text{ Btu}} \times \frac{1 \text{ hr}}{3600 \text{ sec}} = 2389 \text{ kcal/s.}$$

This illustrates the use of a known intermediate unit (Btu) to convert between units for which no factor is readily available.

---

† *Small pressures are often expressed in* **wc** *for water column, which is synonymous with* **wg** *for water gauge or* **H₂O**. *Larger pressures are often measured with a mercury column and expressed as* "Hg *or* mm Hg.

Table C.3. Heat, energy, work equivalents – SI unit = Joule

1 gigajoule = 10^9 Newton·metre	= 1 GJ	= 238 800 kcal	= 0.9479 kk Btu	= 277.8 kWh	= 9.479 therm	238.9 thermie	= 0.02464 toe
1 kcal = 1000 calories	= 4.187×10^{-6} GJ	= 1 kcal	= 3.968×10^{-6} kk Btu	= 0.001163 kWh	= 3.968×10^{-5} therm	0.0010 thermie	= 1.03×10^{-7} toe
1 kk Btu = 1 000 000 Btu	= 1.055 GJ	= 252 000 kcal	= 1 kk Btu	= 293.1 kWh	= 10 therm	252.1 thermie	= 0.02598 toe
1 kWh = 1000 watt·hours	= 0.0036 GJ	= 859.1 kcal	= 0.003412 kk Btu	= 1 kWh	= 0.03412 therm	0.8591 thermie	= 8.86×10^{-5} toe
1 therm	= 0.1055 GJ	= 25 200 kcal	= 0.1 kk Btu	= 29.31 kWh	= 1 therm	25.21 thermie	= 0.002598 toe
1 thermie	= 0.004186 GJ	= 999.7 kcal	= 0.003967 kk Btu	= 1.164 kWh	= 0.03967 therm	1 thermie	9700 toe
1 toe (ton of oil equivalent)	= 40.61 GJ	= 9.71×10^6 kcal	= 38.49 kk Btu	= 11 290 kWh	= 384.9 therm	0.0001031 thermie	= 1 toe
1 hp·hr (horsepower·hour)	= 0.002684 GJ	= 641.6 kcal	= 2546 kk Btu	= 0.7457 kWh	= 0.02546 therm	0.6418 thermie	= 6.609×10^{-5} toe

Table C.4. Volume equivalents, rounded to 4 significant figures. SI unit = metre.

1 cm³ (cc)	= 0.000 001 00 m³						
1 in.³	= 0.000 016 39 m³ =	1 in.³					
1 litre	= 0.001 000 m³ =	61.02 in.³ =	1 L (dm³)				
1 USgal	= 0.003 785 m³ =	231.0 in.³ =	3.785 L =	1 USgal			
1 Br gal	= 0.004 546 m³ =	277.4 in.³ =	4.546 L =	1.201 USgal =	1 Br gal		
1 ft³	= 0.028 32 m³ =	1 728 in.³ =	28.32 L =	7.481 USgal =	6.229 Br gal =	1 ft³	
1 bbl, oil	= 0.159 0 m³ =	9 702 in.³ =	159.0 L =	42.00 USgal =	34.97 Br gal =	5.615 ft³ =	1 bbl
1 m³	= 1 m³ =	61 020 in.³ =	1000 L =	264.2 USgal =	220.0 Br gal =	35.31 ft³ =	6.290 bbl

Table C.5. Pressure equivalents, rounded to 4 significant figures. SI unit = pascal.

1 N/m²	= 0.001 kPa (kilopascal)							
1 mm H₂O	= 0.0098 kPa =	1 mm H₂O						
1 mm Hg	= 0.1333 kPa =	13.60 mm H₂O =	1 mm Hg (torr)					
1" H₂O	= 0.2488 kPa =	25.40 mm H₂O =	1.866 mm Hg =	1 "H₂O				
1 osi	= 0.4309 kPa =	43.94 mm H₂O =	3.232 mm Hg =	1.732"H₂O =	1 osi			
1" Hg	= 3.386 kPa =	345.3 mm H₂O =	25.40 mm Hg =	13.61 "H₂O =	7.858 osi =	1 "Hg		
1 psi	= 6.895 kPa =	703.1 mm H₂O =	51.72 mm Hg =	27.71 "H₂O =	16.00 osi =	2.036 "Hg =	1 psi	
1 kg/cm²	= 98.07 kPa = 10 000	mm H₂O =	735.6 mm Hg =	394. "H₂O =	227.6 osi =	28.96 "Hg =	14.22 psi =	1 kg/cm²
1 bar	= 100.0 kPa = 10 200	mm H₂O =	750.1 mm Hg =	401.5 "H₂O =	232.1 osi =	29.53 "Hg =	14.50 psi =	1.020 kg/cm²
1 atm*	= 101.3 kPa = 10 330	mm H₂O =	760.0 mm Hg =	407.5 "H₂O =	235.1 osi =	29.92 "Hg =	14.70 psi =	1.033 kg/cm²

* normal atmosphere = 760 torr (mm Hg, 0 C); a "technical atmosphere" = 1 kg/cm² = 736 torr

Table C.6. Unit equivalents *(continued)*

SI Metric To American Metric To Metric	American To SI Metric American To American

HEAT CONTENT and SPECIFIC HEAT:

1 cal/g   = 1.80 Btu/lb = 4187 J/kg
1 cal/cm^3 = 112.4 Btu/ft^3
1 kcal/m^3 = 0.1124 Btu/ft^3 = 4187 J/m^3
1 kJ/kg   = 0.43 Btu/lb
1 MJ/m^3 = 26.81 Btu/ft^3 = 3.59 Btu/US gal

1 cal/g·°C = 1 Btu/lb·°F = 4187 J/kg·°K
1 kcal/kg = 0.004 187 MJ/kg = 1.8 Btu/lb
1 joule/m^3 = 3.723 × 10^4 Btu/scf

1 Btu/lb = 0.5556 cal/g = 0.5556 kcal/kg
     = 2326 J/kg = 0.002 326 MJ/kg
1 Btu/ft^3 = 0.008 90 cal/cm^3
     = 8.899 kcal/m^3 = 0.037 30 MJ/m^3
1 Btu/USgal = 0.0666 kcal/L = 66.6 kcal/m^3
     = 0.2787 MJ/m^3

1 Btu/lb·°F = 1 cal/g·°C = 4187 J/kg·°K

HEAT FLOW, POWER:

1 N·m/s = 1 W = 1 J/s
     = 0.001 341 hp = 0.7376 ft·lb/sec
1 kcal/h = 1.162 J/s = 1.162 W
     = 3.968 Btu/hr
     = 0.001 thermies/h
1 kW   = 1000 J/s = 3412 Btu/hr = 1.341 hp
     = 859.8 kcal/hr
1 MW   = 3 412 000 Btu/hr = 3.600 GJ/h

1 Btu/hr = 0.2520 kcal/h
     = 0.000 393 1 hp
     = 0.2931 W = 0.2931 J/s
1 million Btu/hr = 1055 MJ/hr
1 hp = 33 000 ft·lb/min = 550 ft·lb/sec
     = 745.7 W = 745.7 J/s
     = 641.4 kcal/h = 2546 Btu/hr
1 boiler hp = 33 475 Btu/hr = 34.5 lb
     steam/hr from and at 212 F

HEAT FLUX and HEAT TRANSFER COEFFICIENT:

1 cal/cm^2·s = 3.687 Btu/ft^2·sec
     = 41.87 kW/m^2
1 cal/cm^2·h = 1.082 W/ft^2 = 11.65 W/m^2
1 kW/m^2 = 317.2 Btu/ft^2·hr

1 kW/m^2·°C = 1 kW/m^2 °K
     = 176.2 Btu/ft^2·hr·°F
     = 860.4 kcal/m^2·hr·°C
1 J/m^2·s·°K = 0.1722 Btu/ft^2·hr·°F

1 Btu/ft^2·sec = 0.2713 cal/cm^2·s = 11.35 kW/m^2
1 Btu/ft^2·hr = 0.003 153 kW/m^2
     = 2.713 kcal/m^2·h
1 kW/ft^2 = 924.2 cal/cm^2·h

1 Btu/ft^2·hr·°F = 4.89 kcal/m^2·h·°C
     = 5.67 W/m^2·°K

LENGTH:

1 mm = 0.10 cm = 0.039 37 in.
     = 0.003 281 ft
1 m   = 100 cm = 1000 mm = 39.37 in.
     = 3.281 ft = 1.094 yard
1 km = 0.6214 mile

1 in. = 25.40 mm = 2.540 cm = 0.025 40 m
1 ft   = 304.8 mm = 30.48 cm = 0.3048 m
1 mile = 5280 ft; 1 nautical mile = 6076 ft
1 micron = 1 μm = 10^{-6} m = 1 micrometer
     = 10 000 Å
1 Angstrom unit = 1 Å = 10^{-10} m = 10^{-4} μm
     (continued)

Table C.6.   Unit equivalents *(continued)*

SI Metric To American Metric To Metric	American To SI Metric American To American

HEAT CONTENT and SPECIFIC HEAT:

1 cal/g = 1.80 Btu/lb = 4187 J/kg	1 Btu/lb = 0.5556 cal/g = 0.5556 kcal/kg
1 cal/cm³ = 112.4 Btu/ft³	= 2326 J/kg = 0.002 326 MJ/kg
1 kcal/m³ = 0.1124 Btu/ft³ = 4187 J/m³	1 Btu/ft³ = 0.008 90 cal/cm³
1 kJ/kg = 0.43 Btu/lb	= 8.899 kcal/m³ = 0.037 30 MJ/m³
1 MJ/m³ = 26.81 Btu/ft³ = 3.59 Btu/US gal	1 Btu/USgal = 0.0666 kcal/L = 66.6 kcal/m³
	= 0.2787 MJ/m³
1 cal/g·°C= 1 Btu/lb·°F = 4187 J/kg·°K	1 Btu/lb·°F = 1 cal/g·°C = 4187 J/kg·°K
1 kcal/kg = 0.004 187 MJ/kg = 1.8 Btu/lb	
1 joule/m³ = 3.723 × 10⁴ Btu/scf	

HEAT FLOW, POWER:

1 N·m/s = 1 W = 1 J/s	1 Btu/hr = 0.2520 kcal/h
= 0.001 341 hp 0.7070 ft lb/sec	= 0.000 393 1 hp
1 kcal/h = 1.162 J/s = 1.162 W	= 0.2931 W = 0.2931 J/s
= 3.968 Btu/hr	1 million Btu/hr = 1055 MJ/hr
= 0.001 thermies/h	1 hp = 33 000 ft·lb/min = 550 ft·lb/sec
1 kW = 1000 J/s = 3412 Btu/hr = 1.341 hp	= 745.7 W = 745.7 J/s
= 859.8 kcal/hr	= 641.4 kcal/h = 2546 Btu/hr
1 MW = 3 412 000 Btu/hr = 3.600 GJ/h	1 boiler hp = 33 475 Btu/hr = 34.5 lb
	steam/hr from and at 212 F

HEAT FLUX and HEAT TRANSFER COEFFICIENT:

1 cal/cm²·s = 3.687 Btu/ft²·sec	1 Btu/ft²·sec = 0.2713 cal/cm²·s = 11.35 kW/m²
= 41.87 kW/m²	1 Btu/ft²·hr = 0.003 153 kW/m²
1 cal/cm²·h= 1.082 W/ft² = 11.65 W/m²	= 2.713 kcal/m²·h
1 kW/m² = 317.2 Btu/ft²·hr	1 kW/ft² = 924.2 cal/cm²·h
	1 Btu/ft²·hr·°F = 4.89 kcal/m²·h·°C
1 kW/m²·°C = 1 kW/m² °K	= 5.67 W/m²·°K
= 176.2 Btu/ft²·hr·°F	
= 860.4 kcal/m²·hr·°C	
1 J/m²·s·°K = 0.1722 Btu/ft²·hr·°F	

LENGTH:

1 mm = 0.10 cm = 0.039 37 in.	1 in. = 25.40 mm = 2.540 cm = 0.025 40 m
= 0.003 281 ft	1 ft = 304.8 mm = 30.48 cm = 0.3048 m
1 m = 100 cm = 1000 mm = 39.37 in.	1 mile = 5280 ft; 1 nautical mile = 6076 ft
= 3.281 ft = 1.094 yard	1 micron = 1 µm = 10⁻⁶ m = 1 micrometer
1 km = 0.6214 mile	= 10 000 Å
	1 Angstrom unit = 1 Å = 10⁻¹⁰ m = 10⁻⁴ µm

(continued)

Table C.6. Unit equivalents *(continued)*

POLLUTANT CONCENTRATION:

$$10\ 000\ \text{ppm}_V = 1.0\%\ \text{by volume}$$
$$1\ \text{ppm}_V\ CH_4 = 676\ \mu g/sm^3$$
$$1\ \text{ppm}_V\ CO = 1183\ \mu g/sm^3$$
$$1\ \text{ppm}_V\ NO = 1272\ \mu g/sm^3$$
$$1\ \text{ppm}_V\ NO_2 = 1948\ \mu g/sm^3$$
$$1\ \text{ppm}_V\ SO_2 = 2707\ \mu g/sm^3$$

$$1\ \text{ppm}_V\ (\text{by volume}) = (M \times 42.25)\ \mu g/sm^3$$
where M = molecular weight
[See also Table C.13 US.]

| Fuel* | Higher heating value, Btu | Stoichiometric | | | # NOx/ million Btu equiv. to 1 ppm | ppm NOx equiv. to 1# NOx/ million Btu |
		air required, cf	wet and dry poc, cf	dry poc, cf		
Natural gas	1000/cf	9.44/cf	10.47/cf	8.52/cf	0.001 21	829
Coke oven gas	530/cf	4.56/cf	5.30/cf	4.12/cf	0.001 10	909
Commercial propane	2499/cf	23.8/cf	25.77/cf	21.8/cf	0.001 23	812
Methanol	64 630/gal	560/gal	608/gal	524/gal	0.001 14	876
#2 fuel oil	137 080/gal	1356/gal	1441/gal	1270/gal	0.001 31	765
#6 fuel oil	153 120/gal	1478/gal	1554/gal	1410/gal	0.001 29	775

Fuel*	# CH_4/ million Btu equiv. to 1 ppm CH_4	ppm CH_4 equiv. to 1# CH_4/ million Btu	# CO/ million Btu equiv. to 1 ppm CO	ppm CO equiv. to 1# CO/ million Btu	# SO_2/ million Btu equiv. to 1 ppm SO_2	ppm SO_2 equiv. to 1# SO_2/ million Btu
Natural gas	0.000 420	2380	0.000 735	1360	0.001 68	595
Coke oven gas	0.000 384	2608	0.000 670	1490	0.001 53	653
Commercial propane	0.000 430	2328	0.000 750	1330	0.001 72	582
Methanol	0.000 397	2520	0.000 696	1440	0.001 59	628
#2 fuel oil	0.000 454	2205	0.000 795	1260	0.001 82	550
#6 fuel oil	0.000 450	2223	0.000 788	1270	0.001 80	555

*For fuel specifications, see Tables 2.1a and 2.1b of Volume I of the North American Combustion Handbook.

(continued)

Table C.6.   Unit equivalents *(continued)*

SI Metric To American
Metric To Metric

American To SI Metric
American To American

PRESSURE:

$1\ N{\cdot}m^2 = 0.001\ kPa = 1.00\ Pa$
$1\ mm\ H_2O = 0.0098\ kPa$
$1\ mm\ Hg = 0.1333\ kPa = 13.60\ mm\ H_2O$
         $= 1\ torr = 0.019\ 33\ lb/in.^2$
$1\ kg/cm^2 = 98.07\ kPa = 10\ 000\ kg/m^2$
         $= 10\ 000\ mm\ H_2O = 394.1\ in.\ H_2O*$
         $= 735.6\ mm\ Hg = 28.96\ in.\ Hg$
         $= 227.6\ oz/in.^2 = 14.22\ lb/in.^2$
         $= 0.9807\ bar$
$1\ bar\quad = 100.0\ kPa = 1.020\ kg/cm^2$
         $= 10\ 200\ mm\ H_2O = 401.9\ in.\ H_2O$
         $= 750.1\ mm\ Hg = 29.53\ in.\ Hg$
         $= 232.1\ oz/in.^2 = 14.50\ lb/in.^2$
         $= 100\ 000\ N/m^2$
$1\ atm\quad = 101.3\ kPa$
$1\ g/cm^2\quad = 0.014\ 22\ lb/in.^2$
         $= 0.2276\ oz/in.^2$
         $= 0.3937\ in.\ H_2O$

(For rough calculations,
$1\ bar = 1\ atm\dagger = 1\ kg/cm^2$
         $= 10\ m\ H_2O = 100\ kPa)$

$1\ in.\ H_2O* = 0.2488\ kPa = 25.40\ mm\ H_2O$
         $= 1.866\ mm\ Hg = 5.198\ lb/ft^2$
         $= 0.002\ 54\ kg/cm^2 = 2.540\ g/cm^2$
$1\ in.\ Hg = 3.386\ kPa = 25.40\ mm\ Hg$
         $= 345.3\ mm\ H_2O = 13.61\ in.\ H_2O$
         $= 7.858\ oz/in.^2 = 0.491\ lb/in.^2\ \mu$
         $= 25.4\ torr$
$1\ lb/in.^2 = 6.895\ kPa = 6895\ N/m^2$
         $= 703.1\ mm\ H_2O = 27.71\ in.\ H_2O$
         $= 51.72\ mm\ Hg = 2.036\ in.\ Hg$
         $= 16.00\ oz/in.^2$
         $= 0.0703\ kg/cm^2 = 70.31\ g/cm^2$
         $= 0.068\ 97\ bar = 0.068\ 07\ atm$
$1\ oz/in.^2 = 0.4309\ kPa$
         $= 43.94\ mm\ H_2O = 1.732\ in.\ H_2O$
         $= 3.232\ mm\ Hg$
         $= 0.004\ 39\ kg/cm^2 = 4.394\ g/cm^2$
$1\ ft\ of\ head\ (water) = 12\ 000\ milinch$

$1\ atm\dagger = 101.3\ kPa = 101\ 325\ N/m^2$
         $= 10\ 330\ mm\ H_2O = 407.3\ in.\ H_2O$
         $= 760.0\ mm\ Hg = 29.92\ in.\ Hg$
         $= 235.1\ oz/in.^2 = 14.70\ lb/in.^2$
         $= 1.033\ kg/cm^2 = 1.013\ bar$

RADIOACTIVITY:

$1\ curie = 3.7 \times 10^{10}$ disintegrations per second
$1\ microcurie = 3.7 \times 10^4$ disintegrations per second

TEMPERATURE:

$F = (9/5)\ C + 32$
$F = (9/5)\ (K - 255.37)$
   $= (9/5)\ K - 459.67$
$R = (9/5)\ C + 491.67$

$R = F + 459.67;\ F = R - 459.67$

$C = (5/9)\ (F - 32)$
$K = (5/9)\ (F + 459.67)$
   $= (5/9)\ F + 255.37$
$C = (5/9)\ R - 273.15$

$K = C + 273.15;\ C = K - 273.15$

                                                  (continued)

---

* $1"wc = 1"wg$
† Normal atmosphere = 760 torr (mm Hg at 0 C)--not a "technical atmosphere", which is 736
   torr or $1\ kg/cm^2$. Subtract about 0.5 $lb/in.^2$ for each 1000 ft above sea level.

Table C.6. Unit equivalents *(continued)*

SI Metric To American
Metric To Metric

American To SI Metric
American To American

THERMAL CONDUCTIVITY:

1 W/m·°K = 0.5778 Btu·ft/ft²·hr·°F
   = 6.934 Btu·in./ft²·hr·°F
1 cal·cm/cm²·s·°C = 241.9 Btu·ft/ft²·hr·°F
   = 2903 Btu·in./ft²·hr·°F
   = 418.7 W/m·°K

1 Btu·ft/ft²·hr·°F = 1.730 W/m·°K
   = 1.488 kcal/m·h·°K
1 Btu·in./ft²·hr·°F = 0.1442 W/m·°K
1 Btu·ft/ft²·hr·°F = 0.004 134 cal·cm/cm²·s·°C
1 Btu·in./ft²·hr·°F = 0.000 344 5 cal·cm/cm²·s·°C

THERMAL DIFFUSIVITY:

1 m²/s = 38 760 ft²/hr
1 m²/h = 10.77 ft²/hr

1 ft²/hr = 0.000 025 8 m²/s = 0.0929 m²/h

TORQUE (BENDING MOMENT):

1 lb (force)·inch = 0.112 98 N·m

1 lb·in. = 1.152 kg·cm
1 lb·ft = 0.1383 kg·m

VELOCITY:

1 cm/s = 0.3937 in./sec = 0.032 81 ft/sec
   = 10.00 mm/s = 1.969 ft/min
1 m/s = 39.37 in./sec = 3.281 ft/sec
   = 196.9 ft/min = 2.237 mph
   = 3.600 km/h = 1.944 knot

1 in./sec= 25.4 mm/s = 0.0254 m/s
   = 0.0568 mph
1 ft/sec = 304.8 mm/s = 0.3048 m/s
   = 0.6818 mph
1 ft/min = 5.08 mm/s = 0.005 08 m/s
   = 0.0183 km/h
1 mph = 0.4470 m/s = 1.609 km/h
   = 1.467 ft/sec
1 knot = 0.5144 m/s
1 rpm = 0.1047 radians/sec

VISCOSITY, absolute, μ: (See also pp 28-30 of Vol. I or Appendix Tables E.1, E.2, E.3 of North
American's Incineration book.)

0.1 Pa·s = 1 dyne·s/cm² = 360 kg/m·h
   = 1 poise = 100 centipoise
   = 242.1 lb mass/ft·hr
   = 0.002 089 lb force·sec/ft²
1 kg/m·h = 0.672 lb/hr·ft = 0.002 78 g/cm·s
   = 0.000 005 81 lb force·sec/ft²

1 lb mass/hr·ft = 0.000 008 634
   lb force·sec/ft²
   = 0.413 centipoise
   = 0.000 413 Pa·s
1 lb force·sec/ft² = 115 800 lb mass/ft·hr
   = 47 880 centipoise
   = 47.88 Pa·s
1 reyn = 1 lb force·sec/in.²
   = 6.890 × 10⁶ centipoise

μ of water† = 1.124 centipoise
   = 2.72 lb mass/hr·ft
   = 2.349 10⁻⁵ lb·sec/ft²

μ of air† = 0.0180 centipoise
   = 0.0436 lb/hr·ft
   = 3.763 × 10⁻⁷ lb·sec/ft²

† At stp (60 F and 14.7 psia or 15.6 C and 760 mm Hg).

(continued)

Table C.6.   Unit equivalents *(continued)*

SI Metric To American Metric To Metric	American To SI Metric American To American

VOLUME:‡

1 cm³ (cc) = 0.000 001 00 m³
          = 0.0610 in.³ = 0.0338 US fluid oz
1 L (dm³) = 0.0010 m³ = 1000 cm³
          = 61.02 in.³ = 0.035 31 ft³
          = 0.2642 USgal

1 m³ = 1000 L = 1 000 000 cm³
     = 61 024 in.³ = 35.31 ft³
     = 220.0 Br gal
     = 6.290 bbl
     = 264.2 USgal
     = 1.308 yd³

1 in.³ = 16.39 cm³ = 0.000 016 39 m³
       = 0.016 39 L
1 ft³ = 1728 in.³ = 7.481 USgal
       = 6.229 Br gal
       = 28 320 cm³ = 0.028 32 m³ = 28.32 L
       = 62.427 lb of 39.4 F (4 C) water
       = 62.344 lb of 60 F (15.6 C) water

1 USgal = 3785 cm³ = 0.003 785 m³
        = 3.785 L = 231.0 in.³
        = 0.8327 Br gal = 0.1337 ft³
        = sp gr × 8.335 lb
        = 8.334 lb of 60 F (15.6 C) water
        = ¹/₄₂ barrel (oil)
1 Br gal = 277.4 in.³
         = 0.004 546 m³ = 4.546 L
         = 1.201 USgal
1 bbl, oil = 9702 in.³ = 5.615 ft³
           = 0.1590 m³ = 159.0 L
           = 42.00 USgal
           = 34.97 Br gal

VOLUME FLOW RATE:‡

1 cm³ (cc)/s = 1 × 10⁻⁶ m³/s
1 L/s = 1 × 10⁻³ m³/s
1 m³/h = 4.403 US gpm (gal/min)
       = 0.5887 ft³/min

1 gpm (gal/min) = 60.0 gph (gal/hr)
                = 0.016 67 gps (gal/sec)
                = 0.002 23 cfs (ft³/sec)
                = 8.022 cfh (ft³/hr)
                = 0.8326 Br gpm
                = 0.227 m³/h = 0.063 08 L/s
                = 1.429 bbl/hr = 34.29 bbl/day
1 gph (gal/hr) = 0.001 05 L/s
                = 0.000 037 1 cfs (ft³/sec)
1 cfm (ft³/min) = 6.18 Br gpm
                = 0.000 471 m³/s
1 cfs (ft³/sec) = 448.8 gpm
                = 22 250 Br gph

(continued)

‡ Volume equivalents are for the same temperature and pressure in both sets of units.
  A standard cubic foot or scfh is measured at 60 F and 29.92" Hg.
  A standard cubic metre or sm³/h is measured at 15 C and 760 mm Hg.
  A normal cubic metre or nm³/h is measured at 0 C and 760 mm Hg.

Table C.6. **Unit equivalents** *(concluded)*

SI Metric To American **Metric To Metric**	**American To SI Metric** **American To American**

WEIGHT, FORCE, MASS:

1 µg = 1.544 × 10⁻⁵ grains
     = 2.205 × 10⁻⁹ lb avdp mass
1 g = 0.035 27 oz avdp mass
1 kg mass = 1000 g mass = 10⁹ µg
       = 35.27 oz avdp mass
       = 2.205 lb avdp mass
1 kg force = 1000 g force = 9.807 N
       = 2.205 lb avdp force
1 metric ton (1 tonne) = 1000 kg = 2205 lb

1 oz avdp mass = 28.35 g = 0.028 35 kg
1 lb avdp mass = 453.6 g = 0.4536 kg
               = 4.536 × 10⁸ µg
1 lb avdp force = 0.4536 kg force
       = 4.448 N
1 lb avdp = 16 ounces (oz) = 7000 grains
1 short ton = 2000 lb = 907.2 kg
1 long ton = 2240 lb = 1015.9 kg

Table C.7. Temperature scale conversions. Formulas are in Appendix C.6. Examples: From the bold type column, 600 Fahrenheit is the same as 315.6 Celsius or 588.7 Kelvin. 600 C = 1112.0 F. 600 K = 620.3 F.

from F to K	from F to C	Convert F, C, or K	from C to F	from K to F	from F to K	from F to C	Convert F, C, or K	from C to F	from K to F
244.3	−28.9	**−20.0**	−4.0	–	599.8	326.7	**620.0**	1148.0	656.3
249.8	−23.3	**−10.0**	14.0	–	610.9	337.8	**640.0**	1184.0	692.3
255.4	−17.8	**0.0**	32.0	−459.7	622.0	348.9	**660.0**	1220.0	728.3
260.9	−12.2	**10.0**	50.0	−441.7	633.2	360.0	**680.0**	1256.0	764.3
266.5	−6.7	**20.0**	68.0	−423.7	644.3	371.1	**700.0**	1292.0	800.3
272.0	−1.1	**30.0**	86.0	−405.7	672.0	398.9	**750.0**	1382.0	890.3
277.6	4.4	**40.0**	104.0	−387.7	699.8	426.7	**800.0**	1472.0	980.3
283.2	10.0	**50.0**	122.0	−369.7	727.6	454.4	**850.0**	1562.0	1070.3
288.7	15.6	**60.0**	140.0	−351.7	755.4	482.2	**900.0**	1652.0	1160.3
294.3	21.1	**70.0**	158.0	−333.7	783.2	510.0	**950.0**	1742.0	1250.3
299.8	26.7	**80.0**	176.0	−315.7	810.9	537.8	**1000.0**	1832.0	1340.3
305.4	32.2	**90.0**	194.0	−297.7	838.7	565.6	**1050.0**	1922.0	1430.3
310.9	37.8	**100.0**	212.0	−279.7	866.5	593.3	**1100.0**	2012.0	1520.3
316.5	43.3	**110.0**	230.0	−261.7	894.3	621.1	**1150.0**	2102.0	1610.3
322.0	48.9	**120.0**	248.0	−243.7	922.0	648.9	**1200.0**	2192.0	1700.3
327.6	54.4	**130.0**	266.0	−225.7	949.8	676.7	**1250.0**	2282.0	1790.3
333.2	60.0	**140.0**	284.0	−207.7	977.6	704.4	**1300.0**	2372.0	1880.3
338.7	65.6	**150.0**	302.0	−189.7	1005.4	732.2	**1350.0**	2462.0	1970.3
344.3	71.1	**160.0**	320.0	−171.7	1033.2	760.0	**1400.0**	2552.0	2060.3
349.8	76.7	**170.0**	338.0	−153.7	1060.9	787.8	**1450.0**	2642.0	2150.3
355.4	82.2	**180.0**	356.0	−135.7	1088.7	815.6	**1500.0**	2732.0	2240.3
360.9	87.8	**190.0**	374.0	−117.7	1116.5	843.3	**1550.0**	2822.0	2330.3
366.5	93.3	**200.0**	392.0	−99.7	1144.3	871.1	**1600.0**	2912.0	2420.3
372.0	98.9	**210.0**	410.0	−81.7	1172.0	898.9	**1650.0**	3002.0	2510.3
373.2	100.0	**212.0**	413.6	−78.1	1199.8	926.7	**1700.0**	3092.0	2600.3
377.6	104.4	**220.0**	428.0	−63.7	1227.6	954.4	**1750.0**	3182.0	2690.3
383.2	110.0	**230.0**	446.0	−45.7	1255.4	982.2	**1800.0**	3272.0	2780.3
388.7	115.6	**240.0**	464.0	−27.7	1283.2	1010.0	**1850.0**	3362.0	2870.3
394.3	121.1	**250.0**	482.0	−9.7	1310.9	1037.8	**1900.0**	3452.0	2960.3
399.8	126.7	**260.0**	500.0	8.3	1338.7	1065.6	**1950.0**	3542.0	3050.3
405.4	132.2	**270.0**	518.0	26.3	1366.5	1093.3	**2000.0**	3632.0	3140.3
410.9	137.8	**280.0**	536.0	44.3	1394.3	1121.1	**2050.0**	3722.0	3230.3
416.5	143.3	**290.0**	554.0	62.3	1422.0	1148.9	**2100.0**	3812.0	3320.3
422.0	148.9	**300.0**	572.0	80.3	1449.8	1176.7	**2150.0**	3902.0	3410.3
427.6	154.4	**310.0**	590.0	98.3	1477.6	1204.4	**2200.0**	3992.0	3500.3
433.2	160.0	**320.0**	608.0	116.3	1505.4	1232.2	**2250.0**	4082.0	3590.3
438.7	165.6	**330.0**	626.0	134.3	1533.2	1260.0	**2300.0**	4172.0	3680.3
444.3	171.1	**340.0**	644.0	152.3	1560.9	1287.8	**2350.0**	4262.0	3770.3
449.8	176.7	**350.0**	662.0	170.3	1588.7	1315.6	**2400.0**	4352.0	3860.3
455.4	182.2	**360.0**	680.0	188.3	1616.5	1343.3	**2450.0**	4442.0	3950.3
460.9	187.8	**370.0**	698.0	206.3	1644.3	1371.1	**2500.0**	4532.0	4040.3
466.5	193.3	**380.0**	716.0	224.3	1699.8	1426.7	**2600.0**	4712.0	4220.3
472.0	198.9	**390.0**	734.0	242.3	1755.4	1482.2	**2700.0**	4892.0	4400.3
477.6	204.4	**400.0**	752.0	260.3	1810.9	1537.8	**2800.0**	5072.0	4580.3
488.7	215.6	**420.0**	788.0	296.3	1866.5	1593.3	**2900.0**	5252.0	4760.3
499.8	226.7	**440.0**	824.0	332.3	1922.0	1648.9	**3000.0**	5432.0	4940.3
510.9	237.8	**460.0**	860.0	368.3	1977.6	1704.4	**3100.0**	5612.0	5120.3
522.0	248.9	**480.0**	896.0	404.3	2033.2	1760.0	**3200.0**	5792.0	5300.3
533.2	260.0	**500.0**	932.0	440.3	2088.7	1815.6	**3300.0**	5972.0	5480.3
544.3	271.1	**520.0**	968.0	476.3	2144.3	1871.1	**3400.0**	6152.0	5660.3
555.4	282.2	**540.0**	1004.0	512.3	2199.8	1926.7	**3500.0**	6332.0	5840.3
566.5	293.3	**560.0**	1040.0	548.3	2255.4	1982.2	**3600.0**	6512.0	6020.3
577.6	304.4	**580.0**	1076.0	584.3	2310.9	2037.8	**3700.0**	6692.0	6200.3
588.7	315.6	**600.0**	1112.0	620.3	2366.5	2093.3	**3800.0**	6872.0	6380.3

Figure C.8.   Temperature scale alignment chart. (The fahrenheit column is duplicated to make it easy to align a straight-edge across the equivalent columns.) Formulas are in Appendix Table C.6.

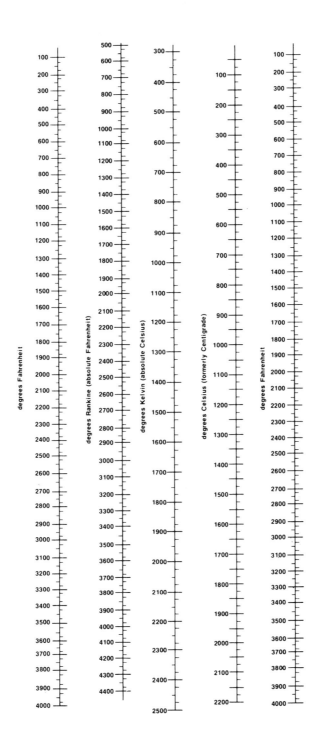

Table C 9. Pyrometric cone equivalents[1] or end points, the temperatures at which the cones tip even with the top of the plaque for Orton Standard Pyrometric Cones when heated at the indicated rates in an air atmosphere.[2] Other heating rates and atmospheres can cause deviations from these values.

The temperatures listed below for number 022 to 20 cones inclusive are for the large or standard cone. Small cones in these numbers have the same composition as the larger cones but require a higher temperature to cause deformation. The temperatures listed for number 12 to 42 cones inclusive are for the small or PCE cone. Large cones in these numbers have the same composition as the small cones but will deform at a lower temperature. Large cones are 2½" long and small cones are 1⅛" long.

End point, temperatures (Cone Number)	Rate of temperature rise[3] 108°F/hr or 60°C/h — F	C	270°F/hr or 150°C/h — F	C
022	1069	576	1086	586
021	1116	602	1137	614
020	1157	625	1175	635
019	1234	668	1261	683
018	1285	696	1290	717
017	1341	727	1377	747
016	1407	764	1458	790
015	1454	790	1479	804
014	1533	834	1540	838
013	1596	869	1566	852
012	1591	866	1623	884
011	1627	886	1641	894
010	1629	887	1641	894
09	1679	915	1693	923
08	1733	945	1751	955
07	1783	973	1803	984
06	1816	991	1830	999
05	1888	1031	1915	1046
04	1922	1050	1940	1060
03	1987	1086	2014	1101
02	2014	1101	2048	1120
01	2043	1117	2079	1137
1	2077	1136	2109	1154
2	2088	1142	2124	1162
3	2106	1152	2134	1168
4	2134	1168	2167	1186
5	2151	1177	2185	1196
6	2194	1201	2232	1222
7	2219	1215	2264	1240
8	2257	1236	2305	1263
9	2300	1260	2336	1280
10	2345	1285	2381	1305
11	2361	1294	2399	1315
12	2383	1306	2419	1326

Left-side groupings: **Soft** (022–011); **Low temperature (iron free · white)[5]** (010–3); **Intermediate temperature** (4–12).

End point, temperatures (Cone Number)	Rate of temperature rise[3] 270°F/hr or 150°C/h — F	C
12	2439	1337
13	2460	1349
14	2548	1398
15	2606	1430
16	2716	1491
17	2754	1512
18	2772	1522
19	2806	1541
20	2847	1564
23	2921	1605
26	2950	1621
27	2984	1640
28	2995	1646
29	3018	1659
30	3029	1665
31	3061	1683
31½	3090	1699
32	3123	1717
32½	3135	1724
33	3169	1743
34	3205	1763
35	3245	1785
36	3279	1804
37	3308	1820
38	3362	1850[4]
39	3389	1865[4]
40	3425	1885[4]
41	3578	1970[4]
42	3659	2015[4]

Right-side grouping: **High temperature, "PCE"** (Cone Number).

[1] Courtesy of Edward Orton, Jr., Ceramic Foundation, 9-64.
[2] Determined at the National Bureau of Standards by H. P. Beerman (see Journal of the American Ceramic Society, vol. 39, 1956), except those marked "4".
[3] During the last several hundred degrees of temperature rise.
[4] Approximate.
[5] Iron-free cones have the same deformation temperatures as the red equivalents when fired at 60°C/h in air.

Table C.10 US. Millivolt-to-Fahrenheit conversions for thermocouples

To convert an observed millivolt reading to a temperature, first find the cold junction correction from *Part a* of this table. Add this to the observed millivolt reading. Opposite this corrected millivolt reading in *Part b* of this table, read the temperature in F. Example: an iron-constantan (J) thermocouple, with its cold junction at 70 F is generating 40 millivolts. Solution: The correction is 1.076 millivolts. The total corrected mv is therefore 40 + 1.076 = 41.076. By interpolation in the J column of *Part b*, the temperature of the hot junction is $1317 + \left(\frac{41.076-40}{45-40}\right) \times (1458-1317) = 1317 + 30.3 = 1347.3\text{F}$.

Part a. **Cold junction correction millivolts**									
	Thermocouple metals and ISA lettter designations								
	copper constantan **T**	chromel constantan **E**	iron constantan **J**	chromel alumel **K**	W/W26 **W**	W5/W26 **W5**	Pt/Pt 13% **R**	Pt/Pt 10% **S**	Pt6/Pt30 **B**
Cold junction temp, F									
32	0	0	0	0	0	0	0	0	0
40	0.173	0.262	0.224	0.176	0.006	0.059	0.024	0.024	−0.001
50	0.391	0.591	0.507	0.397			0.054	0.055	−0.002
60	0.611	0.924	0.791	0.619	0.026	0.211	0.086	0.087	−0.002
70	0.834	1.259	1.076	0.843			0.118	0.119	−0.003
80	1.060	1.597	1.363	1.068	0.050	0.365	0.150	0.152	−0.002
90	1.288	1.937	1.652	1.294			0.184	0.186	−0.002
100	1.518	2.281	1.942	1.520	0.079	0.522	0.218	0.221	−0.001

Part b. **Temperature F**									
	Thermocouple metals and ISA lettter designations								
	copper constantan **T**	chromel constantan **E**	iron constantan **J**	chromel alumel **K**	W/W26 **W**	W5/W26 **W5**	Pt/Pt 13% **R**	Pt/Pt 10% **S**	Pt6/Pt30 **B**
Total millivolts, including correction for cold junction									
0	32.0	32.0	32.0	32.0	32.0	32.0	32.0	32.0	32.0
1	77.3	62.3	67.3	77.0	385.6	158.8	293.0	295.5	841.2
2	120.5	91.8	102.0	121.0	573.4	274.3	496.8	507.8	1176
3	161.7	120.7	136.1	164.5	707.6	382.6	680.5	702.9	1435
4	201.3	148.9	169.8	207.9	834.1	485.9	853.4	889.3	1661
5	239.5	176.4	203.1	251.6	949.4	585.4	1019	1070	1869
6	276.6	203.5	236.2	295.9	1057	682.4	1178	1245	2053
7	312.6	230.1	269.0	340.7	1160	778.3	1331	1415	2230
8	348.2	256.2	301.7	385.9	1258	871.1	1479	1580	2398
9	382.3	282.0	334.3	430.8	1353	964.0	1623	1740	2260
10	416.0	307.4	367.0	475.3	1445	1056	1763	1897	2717
11	449.1	332.5	399.3	519.3	1535	1148	1899	2049	2872
12	481.6	357.4	431.7	563.0	1624	1241	2032	2200	3026
13	513.4	381.8	464.1	603.7	1712	1333	2163	2349	3181
14	545.2	406.1	496.5	649.4	1798	1426	2292	2497	
15	576.3	430.2	529.0	692.4	1884	1520	2420	2646	
16	607.1	454.1	561.4	735.1	1990	1614	2548	2796	
17	637.5	477.8	593.9	777.7	2055	1710	2675	2947	
18	667.5	501.4	626.5	820.3	2140	1807	2803	3101	
19	697.3	524.8	659.1	862.6	2226	1905	2933		
20	726.6	548.1	691.7	904.9	2311	2005	3063		
22		594.3	757.0	989.4	2483	2211			
24		640.1	822.2	1074	2658	2425			
26		685.6	887.2	1159	2837	2649			
28		730.8	951.6	1244	3022	2884			
30		775.7	1015	1330		3132			
32		820.5	1078	1416					
34		865.2	1140	1504					
36		909.8	1200	1593					
38		954.3	1259	1683					
40		998.8	1317	1774					
45		1110	1458	2007					
50		1222	1598	2250					
55		1335	1744						
60		1449	1895						
65		1564	2050						
70		1681							
75		1799							

Data checked 6-94 and found in agreement with ASTM E 230-93.

To find the electromotive force equivalent to a known temperature, it may be more convenient to interpolate from the tables of North American's Mfg. Company's Handbook Supplement 175.

Table C.11 Metric.   Millivolt-to-Celsius conversions for thermocouples

To convert an observed millivolt reading to a temperature, first find the cold junction correction from *Part a* of this table. Add this to the observed millivolt reading. Opposite this corrected millivolt reading in *Part b* of this table, read the temperature in C. Example: an iron-constantan (J) thermocouple, with its cold junction at 30 C is generating 30 millivolts. Solution: The correction is 1.536 millivolts. The total corrected mv is therefore 30 + 1.536 = 31.536. By interpolation in the J column of *Part b*, the temperature of the hot junction is $546.3 + \left(\dfrac{31.536 - 30}{32 - 30} \right) \times (581.2 - 546.3) = 546.3 + 26.8 = 573.1$ C.

Part a. Cold junction correction millivolts	copper constantan T	chromel constantan E	iron constantan J	chromel alumel K	Pt/Pt 13% R	Pt/Pt 10% S	Pt6/Pt30 B
0	0	0	0	0	0	0	0
10	0.391	0.591	0.507	0.397	0.054	0.0552	−0.002
20	0.789	1.192	1.019	0.798	0.111	0.1128	−0.003
30	1.196	1.801	1.536	1.203	0.171	0.1727	−0.002
40	1.611	2.419	2.058	1.611	0.232	0.0010	−0.000
50	2.035	3.047	2.585	2.022	0.296	0.2986	−0.002
60	2.467	3.683	3.115	2.436	0.363	0.3045	−0.006
70	2.909	4.200	0.010	0.000	0.101	0.4525	−0.011

Cold junction temp, C (left label for Part a rows)

Part b. Temperature C	copper constantan T	chromel constantan E	iron constantan J	chromel alumel K	Pt/Pt 13% R	Pt/Pt 10% S	Pt6/Pt30 B
0	0.00	0.00	0.00	0.00	0.00	0.00	42.15
1	25.20	16.83	19.63	25.00	145.0	146.4	449.6
2	49.18	33.24	38.88	49.46	258.2	264.3	634.2
3	72.06	49.26	57.84	73.60	360.3	372.7	779.6
4	94.04	64.92	76.55	97.70	456.3	476.3	905.2
5	115.3	80.26	95.07	122.0	548.2	576.6	1018
6	135.9	95.29	113.4	146.6	636.6	673.9	1123
7	155.9	110.1	131.7	171.5	721.8	768.1	1221
8	175.5	124.6	149.9	196.6	804.2	859.7	1314
9	194.6	138.9	168.0	221.6	884.0	948.9	1404
10	213.3	153.0	186.0	246.3	961.7	1036	1492
11	231.7	166.9	204.0	270.8	1037	1121	1578
12	249.8	180.7	222.0	295.0	1111	1204	1663
13	267.6	194.3	240.0	319.1	1184	1287	1749
14	285.1	207.8	258.1	343.0	1256	1370	
15	302.4	221.2	276.1	366.9	1327	1452	
16	319.5	234.5	294.1	390.6	1398	1535	
17	336.4	247.7	312.2	416.3	1468	1619	
18	353.1	260.8	330.3	437.9	1540	1705	
19	369.6	273.8	348.4	461.5	1611		
20	385.9	286.7	366.5	485.0	1685		
22			402.8	531.9			
24			439.0	578.8			
26			475.1	625.9			
28			510.9	673.2			
30			546.3	720.8			
32			581.2	769.0			
34			615.4	817.7			
36			648.9	867.0			
38			681.7	916.9			
40			713.9	967.0			

Total millivolts, including correction for cold junction (left label for Part b rows)

Table C.12. Equivalent air/fuel ratio specifications

Confusion in combustion communications may result because some people think in terms of air/fuel ratios, others fuel/air ratios; some in weight ratios, others in volume ratios; some in mixed units (such as normal cubic metres of air per metric tonne of coal or ft³ air/gal oil). Unitless ratios such as % air (% aeration), % excess air, % deficiency of air, or equivalence ratio are preferred.

"Stoichiometric" ratio (also called correct or ideal ratio or "on-ratio") refers to the chemically correct proportion of air to fuel to completely burn the fuel with no excess air (typically 10:1 for natural gas). Tables 2.1b and 2.12c (Volume I) list correct air/fuel ratios for a number of common fuels. *"100% air"* is the correct (stoichiometric) amount. 200% air is twice as much as necessary, or 100% excess air.

Equivalence ratio, ϕ, is the actual amount of fuel expressed as a fraction percent of the stoichiometrically correct amount of fuel. The $\phi = 0.9$ is lean; $\phi = 1.1$ is rich; $\phi = 1.0$ is "on-ratio."

In Europe, "stoichiometric ratio" = $1/\phi$ = "S.R."

$$\phi = \frac{100}{\%XS + 100} = \frac{1}{1 - \%def}$$

% air = $100/\phi$ = %XS + 100 = 100 – %def

%XS = %air – 100 = $\frac{1-\phi}{\phi} \times 100$

%def = $100 - \%air = \frac{\phi - 1}{\phi} \times 100$

% rich = $(\phi - 1) \times 100$

(for rich conditions only)

The table below lists a number of equivalent terms for convenience to save the above calculations in converting values from one "language" to another.

	ϕ	% air	% def	% rich	%XS air		ϕ	% air	%XS
Fuel Rich	2.50	40	60	150	–60		*(continued from table at left)*		
(air lean)	1.67	60	40	66.7	–40		0.62	160	60
	1.25	80	20	25	–20		0.56	180	80
	1.11	90	10	11.1	–10		0.50	200	100
	1.056	95	5	5.56	–5		0.40	250	150
Stoichiometric	1.000	100	0	0	0		0.33	300	200
Fuel Lean	0.952	105	—		5	**Fuel Lean (air rich)**	0.25	400	300
(air rich)	0.909	110	—		10		0.20	500	400
	0.833	120	—		20		0.167	600	500
	0.769	130	—		30		0.091	1100	1000
	0.714	140	—		40		0.048	2100	2000

Examples: If $\phi = 0.9$, then $^9/_{10}$ of the stoichiometrically correct amount of fuel is supplied or %XS = $(1 - 0.9)/0.9 = 0.1/0.9 = 11\%$; or %air = $100/0.9 = 111\%$. For a *natural gas* requiring 10.00 ft³ air/ft³ gas for stoichiometric combustion, this means an air/fuel ratio of 111% of 10.00 = 11.10 ft³ air/ft³ gas, or a fuel/air ratio of 1/11.10 = 0.0901 ft³ gas/ft³ air. For a *heavy oil* requiring 1492 ft³ air/gal for stoichiometric combustion, this means an air/fuel ratio 111% of 1492 or 1656 ft³/gal.

If $\phi = 1.25$, then 125% of the stoichiometrically correct amount of fuel supplied, 100/1.25 = 80% aeration, or % deficiency of air (1.25 – 1.0)/1.25 = 20%. For a *producer gas* requiring 1.18 m³ air/kg gas for proper combustion, this means 1.18/1.25 = 0.944 m³ air/kg fuel.

Table C.13 US.   Pipe capacities, velocities, pressure losses

Pipe flow, scfh			Mean pipe velocity			Pressure loss "wc/100 ft w/40 fps*
Nominal pipe size, inches	Pipe schedule	Cross-sectional area of pipe, ft²	40 fps	60 fps	80 fps	
½	40	0.0021	303	456	608	32.8
¾	40	0.0037	534	801	1 068	25.0
1	40	0.0060	864	1 300	1 730	19.5
1¼	40	0.0104	1 500	2 250	3 000	13.7
1½	40	0.0141	2 040	3 050	4 070	10.6
2	40	0.0233	3 360	5 030	6 710	6.8
2½	40	0.0332	4 780	.7 160	9 570	5.7
3	40	0.0513	7 390	11 100	14 800	3.9
4	40	0.0884	12 700	19 100	25 500	2.8
6	40	0.201	28 900	43 300	57 800	1.8
8	30	0.355	51 000	76 700	102 000	1.4
10	30	0.560	80 700	121 000	161 000	1.1
12	30	0.797	115 000	172 000	230 000	0.95
14	10	0.994	143 000	215 000	286 000	0.80
16	10	1.31	189 000	283 000	378 000	0.70
18	10	1.67	240 000	361 000	481 000	0.62
20	10	2.07	299 000	447 000	596 000	0.56
22	10	2.64	380 000	570 000	760 000	0.51
24	10	3.01	434 000	650 000	867 000	0.47
30	10	4.71	678 000	1 020 000	1 360 000	0.376
36	10	6.88	990 000	1 490 000	1 980 000	0.312
42	10	9.38	1 350 000	2 020 000	2 700 000	0.268
48	10	12.5	1 800 000	2 700 000	3 600 000	0.245

For **metering** with 4"wc differential and β of 0.7, use the following for maximum capacity:

cold air .... 40 fps
natural gas .... slightly less than 60 fps
coke oven gas .... slightly more than 60 fps

---

*   Multiplying factors for pipe velocities other than 40 ft/second

	Velocities in ft/sec. (COLD AIR)								
	20	30	40	50	60	70	80	90	100
Volume factor	0.5	0.75	1.0	1.25	1.5	1.75	2.0	2.25	2.5
Pressure drop factor	1.25	0.56	1.0	1.56	2.25	3.06	4.00	5.06	6.25

Table D.1. Dimensions of ANSI malleable threaded pipe fittings, Class 150 and 300 (per ANSI B16.3-1977)

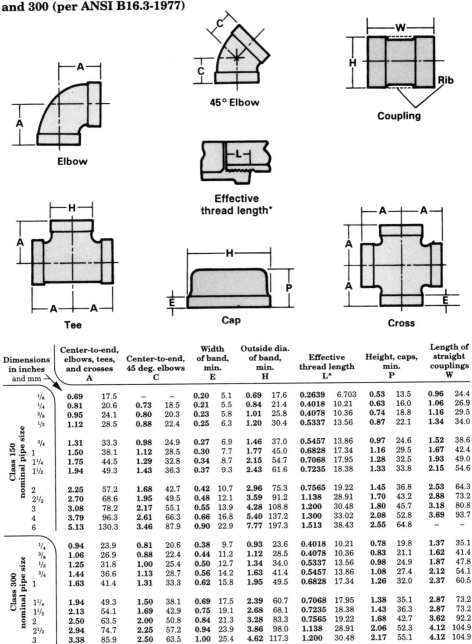

Dimensions in inches and mm	Center-to-end, elbows, tees, and crosses A		Center-to-end, 45 deg. elbows C		Width of band, min. E		Outside dia. of band, min. H		Effective thread length L*		Height, caps, min. P		Length of straight couplings W	
Class 150 nominal pipe size														
1/8	0.69	17.5	–	–	0.20	5.1	0.69	17.6	0.2639	6.703	0.53	13.5	0.96	24.4
1/4	0.81	20.6	0.73	18.5	0.21	5.5	0.84	21.4	0.4018	10.21	0.63	16.0	1.06	26.9
3/8	0.95	24.1	0.80	20.3	0.23	5.8	1.01	25.8	0.4078	10.36	0.74	18.8	1.16	29.5
1/2	1.12	28.5	0.88	22.4	0.25	6.3	1.20	30.4	0.5337	13.56	0.87	22.1	1.34	34.0
3/4	1.31	33.3	0.98	24.9	0.27	6.9	1.46	37.0	0.5457	13.86	0.97	24.6	1.52	38.6
1	1.50	38.1	1.12	28.5	0.30	7.7	1.77	45.0	0.6828	17.34	1.16	29.5	1.67	42.4
1 1/4	1.75	44.5	1.29	32.8	0.34	8.7	2.15	54.7	0.7068	17.95	1.28	32.5	1.93	49.0
1 1/2	1.94	49.3	1.43	36.3	0.37	9.3	2.43	61.6	0.7235	18.38	1.33	33.8	2.15	54.6
2	2.25	57.2	1.68	42.7	0.42	10.7	2.96	75.3	0.7565	19.22	1.45	36.8	2.53	64.3
2 1/2	2.70	68.6	1.95	49.5	0.48	12.1	3.59	91.2	1.138	28.91	1.70	43.2	2.88	73.2
3	3.08	78.2	2.17	55.1	0.55	13.9	4.28	108.8	1.200	30.48	1.80	45.7	3.18	80.8
4	3.79	96.3	2.61	66.3	0.66	16.8	5.40	137.2	1.300	33.02	2.08	52.8	3.69	93.7
6	5.13	130.3	3.46	87.9	0.90	22.9	7.77	197.3	1.513	38.43	2.55	64.8	–	–
Class 300 nominal pipe size														
1/4	0.94	23.9	0.81	20.6	0.38	9.7	0.93	23.6	0.4018	10.21	0.78	19.8	1.37	35.1
3/8	1.06	26.9	0.88	22.4	0.44	11.2	1.12	28.5	0.4078	10.36	0.83	21.1	1.62	41.4
1/2	1.25	31.8	1.00	25.4	0.50	12.7	1.34	34.0	0.5337	13.56	0.98	24.9	1.87	47.8
3/4	1.44	36.6	1.13	28.7	0.56	14.2	1.63	41.4	0.5457	13.86	1.08	27.4	2.12	54.1
1	1.63	41.4	1.31	33.3	0.62	15.8	1.95	49.5	0.6828	17.34	1.26	32.0	2.37	60.5
1 1/4	1.94	49.3	1.50	38.1	0.69	17.5	2.39	60.7	0.7068	17.95	1.38	35.1	2.87	73.2
1 1/2	2.13	54.1	1.69	42.9	0.75	19.1	2.68	68.1	0.7235	18.38	1.43	36.3	2.87	73.2
2	2.50	63.5	2.00	50.8	0.84	21.3	3.28	83.3	0.7565	19.22	1.68	42.7	3.62	92.2
2 1/2	2.94	74.7	2.25	57.2	0.94	23.9	3.86	98.0	1.138	28.91	2.06	52.3	4.12	104.9
3	3.38	85.9	2.50	63.5	1.00	25.4	4.62	117.3	1.200	30.48	2.17	55.1	4.12	104.9

* See Table D.3 for normal thread engagements for tight joints.

Table D.2. Dimensions of butt-welding pipe fittings

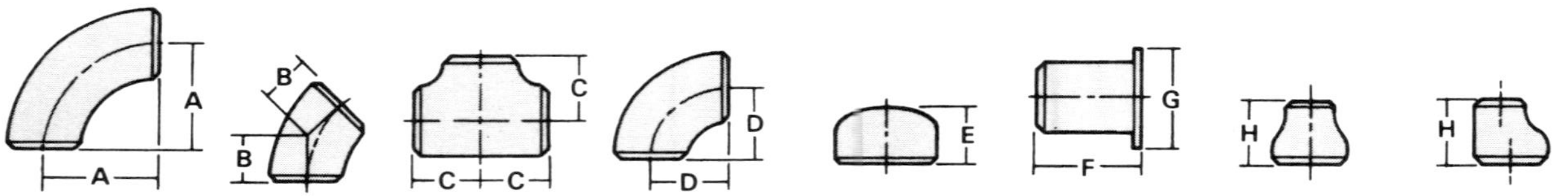

Nominal pipe size — Dimensions in inches and millimetres	90° long radius ell		45° long radius ell		Straight tee		90° short radius ell		Cap		Lap joint stub end		Concentric reducer		Eccentric reducer	
	A		**B**		**C**		**D**		**E**		**F**		**G**		**H**	
1	1½	38.10	⅞	22.23	1½	38.10	1	25.40	1½	38.10	4	101.6	2	50.80	2	50.80
1¼	1⅞	47.63	1	25.40	1⅞	47.63	1¼	31.75	1½	38.10	4	101.6	2½	63.50	2	50.80
1½	2¼	57.15	1⅛	28.58	2¼	57.15	1½	38.10	1½	38.10	4	101.6	2⅞	73.02	2½	63.50
2	3	76.20	1⅜	34.93	2½	63.50	2	50.80	1½	38.10	6	152.4	3⅝	92.08	3	76.20
2½	3¾	95.25	1¾	44.45	3	76.20	2½	63.50	1½	38.10	6	152.4	4⅛	104.8	3½	88.90
3	4½	114.3	2	50.80	3⅜	85.73	3	76.20	2	50.80	6	152.4	5	127.0	3½	88.90
4	6	152.4	2½	63.50	4⅛	104.8	4	101.6	2½	63.50	6	152.4	6³/₁₆	157.2	4	101.6
6	9	228.6	3¾	95.25	5⅝	142.9	6	152.4	3½	88.90	8	203.2	8½	215.9	5½	139.7
8	12	304.8	5	127.0	7	177.8	8	203.2	4	101.6	8	203.2	10⅝	269.9	6	152.4
10	15	381.0	6¼	158.8	8½	215.9	10	254.0	5	127.0	10	254.0	12¾	323.9	7	177.8
12	18	457.2	7½	190.5	10	254.0	12	304.8	6	152.4	10	254.0	15	381.0	8	203.2
14	21	533.4	8¾	222.3	11	279.4	14	355.6	6½	165.1	12	304.8	16¼	412.8	13	330.2

(continued)

Table D.2. Dimensions of butt-welding pipe fittings *(concluded)*

Nominal pipe size	90° long radius ell A	45° long radius ell B	Straight tee C	90° short radius ell D	Cap E	Lap joint stub end F	Concentric reducer G	Eccentric reducer H
16	24 / 609.6	10 / 254.0	12 / 304.8	16 / 406.4	7 / 177.8	12 / 304.8	18½ / 469.9	14 / 355.6
18	27 / 685.8	11¼ / 285.8	13½ / 342.9	18 / 457.2	8 / 203.2	12 / 304.8	21 / 533.4	15 / 381.0
20	30 / 762.0	12½ / 317.5	15 / 381.0	20 / 508.0	9 / 228.6	12 / 304.8	23 / 584.2	20 / 508.0
22	33 / 838.2	—	16½ / 419.1	22 / 558.8	10 / 254.0	—	—	20 / 508.0
24	36 / 914.4	15 / 381.0	17 / 431.8	24 / 609.6	10½ / 266.7	12 / 304.8	27½ / 698.5	20 / 508.0
26	39 / 990.6	—	19½ / 495.3	26 / 660.4	10½ / 266.7	—	—	—
28	42 / 1067	—	20½ / 520.7	28 / 711.2	10½ / 266.7	—	—	—
30	45 / 1143	18½ / 469.9	22 / 558.8	30 / 762.0	10½ / 266.7	—	—	24 / 609.6
32	48 / 1219	—	23½ / 596.9	32 / 812.8	10½ / 266.7	—	—	—
34	51 / 1296	—	25 / 635.0	34 / 863.6	10½ / 266.7	—	—	—
36	54 / 1372	22¼ / 565.2	26½ / 673.1	36 / 914.4	10½ / 266.7	—	—	24 / 609.6
42	63 / 1600	26 / 680.4	30 / 762.0	—	12 / 304.8	—	—	24 / 609.6

Dimensions in inches and millimetres

Table D.3. Thread engagements and lengths of pipe nipples. For all other nipples, specify length × pipe size.

Nominal pipe size		Normal thread engagement for a tight joint†		Nipple lengths			
				Close nipple		Short nipple	
$1/8$"	6mm	0.250"	6.35mm	$3/4$"	20mm	$1^1/2$"	40mm
$1/4$	8	0.375	9.56	$7/8$	22	$1^1/2$	40
$3/8$	10	0.375	9.56	1	25	$1^1/2$	40
$1/2$	15	0.500	12.7	$1^1/8$	29	$1^1/2$	40
$3/4$	20	0.562	14.3	$1^3/8$	35	$1^1/2$	40
1	25	0.687	17.5	$1^1/2$	40	2	50
$1^1/4$	32	0.687	17.5	$1^5/8$	41	2	50
$1^1/2$	40	0.687	17.5	$1^3/4$	44	2	50
2	50	0.750	19.1	?	50	$2^1/2$	65
$2^1/2$	65	0.937	23.8	$2^1/2$	65	3	80
3	80	1.000	25.4	$2^5/8$	67	0	00
4	100	1.125	28.6	$2^7/8$	73	4	100
6	150	1.312	33.3	$3^1/8$	79	$4^1/2$	114
8	200	1.437	36.5	$3^1/2$	89	5	127
10	250	1.625	41.3	$3^7/8$	98	5	127
12	300	1.750	44.5	$4^1/2$	114	6	150

† Not the same as dimension L in Table D.1.

Table D.4. Thermal expansion and contraction of metals[a]

Length change, inches per 100 ft[f]	−325F −198C	−150F −101C	−50F −46C	70F 21	Temperature range,[b] 70 F (21 C) to 200F 93	300F 149	400F 204	500F 260	600F 316	700F 371	800F 427	900F 482	1000F 538	1100F 593	1200F 649	1300F 704	1400F 760C
Aluminum	−4.68	−2.88	−1.67	0	2.00	3.66	5.39	7.17	9.03	–	–	–	–	–	–	–	–
Brass	−3.88	−2.24	−1.29	0	1.52	2.76	4.05	5.40	6.80	8.26	9.78	11.35	12.98	14.65	16.39	–	–
Bronze	−3.98	−2.31	−1.32	0	1.56	2.79	4.05	5.33	6.64	7.95	9.30	10.68	12.05	13.47	14.92	–	–
Copper	–	–	–	0	1.51	2.67	3.88	5.11	6.40	7.67	9.00	–	–	–	–	–	–
Copper-nickel	−3.15	−1.95	−1.13	0	1.33	2.40	3.52	–	–	–	–	–	–	–	–	–	–
Monel (67Ni, 30Cu)	−2.62	−1.79	–	0	1.22	2.21	3.25	4.33	5.46	6.64	7.85	9.12	10.42	11.77	13.15	14.58	16.02
(66Ni, 29CuAl)	−2.53	−1.70	−0.98	0	1.17	2.12	3.13	4.17	5.28	6.43	7.62	8.86	10.16	11.50	13.00	14.32	15.78
Iron, cast gray	–	–	–	0	0.90	1.64	2.42	3.24	4.11	5.03	5.98	6.97	8.02	–	–	–	–
wrought	−2.70	−1.67	−0.96	0	1.14	2.06	3.01	3.99	5.01	6.06	7.12	8.26	9.36	–	–	–	–
Steel, carbon, carbon-moly, low-Cr (through 3% Cr)	−2.37	−1.45	−0.84	0	0.99	1.82	2.70	3.62	4.60	5.63	6.70	7.81	8.89	10.04	11.10	12.22	13.34
intermed. alloys (5 CrMo through 9 CrMo)	−2.22	−1.37	−0.79	0	0.94	1.71	2.50	3.35	4.24	5.14	6.10	7.07	8.06	9.05	10.00	11.06	12.05
austenitic stainless steels[c] (18 Cr, 8 Ni)	−3.85	−2.27	–	0	1.46	2.61	3.80	5.01	6.24	7.50	8.80	10.12	11.48	12.84	14.20	15.56	16.92
25 Cr, 20 Ni[d]	−2.04	−1.24	−0.98	0	1.21	2.18	3.20	4.24	5.33	6.44	7.60	8.78	9.95	11.12	12.31	13.46	14.65
straight Cr stainless steels[e] (12Cr, 17Cr, 27Cr)	−3.00	−1.81	−0.72	0	0.86	1.56	2.30	3.08	3.90	4.73	5.60	6.49	7.40	8.31	9.20	10.11	11.01

[a] Partially abstracted from ANSI B31.3–"Chemical Plant and Petroleum Refinery Piping".
[b] This does not imply that the materials are suitable for all the temperature ranges shown—see Table E.5.
[c] 300 series stainless steels.
[d] 310 stainless.
[e] 400 series stainless steels.
[f] or multiply (inches expansion per 100 ft) by 0.833 to obtain (mm expansion per metre).

Figure D.5.   Approximate reversible thermal expansion of refractories
(Courtesy of G. Turton of K.T.G. Glassworks Technology, Pittsburgh, PA from his article in the Sept. 1980 issue of GLASS INDUSTRY.)

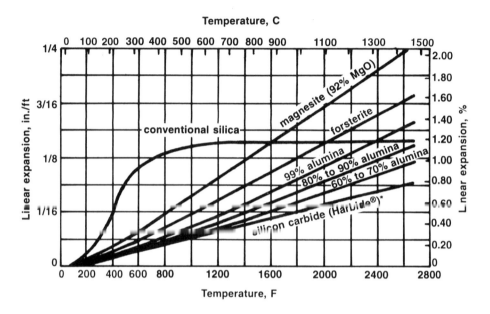

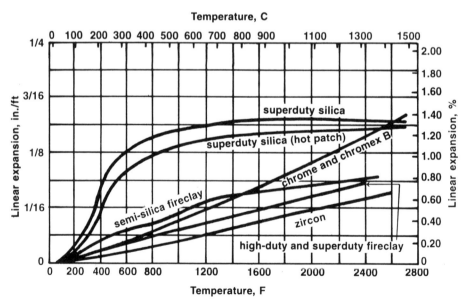

* Harbide® is a registered trademark of Harbison-Walker Refractories division of Dresser Industries.

Table D.6 US. Pipe flange template data

Nominal pipe size, in.	ANSI Class 150 flanges per B16.5-1977 (**bold**) Old ASA 125 psi (*italics*)					Riveted Pipe Manufacturers (RPM) flanges			
	od of flange, in.	Diameter of bolt circle, in.	Diameter of bolt holes, in.	Diameter of bolts, in.	No. of bolts	od of flange, in.	Diameter of bolt circle, in.	Diameter of bolts, in.	No. of bolts
3	**7 1/2**	**6**	**0.75**	5/8	**4**	6	4 3/4	1/2	4
4	**9**	**7 1/2**	**0.75**	5/8	**8**	7	5 15/16	1/2	8
6	**11**	**9 1/2**	**0.88**	3/4	**8**	9	7 7/8	1/2	8
8	**13 1/2**	**11 3/4**	**0.88**	3/4	**8**	11	10	1/2	8
10	**16**	**14 1/4**	**1.00**	7/8	**12**	14	12 1/4	5/8	12
12	**19**	**17**	**1.00**	7/8	**12**	16	14 1/4	5/8	12
14	**21**	**18 3/4**	**1.12**	1	**12**	18	16 1/4	5/8	12
16	**23 1/2**	**21 1/4**	**1.12**	1	**16**	21 1/4	19 1/4	5/8	16
18	**25**	**22 3/4**	**1.25**	1 1/8	**16**	23 1/4	21 1/4	5/8	16
20	**27 1/2**	**25**	**1.25**	1 1/8	**20**	25 1/4	23 1/8	5/8	20
22	–	–	–	–	–	28 1/4	26	5/8	20
24	**32**	**29 1/2**	**1.38**	1 1/4	**20**	30	27 3/4	5/8	20
26	*34 1/4*	*31 3/4*		*1 1/4*	*24*	32	29 3/4	3/4	24
28	*36 1/2*	*34*		*1 1/4*	*28*	34	31 3/4	3/4	28
30	*38 3/4*	*36*		*1 1/4*	*28*	36	33 3/4	3/4	28
32	*41 3/4*	*38 1/2*		*1 1/2*	*28*	38	35 3/4	3/4	28
34	*43 3/4*	*40 1/2*		*1 1/2*	*32*	40	37 3/4	3/4	32
36	*46*	*42 3/4*		*1 1/2*	*32*	42	39 3/4	3/4	32
38	*48 3/4*	*45 1/4*		*1 1/2*	*32*	44	41 3/4	3/4	32
40	*50 3/4*	*47 1/4*		*1 1/2*	*36*	46	43 3/4	3/4	36
42	*53*	*49 1/2*		*1 1/2*	*36*	48	45 3/4	3/4	36
44	*55 1/4*	*51 3/4*		*1 1/2*	*40*	50	47 3/4	3/4	40
46	*57 1/4*	*53 3/4*		*1 1/2*	*40*	52	49 3/4	3/4	40
48	*59 1/2*	*56*		*1 1/2*	*44*	54	51 3/4	3/4	44

Figure D.7.   Schematic piping symbols for combustion systems

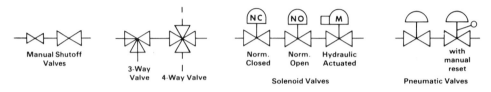

Manual Shutoff Valves

3-Way Valve   4-Way Valve

NC Norm. Closed   NO Norm. Open   M Hydraulic Actuated

Solenoid Valves

with manual reset

Pneumatic Valves

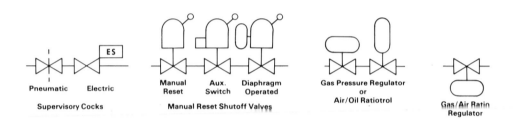

Pneumatic   Electric   ES

Supervisory Cocks

Manual Reset   Aux. Switch   Diaphragm Operated

Manual Reset Shutoff Valves

Gas Pressure Regulator or Air/Oil Ratiotrol

Gas/Air Ratio Regulator

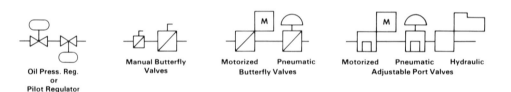

Oil Press. Reg. or Pilot Regulator

Manual Butterfly Valves

M   Pneumatic

Motorized Butterfly Valves

M   Pneumatic   Hydraulic

Motorized   Adjustable Port Valves

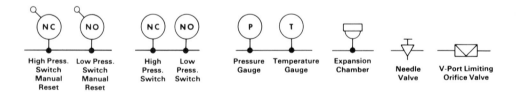

NC High Press. Switch Manual Reset   NO Low Press. Switch Manual Reset

NC High Press. Switch   NO Low Press. Switch

P Pressure Gauge   T Temperature Gauge

Expansion Chamber

Needle Valve

V-Port Limiting Orifice Valve

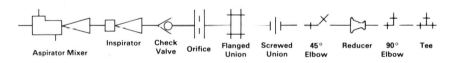

Aspirator Mixer

Inspirator   Check Valve   Orifice   Flanged Union   Screwed Union   45° Elbow   Reducer   90° Elbow   Tee

(continued)

Figure D.7. Schematic piping symbols for combustion systems *(concluded)*

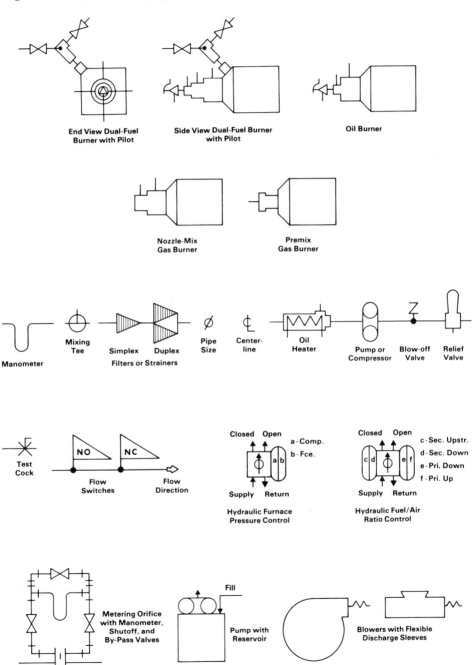

End View Dual-Fuel Burner with Pilot

Side View Dual-Fuel Burner with Pilot

Oil Burner

Nozzle-Mix Gas Burner

Premix Gas Burner

Manometer

Mixing Tee

Simplex   Duplex
Filters or Strainers

Pipe Size

Center-line

Oil Heater

Pump or Compressor

Blow-off Valve

Relief Valve

Test Cock

NO   NC
Flow Switches

Flow Direction

Closed   Open
a - Comp.
b - Fce.
Supply   Return
Hydraulic Furnace Pressure Control

Closed   Open
c - Sec. Upstr.
d - Sec. Down
e - Pri. Down
f - Pri. Up
Supply   Return
Hydraulic Fuel/Air Ratio Control

Metering Orifice with Manometer, Shutoff, and By-Pass Valves

Fill

Pump with Reservoir

Blowers with Flexible Discharge Sleeves

Table D.8 US.   Pipe color codes and labels

There is general agreement between ANSI Standard A13.1 and the MCAA (Mechanical Contractors Association of America) Engineering Standard Part V on the following *general* pipe color code:

Classification	Color of field	Color of letters
Fire quenching materials		
water, foam, CO_2	RED	WHITE
Inherently hazardous materials		
flammable or explosive	YELLOW	BLACK
chemically active or toxic	YELLOW	BLACK
extreme temperature or pressure	YELLOW	BLACK
radioactive	YELLOW	BLACK
Inherently low hazard materials		
liquid or liquid admixture	GREEN	WHITE
gas or gaseous admixture	BLUE	WHITE

The above codes are so general that it is difficult to distinquish the many fluids involved in combustion systems; so North American Mfg. Co. has established the following specific pipe color code for its own laboratories:

Fire protection materials and equipment	RED
High pressure fuel gas	ORANGE
Low pressure fuel gas	YELLOW
Fuel oils	TAN or BROWN
Low pressure air	LIGHT BLUE
Compressed air	PINK
Steam	WHITE
Water	GREEN
Oxygen gas	Bare stainless steel with frequent GREEN LABELS WITH WHITE LETTERING

Even with pipe color coding, FREQUENT USE OF STICK-ON LABELS, with large easy-to-read type, is recommended to minimize the chance that someone will open or close the wrong valve. Multiple applications of stick-on FLOW DIRECTION ARROWS are also advised.

The multitude of fluids piped around process plants may necessitate use of different shades of the above colors or spiral striping, for example, to distinguish cooling water from boiler feedwater, producer gas from natural gas, recirculated flue gas from air, hot air from cold air, main loop oil from individual furnace loop oil.

In all cases, pipe color coding should be accompanied by many stick-on labels identifying the fluid so that all personnel can easily learn the code.

Table E.1 US. Sheet metal and wire gauges.[a] Preferred practice is to specify sheet thickness or wire diameter in decimal parts of an inch or millimetre. ANSI preferred thickness for uncoated sheet, strip, and plate (0.25 in.) are: 0.224, .220, .180, .160, .140, .125, .112, .100, .090, .080, .071, .063, .056, .050, .045, .040, .036, .032, .028, .025, .022, .020, .018, .016, .014, .012, .011, .010, .009, .008, .007, .006, .005, .004.

Gauge number	AWG, American wire gauge; or B & S, Brown & Sharpe (nonferrous sheet, rod, wire)	US Steel wire gauge, AS & W, Washburn & Moen, Roebling (steel wire except music)	USSG, United States standard gauge (old) (uncoated carbon steel sheet, tin plate)	Manufacturers' standard gauge, revised US standard (uncoated steel sheets)[b]	BWG, Birmingham wire gauge; or WWG, Warrington Stubs iron wire (telephone and telegraph wire)	WG, BWG, ISWG, SWG, British Imperial standard wire gauge	BG, British standard gauge for iron & steel sheets & hoops, 1914[c]
7/0's	—	0.4900	0.4902	—	—	0.500	0.6666
6/0's	0.580 000	.4615	.4596	—	—	.464	.6250
5/0's	.516 500	.4305	.4289	—	—	.432	.5883
4/0's	.460 000	.3938	.3983	—	0.454	.400	.5416
3/0's	.409 642	.3625	.3676	—	.425	.372	.5000
2/0's	.364 796	.3310	.3370	—	.380	.348	.4452
0	.324 861	.3065	.3064	—	.340	.324	.3964
1	.289 297	.2830	.2757	—	.300	.300	.3532
2	.257 627	.2625	.2604	—	.284	.276	.3147
3	.229 423	.2437	.2451	0.2391	.259	.252	.2804
4	.204 307	.2253	.2298	.2242	.238	.232	.2500
5	.181 940	.2070	.2145	.2092	.220	.212	.2225
6	.162 023	.1920	.1991	.1943	.203	.192	.1981
7	.144 285	.1770	.1838	.1793	.180	.176	.1764
8	.128 490	.1620	.1685	.1644	.165	.160	.1570
9	.114 423	.1483	.1532	.1495	.148	.144	.1398
10	.101 897	.1350	.1379	.1345	.134	.128	.1250
11	.090 742	.1205	.1225	.1196	.120	.116	.1113
12	.080 808	.1055	.1072	.1046	.109	.104	.0991
13	.071 962	.0915	.0919	.0897	.095	.092	.0882
14	.064 084	.0800	.0766	.0747	.083	.080	.0785
15	.057 068	.0720	.0689	.0673	.072	.072	.0699
16	.050 821	.0625	.0613	.0598	.065	.064	.0625
17	.045 257	.0540	.0551	.0538	.058	.056	.0556
18	.040 303	.0475	.0490	.0478	.049	.048	.0495
19	.035 890	.0410	.0429	.0418	.042	.040	.0440
20	.031 961	.0348	.0368	.0359	.035	.036	.0392

(continued)

See footnotes on next page.

Table E.1 US. Sheet metal and wire gauges. [a] (concluded)

Gauge number / Thickness or diameter in decimal parts of an inch	AWG, American wire gauge; or B & S, Brown & Sharpe (non-ferrous sheet, rod, wire)	US Steel wire gauge, AS & W, Washburn & Moen, Roebling (steel wire except music)	USSG, United States standard gauge (old) (un-coated carbon steel sheet, tin plate)	Manufacturers' standard gauge, revised US standard (uncoated steel sheets) [b]	BWG, Birmingham wire gauge; or WWG, Warrington Stubs iron wire (telephone and telegraph wire)	WG, BWG, ISWG, SWG, British Imperial standard wire gauge	BG, British standard gauge for iron & steel sheets & hoops, 1914 [c]
21	0.028 462	0.0317	0.0337	0.0329	0.032	0.032	0.0349
22	.025 346	.0286	.0306	.0299	.028	.028	.0313
23	.022 572	.0258	.0276	.0269	.025	.024	.0278
24	.020 101	.0230	.0245	.0239	.022	.022	.0248
25	.017 900	.0204	.0214	.0209	.020	.020	.0220
26	.015 941	.0181	.0184	.0179	.018	.018	.0196
27	.014 195	.0173	.0169	.0164	.016	.0164	.0175
28	.012 641	.0162	.0153	.0149	.014	.0148	.0156
29	.011 257	.0150	.0138	.0135	.013	.0136	.0139
30	.010 025	.0140	.0123	.0120	.012	.0124	.0123
31	.008 928	.0132	.0107	.0105	.010	.0116	.0110
32	.007 950	.0128	.0100	.0097	.009	.0108	.0098
33	.007 080	.0118	.0092	.0090	.008	.0100	.0087
34	.006 305	.0104	.0084	.0082	.007	.0092	.0077
35	.005 615	.0095	.0077	.0075	.005	.0084	.0069
36	.005 000	.0090	.0069	.0067	.004	.0076	.0061
37	.004 453	.0085	.0065	.0064	—	.0068	.0054
38	.003 965	.0080	.0061	.0060	—	.0060	.0048
39	.003 531	.0075	.0057	—	—	.0052	.0043
40	.003 144	.0070	.0054	—	—	.0048	.0039
41	—	.0066	.0052	—	—	.0044	.0034
42	—	.0062	.0050	—	—	.0040	.0031
43	—	.0060	.0048	—	—	.0036	.0027
44	—	.0058	.0046	—	—	.0032	.0024
45	—	.0055	—	—	—	.0028	.0022
46	—	.0052	—	—	—	.0024	.0019
47	—	.0050	—	—	—	.0020	.0017
48	—	.0048	—	—	—	.0016	.0015
49	—	.0046	—	—	—	.0012	.0014
50	—	.0044	—	—	—	.0010	.0012

[a] Abstracted from:
Baumeister, R. (ed.): "Marks' Standard Handbook for Mechanical Engineers", 7th ed., McGraw-Hill Book Co, New York, NY, 1967.
Bolz, R.E. and Tuve, G. L.: "Handbook of Tables for Applied Engineering Science", 2nd ed., CRC Press, Cleveland, OH, 1973.
McGannon, H. E. (ed.): "The Making, Shaping and Treating of Steel", 9th ed., United States Steel Corp., Pittsburgh, PA, 1971.

[b] Current US standard, but see subtitle on page 343. For converting from thickness to weight *for steel sheets*, use 41.82 lb/ft² in. This accounts for a slight extra thickness in the center, due to spring in the rolls, and for the fact that sheets are usually gauged near the edge.

[c] Last British standard before metrication. See subtitle on page 343.

Table E.2. Squares and square roots. Within limited ranges, Bernoulli's equation (formula 5/1 or 5/2 in Volume I) can be modified into the "Square Root Law" for flow and pressure drop calculations, as follows:

$$\frac{flow_2}{flow_1} = \frac{\sqrt{\Delta P_2}}{\sqrt{\Delta P_1}} \quad \text{or} \quad \frac{(flow_2)^2}{(flow_1)^2} = \frac{\Delta P_2}{\Delta P_1}.$$

The table below is an accurate alternate to Tables 5.4a, b, and c in Volume I. Use any units, but they must be consistent through any one problem.

Example A: A gas train has a measured capacity of 3000 cfh with 7"wc total pressure drop. How much pressure drop would be needed for 3500 cfh of gas?

Opposite 3.0 (for 3000) in the bold face column, read 9.00 in the light face column. Opposite 3.5, read 12.3.

Solve the second formula above for $\Delta P_2 = \Delta P_1 \times \left(\dfrac{flow_2}{flow_1}\right)^2 = 7\text{"wc} \times \left(\dfrac{3500}{3000}\right)^2 = 9.57\text{"wc; or } 7\text{"wc} \times \left(\dfrac{12.3}{9.00}\right) = 9.57\text{"wc.}$

Example B: An air orifice passes 991 nm³/h with 3.03 kPa pressure drop. How much air will it pass with 5.02 kPa drop?

Opposite 3.03 in the light face column, read 1.74 in the bold face column. Opposite 5.02, read 2.24.

Solve the first formula above for $flow_2 = flow_1 \times \sqrt{\dfrac{\Delta P_2}{\Delta P_1}} = 991 \times \sqrt{\dfrac{5.02}{3.03}} = 1276 \text{ nm}^3/\text{h; or } 991 \times \left(\dfrac{2.24}{1.74}\right) = 1276 \text{ nm}^3/\text{h.}$

flow or √ΔP	flow² or ΔP	flow or √ΔP	flow² or ΔP	flow or √ΔP	flow² or ΔP	flow or √ΔP	flow² or ΔP	flow or √ΔP	flow² or ΔP	flow or √ΔP	flow² or ΔP	flow or √ΔP	flow² or ΔP	flow or √ΔP	flow² or ΔP	flow or √ΔP	flow² or ΔP	flow or √ΔP	flow² or ΔP
1.01	1.02	1.45	2.10	1.89	3.57	2.33	5.43	2.77	7.67	3.42	11.7	4.30	18.5	5.27	27.8	6.60	43.6	8.24	67.9
1.02	1.04	1.46	2.13	1.90	3.61	2.34	5.48	2.78	7.73	3.44	11.8	4.32	18.7	5.30	28.1	6.63	44.0	8.28	68.6
1.03	1.06	1.47	2.16	1.91	3.65	2.35	5.52	2.79	7.78	3.46	12.0	4.34	18.8	5.33	28.4	6.66	44.4	8.32	69.2
1.04	1.08	1.48	2.19	1.92	3.69	2.36	5.57	2.80	7.84	3.48	12.1	4.36	19.0	5.36	28.7	6.69	44.8	8.36	70.0
1.05	1.10	1.49	2.22	1.93	3.72	2.37	5.62	2.81	7.90	3.50	12.3	4.38	19.2	5.39	29.1	6.72	45.2	8.40	70.6
1.06	1.12	1.50	2.25	1.94	3.76	2.38	5.66	2.82	7.95	3.52	12.4	4.40	19.4	5.42	29.4	6.75	45.6	8.44	71.2
1.07	1.14	1.51	2.28	1.95	3.80	2.39	5.71	2.83	8.01	3.54	12.5	4.42	19.5	5.45	29.7	6.78	46.0	8.48	71.9
1.08	1.17	1.52	2.31	1.96	3.84	2.40	5.76	2.84	8.07	3.56	12.7	4.44	19.7	5.48	30.0	6.81	46.4	8.52	72.6
1.09	1.19	1.53	2.34	1.97	3.88	2.41	5.81	2.85	8.12	3.58	12.8	4.46	19.9	5.51	30.4	6.84	46.8	8.56	73.3
1.10	1.21	1.54	2.37	1.98	3.92	2.42	5.86	2.86	8.18	3.60	13.0	4.48	20.0	5.54	30.7	6.87	47.2	8.60	74.0

Table E.2.   Squares and square roots (concluded)

flow or $\sqrt{\Delta P}$	flow^2 or ΔP	flow or $\sqrt{\Delta P}$	flow^2 or ΔP	flow or $\sqrt{\Delta P}$	flow^2 or ΔP	flow or $\sqrt{\Delta P}$	flow^2 or ΔP	flow or $\sqrt{\Delta P}$	flow^2 or ΔP	flow or $\sqrt{\Delta P}$	flow^2 or ΔP	flow or $\sqrt{\Delta P}$	flow^2 or ΔP	flow or $\sqrt{\Delta P}$	flow^2 or ΔP	flow or $\sqrt{\Delta P}$	flow^2 or ΔP	flow or $\sqrt{\Delta P}$	flow^2 or ΔP
1.11	1.23	1.55	2.40	1.99	3.96	2.43	5.90	2.87	8.24	3.62	13.1	4.50	20.3	5.57	31.0	6.90	47.6	8.64	74.6
1.12	1.25	1.56	2.43	2.00	4.00	2.44	5.95	2.88	8.29	3.64	13.2	4.52	20.4	5.60	31.4	6.93	48.0	8.68	75.3
1.13	1.28	1.57	2.46	2.01	4.04	2.45	6.00	2.89	8.35	3.66	13.4	4.54	20.6	5.63	31.7	6.96	48.4	8.72	76.0
1.14	1.30	1.58	2.50	2.02	4.08	2.46	6.05	2.90	8.41	3.68	13.5	4.55	20.8	5.66	32.0	7.00	49.0	8.76	76.7
1.15	1.32	1.59	2.53	2.03	4.12	2.47	6.10	2.91	8.47	3.70	13.7	4.58	21.0	5.69	32.4	7.04	49.6	8.80	77.4
1.16	1.35	1.60	2.56	2.04	4.16	2.48	6.15	2.92	8.53	3.72	13.8	4.60	21.2	5.72	32.7	7.08	50.1	8.84	78.1
1.17	1.37	1.61	2.59	2.05	4.20	2.49	6.20	2.93	8.58	3.74	14.0	4.62	21.3	5.75	33.1	7.12	50.7	8.88	78.9
1.18	1.39	1.62	2.62	2.06	4.24	2.50	6.25	2.94	8.64	3.76	14.1	4.64	21.5	5.78	33.4	7.16	51.3	8.92	79.6
1.19	1.42	1.63	2.66	2.07	4.28	2.51	6.30	2.95	8.70	3.78	14.3	4.66	21.7	5.81	33.8	7.20	51.8	8.96	80.3
1.20	1.44	1.64	2.69	2.08	4.33	2.52	6.35	2.96	8.76	3.80	14.4	4.68	21.9	5.84	34.1	7.24	52.4	9.00	81.0
1.21	1.46	1.65	2.72	2.09	4.37	2.53	6.40	2.97	8.82	3.82	14.6	4.70	22.1	5.87	34.5	7.28	53.0	9.04	81.7
1.22	1.49	1.66	2.76	2.10	4.41	2.54	6.45	2.98	8.88	3.84	14.7	4.72	22.3	5.90	34.8	7.32	53.6	9.08	82.4
1.23	1.51	1.67	2.79	2.11	4.45	2.55	6.50	2.99	8.94	3.86	14.9	4.74	22.5	5.93	35.2	7.36	54.2	9.12	83.2
1.24	1.54	1.68	2.82	2.12	4.49	2.56	6.55	3.00	9.00	3.88	15.1	4.76	22.7	5.96	35.5	7.40	54.8	9.16	83.9
1.25	1.56	1.69	2.86	2.13	4.54	2.57	6.60	3.02	9.12	3.90	15.2	4.78	22.8	6.00	36.0	7.44	55.4	9.20	84.6
1.26	1.59	1.70	2.89	2.14	4.58	2.58	6.66	3.04	9.24	3.92	15.4	4.80	23.0	6.03	36.4	7.48	56.0	9.24	85.4
1.27	1.61	1.71	2.92	2.15	4.62	2.59	6.71	3.06	9.36	3.94	15.5	4.82	23.2	6.06	36.7	7.52	56.6	9.28	86.1
1.28	1.64	1.72	2.96	2.16	4.67	2.60	6.76	3.08	9.49	3.96	15.7	4.84	23.4	6.09	37.1	7.56	57.2	9.32	86.9
1.29	1.66	1.73	3.00	2.17	4.71	2.61	6.81	3.10	9.61	3.98	15.8	4.85	23.6	6.12	37.5	7.60	57.8	9.36	87.6
1.30	1.69	1.74	3.03	2.18	4.75	2.62	6.86	3.12	9.73	4.00	16.0	4.88	23.8	6.15	37.8	7.64	58.4	9.40	88.4
1.31	1.72	1.75	3.06	2.19	4.80	2.63	6.92	3.14	9.86	4.02	16.2	4.90	24.0	6.18	38.2	7.68	59.0	9.44	89.1
1.32	1.74	1.76	3.10	2.20	4.84	2.64	6.97	3.16	10.0	4.04	16.3	4.92	24.2	6.21	38.6	7.72	59.6	9.48	89.9
1.33	1.77	1.77	3.13	2.21	4.88	2.65	7.02	3.18	10.1	4.06	16.5	4.94	24.4	6.24	38.9	7.76	60.2	9.52	90.6
1.34	1.80	1.78	3.17	2.22	4.93	2.66	7.08	3.20	10.2	4.08	16.6	4.96	24.6	6.27	39.3	7.80	60.8	9.56	91.4
1.35	1.82	1.79	3.20	2.23	4.97	2.67	7.13	3.22	10.4	4.10	16.8	4.98	24.8	6.30	39.7	7.84	61.5	9.60	92.2
1.36	1.85	1.80	3.24	2.24	5.02	2.68	7.18	3.24	10.5	4.12	17.0	5.00	25.0	6.33	40.1	7.88	62.1	9.64	92.9
1.37	1.88	1.81	3.28	2.25	5.06	2.69	7.24	3.26	10.6	4.14	17.1	5.04	25.3	6.36	40.4	7.92	62.7	9.68	93.7
1.38	1.90	1.82	3.31	2.26	5.11	2.70	7.29	3.28	10.8	4.16	17.3	5.06	25.6	6.39	40.8	7.96	63.4	9.72	94.5
1.39	1.93	1.83	3.35	2.27	5.15	2.71	7.34	3.30	10.9	4.18	17.5	5.09	25.9	6.42	41.2	8.00	64.0	9.76	95.3
1.40	1.96	1.84	3.39	2.28	5.20	2.72	7.40	3.32	11.0	4.20	17.6	5.12	26.2	6.45	41.6	8.04	64.6	9.80	96.0
1.41	1.98	1.85	3.42	2.29	5.24	2.73	7.45	3.34	11.2	4.22	17.8	5.15	26.5	6.48	42.0	8.08	65.3	9.84	96.8
1.42	2.02	1.86	3.46	2.30	5.29	2.74	7.51	3.36	11.3	4.24	18.0	5.18	26.8	6.51	42.4	8.12	65.9	9.88	97.6
1.43	2.04	1.87	3.50	2.31	5.34	2.75	7.54	3.38	11.4	4.26	18.1	5.21	27.1	6.54	42.8	8.16	66.6	9.92	98.4
1.44	2.07	1.88	3.53	2.32	5.38	2.76	7.62	3.40	11.6	4.28	18.3	5.24	27.5	6.57	43.2	8.20	67.2	9.96	99.2

Table E.3 US. Areas, circumferences, and flow capacities of circles and drill sizes.

Drill size or diameter	Diameter, inches	Circumference, inches	Area, in.²	Area, ft²	Flow, cfh with $\Delta p = 1''wc$, $\overline{K} = 1.0$ air	natl gas
80	0.0135	0.042 41	0.000 143	0.000 000 9	0.238	0.306
79	0.0145	0.045 55	0.000 165	0.000 001 1	0.274	0.354
1/64"	0.0156	0.049 09	0.000 191	0.000 001 3	0.317	0.409
78	0.0160	0.050 27	0.000 201	0.000 001 4	0.334	0.430
77	0.0180	0.056 55	0.000 254	0.000 001 8	0.422	0.545
76	0.0200	0.062 83	0.000 314	0.000 002 2	0.522	0.673
75	0.0210	0.065 97	0.000 346	0.000 002 4	0.575	0.742
74	0.0225	0.070 69	0.000 398	0.000 002 8	0.660	0.851
73	0.0240	0.075 40	0.000 452	0.000 003 1	0.751	0.969
72	0.0250	0.098 54	0.000 491	0.000 003 4	0.815	1.051
71	0.0260	0.081 68	0.000 531	0.000 003 7	0.881	1.137
70	0.0280	0.087 96	0.000 616	0.000 004 3	1.022	1.318
69	0.0292	0.091 73	0.000 670	0.000 004 7	1.112	1.434
68	0.0310	0.097 39	0.000 755	0.000 005 2	1.253	1.616
1/12"	0.0313	0.098 18	0.000 765	0.000 005 3	1.277	1.647
67	0.0320	0 100 53	0.000 804	0.000 005 6	1.335	1.722
66	0.0330	0.103 67	0.000 855	0.000 005 9	1.420	1.831
65	0.0350	0.109 96	0.000 962	0.000 006 7	1.597	2.060
64	0.0360	0.113 10	0.001 018	0.000 007 1	1.690	2.179
63	0.0370	0.116 24	0.001 075	0.000 007 5	1.785	2.302
62	0.0380	0.119 38	0.001 134	0.000 007 9	1.883	2.428
61	0.0390	0.122 52	0.001 195	0.000 008 3	1.983	2.558
60	0.0400	0.125 66	0.001 257	0.000 008 7	2.086	2.691
59	0.0410	0.128 81	0.001 320	0.000 009 2	2.192	2.827
58	0.0420	0.131 95	0.001 385	0.000 009 6	2.300	2.966
57	0.0430	0.135 09	0.001 452	0.000 010 1	2.411	3.109
56	0.0465	0.146 08	0.001 698	0.000 011 8	2.819	3.636
3/64"	0.0469	0.147 26	0.001 73	0.000 012 0	2.868	3.699
55	0.0520	0.163 36	0.002 12	0.000 014 7	3.525	4.547
54	0.0550	0.172 79	0.002 38	0.000 016 5	3.944	5.087
53	0.0595	0.186 93	0.002 78	0.000 019 3	4.616	5.953
1/16"	0.0625	0.196 35	0.003 07	0.000 021 3	5.093	6.569
52	0.0635	0.199 49	0.003 17	0.000 022 0	5.257	6.780
51	0.0670	0.210 49	0.003 53	0.000 024 5	5.853	7.548
50	0.0700	0.219 91	0.003 85	0.000 026 7	6.388	8.240
49	0.0730	0.229 34	0.004 19	0.000 029 1	6.948	8.961
48	0.0760	0.238 76	0.004 54	0.000 031 5	7.531	9.713
5/64"	0.0781	0.245 44	0.004 79	0.000 033 3	7.952	10.257
47	0.0785	0.246 62	0.004 84	0.000 033 6	8.034	10.362
46	0.0810	0.254 47	0.005 15	0.000 035 8	8.554	11.033
45	0.0820	0.257 61	0.005 28	0.000 036 7	8.767	11.307
44	0.0860	0.270 18	0.005 81	0.000 040 3	9.643	12.437
43	0.0890	0.279 60	0.006 22	0.000 043 2	10.398	13.320
42	0.0935	0.293 74	0.006 87	0.000 047 7	11.398	14.700

(continued)

Table E.3 US. *(continued)*

Drill size or diameter	Diameter, inches	Circumference, inches	Area, in.²	Area, ft²	Flow, cfh with Δp = 1"wc, K̄ = 1.0 air	natl gas
³/₃₂"	0.0937	0.294 52	0.006 90	0.000 047 9	11.447	14.763
41	0.0960	0.301 59	0.007 24	0.000 050 3	12.016	15.497
40	0.0980	0.307 88	0.007 54	0.000 052 4	12.521	16.150
39	0.0995	0.312 59	0.007 78	0.000 054 0	12.908	16.648
38	0.1015	0.318 87	0.008 09	0.000 056 2	13.432	17.324
37	0.1040	0.326 73	0.008 49	0.000 059 0	14.102	18.188
36	0.1065	0.334 58	0.008 91	0.000 061 9	14.788	19.072
⁷/₆₄"	0.1094	0.343 61	0.009 40	0.000 065 2	15.604	20.125
35	0.1100	0.345 58	0.009 50	0.000 066 0	15.776	20.347
34	0.1110	0.348 72	0.009 68	0.000 067 2	16.064	20.718
33	0.1130	0.355 00	0.010 03	0.000 069 6	16.648	21.472
32	0.1160	0.364 43	0.010 57	0.000 073 4	17.543	22.627
31	0.1200	0.376 99	0.011 31	0.000 078 5	18.774	24.214
¹/₈	0.1250	0.392 70	0.012 27	0.000 085 2	20.371	26.274
30	0.1285	0.403 70	0.012 96	0.000 090 1	21.528	27.766
29	0.1360	0.427 26	0.014 53	0.000 100 9	24.114	31.102
28	0.1405	0.441 39	0.015 49	0.000 107 7	25.737	33.194
⁹/₆₄"	0.1406	0.441 79	0.015 53	0.000 107 9	25.773	33.241
27	0.1440	0.442 39	0.016 29	0.000 113 1	27.035	34.868
26	0.1470	0.461 82	0.016 97	0.000 117 9	28.173	36.336
25	0.1495	0.469 67	0.017 55	0.000 121 9	29.139	37.583
24	0.1520	0.477 52	0.018 15	0.000 126 0	30.122	38.850
23	0.1540	0.483 81	0.018 63	0.000 129 4	30.920	39.879
⁵/₃₂	0.1562	0.490 87	0.019.17	0.000 133 1	31.810	41.027
22	0.1570	0.493 23	0.019 36	0.000 134 4	32.136	41.448
21	0.1590	0.499 51	0.019 86	0.000 137 9	32.960	42.511
20	0.1610	0.505 80	0.020 36	0.000 141 4	33.794	43.587
19	0.1660	0.521 51	0.021 64	0.000 150 3	35.926	46.336
18	0.1695	0.532 50	0.022 56	0.000 156 7	37.457	48.311
¹¹/₆₄"	0.1719	0.539 96	0.023 20	0.000 161 1	38.526	49.689
17	0.1730	0.543 50	0.023 51	0.000 163 2	39.020	50.327
16	0.1770	0.556 06	0.024 61	0.000 170 9	40.846	52.681
15	0.1800	0.565 49	0.025 45	0.000 176 7	42.242	54.482
14	0.1820	0.571 77	0.026 02	0.000 180 7	43.186	55.699
13	0.1850	0.581 20	0.026 88	0.000 186 7	44.621	57.551
³/₁₆"	0.1875	0.589 05	0.027 61	0.000 191 7	45.835	59.117
12	0.1890	0.593 76	0.028 06	0.000 194 8	46.572	60.066
11	0.1910	0.600 05	0.028 65	0.000 199 0	47.563	61.344
10	0.1930	0.606 33	0.029 40	0.000 203 2	48.564	62.636
9	0.1960	0.615 75	0.030 17	0.000 209 5	50.085	64.598
8	0.1990	0.625 18	0.031 10	0.000 216 0	51.630	66.591
7	0.2010	0.631 46	0.031 73	0.000 220 4	52.673	67.939
¹³/₆₄"	0.2031	0.638 14	0.032 41	0.000 224 8	53.780	69.363
6	0.2040	0.640 89	0.032 69	0.000 227 0	54.257	69.979

(continued)

Table E.3 US. *(continued)*

Drill size or diameter	Diameter, inches	Circumference, inches	Area, in.²	Area, ft²	Flow, cfh with $\Delta p = 1''wc$, $\overline{K} = 1.0$ air	natl gas
5	0.2055	0.645 60	0.033 17	0.000 230 3	55.058	71.012
4	0.2090	0.656 59	0.034 31	0.000 238 2	56.950	73.451
3	0.2130	0.669 16	0.035 63	0.000 247 5	59.150	76.290
$7/32''$	0.2187	0.687 22	0.037 58	0.000 261 0	62.358	80.427
2	0.2210	0.694 29	0.038 36	0.000 266 4	63.128	82.128
1	0.2280	0.716 28	0.040 83	0.000 283 5	67.775	87.413
A	0.2340	0.735 13	0.043 01	0.000 298 7	71.389	92.074
$15/64''$	0.2344	0.736 31	0.043 14	0.000 299 6	71.633	92.389
B	0.2380	0.747 70	0.044 49	0.000 308 9	73.850	95.249
C	0.2420	0.760 27	0.046 00	0.000 319 4	76.353	98.478
D	0.2460	0.772 83	0.047 53	0.000 330 1	78.898	101.76
$E = 1/4''$	0.2500	0.785 40	0.049 09	0.000 340 9	81.485	105.10
F	0.2570	0.807 39	0.051 87	0.000 360 2	86.112	111.06
G	0.2610	0.819 96	0.053 50	0.000 371 5	88.813	114.55
$17/64''$	0.2656	0.834 41	0.055 42	0.000 384 9	91.972	118.62
H	0.2660	0.835 67	0.055 57	0.000 385 9	92.249	118.98
I	0.2720	0.854 52	0.058 11	0.000 403 5	96.457	124.40
J	0.2770	0.870 22	0.060 26	0.000 418 5	100.04	129.02
K	0.2810	0.882 79	0.062 02	0.000 430 7	102.95	132.78
$9/32''$	0.2812	0.883 57	0.062 13	0.000 431 5	103.09	132.97
L	0.2900	0.911 06	0.066 05	0.000 458 7	109.65	141.42
M	0.2950	0.926 77	0.068 35	0.000 474 7	113.46	146.37
$19/64''$	0.2969	0.932 66	0.069 22	0.000 480 7	114.93	148.23
N	0.3030	0.951 90	0.071 63	0.000 500 7	118.91	153.36
$5/16''$	0.3125	0.981 75	0.076 70	0.000 532 6	127.32	164.21
O	0.3160	0.992 75	0.078 43	0.000 544 6	130.19	167.91
P	0.3230	1.014 74	0.081 94	0.000 569 0	136.02	175.43
$21/64''$	0.3281	1.030 8	0.084 56	0.000 587 2	140.35	181.02
Q	0.3320	1.043 0	0.086 57	0.000 601 2	143.71	185.35
R	0.3390	1.065 0	0.090 26	0.000 626 8	149.83	193.24
$11/32''$	0.3437	1.079 8	0.092 81	0.000 644 5	154.01	198.64
S	0.3480	1.093 3	0.095 11	0.000 660 5	157.89	203.64
T	0.3580	1.124 7	0.100 6	0.000 699 0	167.10	215.51
$23/64''$	0.3594	1.129 0	0.101 4	0.000 704 4	168.41	217.20
U	0.3680	1.156 1	0.106 4	0.000 738 6	176.56	227.72
$3/8''$	0.3750	1.178 1	0.110 5	0.000 767 0	183.34	236.47
V	0.3770	1.184 4	0.111 6	0.000 775 2	185.30	239.00
W	0.3860	1.212 7	0.117 0	0.000 812 7	194.26	250.54
$25/64''$	0.3906	1.227 2	0.119 8	0.000 832 2	198.91	256.55
X	0.3970	1.247 2	0.123 8	0.000 859 6	205.48	265.03
Y	0.4040	1.269 2	0.128 2	0.000 890 2	212.80	274.45
$13/32''$	0.4062	1.276 3	0.129 6	0.000 900 1	215.12	277.45
Z	0.4130	1.297 5	0.134 0	0.000 930 3	222.38	286.82
$27/64''$	0.4219	1.325 4	0.139 8	0.000 970 8	232.07	299.31

(continued)

Table E.3 US. *(continued)*

Drill size or diameter	Diameter, inches	Circumference, inches	Area, in.²	Area, ft²	Flow, cfh with Δp = 1"wc, K̄ = 1.0 air	natl gas
7/16"	0.4375	1.3745	0.1503	0.001 044	249.55	321.86
29/64"	0.4531	1.4235	0.1613	0.001 120	267.66	345.22
15/32"	0.4687	1.4726	0.1726	0.001 198	286.41	369.40
31/64"	0.4844	1.5217	0.1843	0.001 280	305.92	394.54
1/2"	0.5000	1.5708	0.1964	0.001 364	325.94	420.39
33/64"	0.5156	1.6199	0.2088	0.001 450	346.60	447.04
17/32"	0.5313	1.6690	0.2217	0.001 539	368.03	474.68
35/64"	0.5469	1.7181	0.2349	0.001 631	389.95	502.95
9/16"	0.5625	1.7672	0.2485	0.001 726	412.52	532.05
37/64"	0.5781	1.8162	0.2625	0.001 823	435.72	561.97
19/32"	0.5938	1.8653	0.2769	0.001 923	459.70	592.91
39/64"	0.6094	1.9144	0.2917	0.002 025	484.18	624.47
5/8"	0.0250	0.9035	0.3008	0.002 131	509.28	656.85
41/64"	0.6406	2.0126	0.3223	0.002 238	535.02	690.05
21/32"	0.6562	2.0617	0.3382	0.002 350	561.40	724.07
43/64"	0.6719	2.1108	0.3545	0.002 462	588.58	759.13
11/16"	0.6875	2.1598	0.3712	0.002 578	616.23	794.79
23/32"	0.7188	2.2580	0.4057	0.002 818	673.62	868.81
3/4"	0.7500	2.3562	0.4418	0.003 068	733.37	945.87
25/32"	0.7812	2.4544	0.4794	0.003 329	795.65	1026.2
13/16"	0.8125	2.5525	0.5185	0.003 601	860.69	1110.1
27/32"	0.8438	2.6507	0.5591	0.003 883	928.28	1197.3
7/8"	0.8750	2.7489	0.6013	0.004 176	998.19	1287.4
29/32"	0.9062	2.8471	0.6450	0.004 479	1070.6	1380.9
15/16"	0.9375	2.9452	0.6903	0.004 794	1145.9	1478.0
31/32"	0.9688	3.0434	0.7371	0.005 119	1223.7	1578.3
1"	1.0000	3.1416	0.7854	0.005 454	1303.8	1681.6
1 1/16"	1.0625	3.3379	0.8866	0.006 157	1471.8	1898.3
1 1/8"	1.1250	3.5343	0.9940	0.006 903	1650.1	2128.2
1 3/16"	1.1875	3.7306	1.1075	0.007 691	1838.5	2371.2
1 1/4"	1.2500	3.9270	1.2272	0.008 522	2037.1	2627.4
1 5/16"	1.3125	4.1233	1.3530	0.009 396	2245.9	2896.7
1 3/8"	1.3750	4.3170	1.4849	0.010 31	2464.9	3179.2
1 7/16"	1.4375	4.5160	1.6230	0.011 27	2694.1	3474.7
1 1/2"	1.5000	4.7124	1.7671	0.012 27	2933.5	3783.5
1 9/16"	1.5625	4.9087	1.9175	0.013 32	3183.0	4105.3
1 5/8"	1.6250	5.1051	2.0739	0.014 40	3442.7	4440.3
1 11/16"	1.6875	5.3014	2.2365	0.015 53	3712.7	4788.4
1 3/4"	1.7500	5.4978	2.4053	0.016 70	3992.8	5149.7
1 13/16"	1.8125	5.6941	2.5802	0.017 92	4342.3	5600.6
1 7/8"	1.8750	5.8905	2.7612	0.019 18	4583.5	5911.7
1 15/16"	1.9375	6.0868	2.9483	0.020 47	4894.2	6312.3
2"	2.0000	6.2832	3.1416	0.021 82	5215.0	6726.1
2 1/16"	2.0625	6.4795	3.3410	0.023 20	5546.1	7153.1

(continued)

Table E.3 US. *(continued)*

Drill size or diameter	Diameter, inches	Circumference, inches	Area, in.²	Area, ft²	Flow, cfh with Δp = 1"wc, K = 1.0 air	natl gas
2¹/₈"	2.1250	6.6759	3.5466	0.024 63	5887.3	7593.2
2³/₁₆"	2.1875	6.8722	3.7583	0.026 10	6238.7	8046.6
2¹/₄"	2.2500	7.0686	3.9761	0.027 61	6600.3	8512.9
2⁵/₁₆"	2.3125	7.2649	4.2000	0.029 17	6972.1	8992.5
2³/₈"	2.3750	7.4613	4.4301	0.030 76	7354.0	9484.9
2⁷/₁₆"	2.4375	7.6576	4.6664	0.032 41	7746.2	9990.7
2¹/₂"	2.5000	7.8540	4.9087	0.034 09	8148.5	10 510
2⁹/₁₆"	2.5625	8.0503	5.1572	0.035 81	8561.0	11 042
2⁵/₈"	2.6250	8.2467	5.4119	0.037 58	8983.7	11 587
2¹¹/₁₆"	2.6875	8.4430	5.6727	0.039 39	9416.6	12 145
2³/₄"	2.7500	8.6394	5.9396	0.041 25	9859.7	12 717
2¹³/₁₆"	2.8125	8.8357	6.2126	0.043 14	10 313	13 301
2⁷/₈"	2.8750	9.0321	6.4918	0.045 08	10 776	13 899
2¹⁵/₁₆"	2.9375	9.2284	6.7771	0.047 06	11 250	14 510
3"	3.0000	9.4248	7.0686	0.049 09	11 734	15 134
3¹/₁₆"	3.0625	9.6211	7.3662	0.051 15	12 228	15 771
3¹/₈"	3.1250	9.8175	7.6699	0.053 26	12 732	16 421
3³/₁₆"	3.1875	10.014	7.9798	0.055 42	13 246	17 085
3¹/₄"	3.2500	10.210	8.2958	0.057 36	13 771	17 761
3⁵/₁₆"	3.3125	10.407	8.6179	0.059 85	14 306	18 451
3³/₈"	3.3750	10.603	8.9462	0.062 13	14 851	19 154
3⁷/₁₆"	3.4375	10.799	9.2806	0.064 45	15 406	19 870
3¹/₂"	3.5000	10.996	9.6211	0.066 81	15 971	20 599
3⁹/₁₆"	3.5625	11.192	9.9678	0.069 22	16 547	21 341
3⁵/₈"	3.6250	11.388	10.321	0.071 67	17 132	22 096
3¹¹/₁₆"	3.6875	11.585	10.680	0.074 17	17 728	22 865
3³/₄"	3.7500	11.781	11.045	0.076 70	18 334	23 647
3¹³/₁₆"	3.8125	11.977	11.416	0.079 28	18 950	24 441
3⁷/₈"	3.8750	12.174	11.793	0.081 90	19 577	25 249
3¹⁵/₁₆"	3.9375	12.370	12.177	0.084 56	20 213	26 070
4"	4.0000	12.566	12.566	0.087 26	20 860	26 905
4¹/₁₆"	4.0625	12.763	12.962	0.090 02	21 517	27 752
4¹/₈"	4.1250	12.959	13.364	0.092 81	22 184	28 612
4³/₁₆"	4.1875	13.155	13.772	0.095 64	22 862	29 486
4¹/₄"	4.2500	13.352	14.186	0.098 52	23 549	30 373
4⁵/₁₆"	4.3125	13.548	14.607	0.101 4	24 247	31 273
4³/₈"	4.3750	13.745	15.033	0.104 3	24 955	32 186
4⁷/₁₆"	4.4375	13.941	15.466	0.107 4	25 673	33 112
4¹/₂"	4.5000	14.137	15.904	0.110 4	26 401	34 051
4⁹/₁₆"	4.5625	14.334	16.349	0.113 5	27 140	35 004
4⁵/₈"	4.6250	14.530	16.800	0.116 7	27 888	35 969
4¹¹/₁₆"	4.6875	14.726	17.257	0.119 8	28 647	36 948
4³/₄"	4.7500	14.923	17.721	0.123 1	29 416	37 940
4¹³/₁₆"	4.8125	15.119	18.190	0.126 3	30 195	38 945

(continued)

Table E.3 US. *(continued)*

Drill size or diameter	Diameter, inches	Circumference, inches	Area, in.²	Area, ft²	Flow, cfh with Δp = 1"wc, K̄ = 1.0 air	natl gas
4⁷/₈"	4.8750	15.315	18.665	0.1296	30 985	39 963
4¹⁵/₁₆"	4.9375	15.512	19.147	0.1330	31 784	40 994
5"	5.0000	15.708	19.635	0.1364	32 594	42 038
5¹/₁₆"	5.0625	15.904	20.129	0.1398	33 414	43 096
5¹/₈"	5.1250	16.101	20.629	0.1433	34 244	44 167
5³/₁₆"	5.1875	16.297	21.135	0.1468	35 084	45 250
5¹/₄"	5.2500	16.493	21.648	0.1503	35 936	46 347
5⁵/₁₆"	5.3125	16.690	22.166	0.1539	36 796	47 457
5³/₈"	5.3750	16.886	22.691	0.1576	37 666	48 581
5⁷/₁₆"	5.4375	17.082	23.221	0.1613	38 548	49 717
5¹/₂"	5.5000	17.279	23.758	0.1650	39 439	50 867
5⁹/₁₆"	5.5625	17.475	24.301	0.1688	40 340	52 029
5⁵/₈"	5.6250	17.671	24.851	0.1726	41 252	53 205
5¹¹/₁₆"	5.6875	17.868	25.406	0.1764	42 174	54 394
5³/₄"	5.7500	18.064	25.967	0.1803	43 106	55 596
5¹³/₁₆"	5.8125	18.261	26.535	0.1843	44 048	56 811
5⁷/₈"	5.8750	18.457	27.109	0.1883	45 000	58 039
5¹⁵/₁₆"	5.9375	18.653	27.688	0.1923	45 963	59 281
6"	6.0000	18.850	28.274	0.1963	46 935	60 535
6¹/₈"	6.1250	19.242	29.465	0.2046	48 911	63 084
6¹/₄"	6.2500	19.649	30.680	0.2131	50 928	65 685
6³/₈"	6.3750	20.028	31.919	0.2217	52 986	68 339
6¹/₂"	6.5000	20.420	33.183	0.2304	55 084	71 045
6⁵/₈"	6.6250	20.813	34.472	0.2394	57 223	73 804
6³/₄"	6.7500	21.206	35.785	0.2485	59 403	76 615
6⁷/₈"	6.8750	21.598	37.122	0.2578	61 623	79 479
7"	7.0000	21.991	38.485	0.2673	63 884	82 395
7¹/₈"	7.1250	22.384	39.871	0.2769	66 186	85 364
7¹/₄"	7.2500	22.777	41.283	0.2867	68 529	88 386
7³/₈"	7.3750	23.169	42.718	0.2967	70 912	91 460
7¹/₂"	7.5000	23.562	44.179	0.3068	73 337	94 586
7⁵/₈"	7.6250	23.955	45.664	0.3171	75 801	97 766
7³/₄"	7.7500	24.347	47.173	0.3276	78 307	101 000
7⁷/₈"	7.8750	24.740	48.707	0.3382	80 854	104 280
8"	8.0000	25.133	50.266	0.3491	83 441	107 620
8¹/₈"	8.1250	25.525	51.849	0.3601	86 069	111 010
8¹/₄"	8.2500	25.918	53.456	0.3712	88 737	114 450
8³/₈"	8.3750	26.301	55.088	0.3826	91 447	117 940
8¹/₂"	8.5000	26.704	56.745	0.3941	94 197	121 490
8⁵/₈"	8.6250	27.096	58.426	0.4057	96 988	125 090
8³/₄"	8.7500	27.489	60.132	0.4176	99 819	128 740
8⁷/₈"	8.8750	27.882	61.862	0.4296	102 690	132 450
9"	9.0000	28.274	63.617	0.4418	105 610	136 210
9¹/₈"	9.1250	28.667	65.397	0.4541	108 560	140 010

(continued)

Table E.3 US. *(continued)*

Drill size or diameter	Diameter, inches	Circumference, inches	Area, in.^{2}	Area, ft^2	Flow, cfh with $\Delta p = 1"wc, \overline{K} = 1.0$ air	natl gas
9¼"	9.2500	29.060	67.201	0.4667	111 550	143 880
9³/₈"	9.3750	29.452	69.029	0.4794	114 590	147 790
9½"	9.5000	29.845	70.882	0.4922	117 660	151 760
9⅝"	9.6250	30.238	72.760	0.5053	120 780	155 780
9¾"	9.7500	30.631	74.662	0.5185	123 940	159 850
9⅞"	9.8750	31.023	76.589	0.5319	127 140	163 980
10"	10.000	31.416	78.540	0.5454	130 380	168 150
10⅛"	10.125	31.809	80.516	0.5591	133 660	172 380
10¼"	10.250	32.201	82.516	0.5730	136 980	176 670
10³/₈"	10.375	32.594	84.541	0.5871	140 340	181 000
10½"	10.500	32.987	86.590	0.6013	143 740	185 390
10⅝"	10.625	33.379	88.664	0.6157	147 180	189 830
10¾"	10.750	33.772	90.763	0.6303	150 670	194 320
10⅞"	10.875	34.165	92.886	0.6450	154 190	198 870
11"	11.000	34.558	95.033	0.6600	157 760	203 470
11⅛"	11.125	34.950	97.205	0.6750	161 360	208 120
11¼"	11.250	35.343	99.402	0.6903	165 010	212 830
11³/₈"	11.375	35.736	101.62	0.7056	168 690	217 580
11½"	11.500	36.128	103.87	0.7213	172 420	222 380
11⅝"	11.625	36.521	106.14	0.7371	176 190	227 240
11¾"	11.750	36.914	108.43	0.7530	180 000	232 160
11⅞"	11.875	37.306	110.75	0.7691	183 850	237 120
12"	12.000	37.699	113.10	0.7854	187 740	242 140
12¼"	12.250	38.485	117.86	0.819	195 650	252 340
12½"	12.500	39.269	122.72	0.851	203 710	262 740
12¾"	12.750	40.055	127.68	0.886	211 940	273 360
13"	13.000	40.841	132.73	0.921	220 340	284 180
13¼"	13.250	41.626	137.89	0.957	228 890	295 220
13½"	13.500	42.412	143.14	0.995	237 610	306 460
13¾"	13.750	43.197	148.49	1.031	246 490	317 920
14"	14.000	43.982	153.94	1.069	255 540	329 580
14¼"	14.250	44.768	159.48	1.109	264 750	341 460
14½"	14.500	45.553	165.13	1.149	274 120	353 540
14¾"	14.750	46.339	170.87	1.185	283 650	365 840
15"	15.000	47.124	176.71	1.228	293 350	378 350
15¼"	15.250	47.909	182.65	1.269	303 210	391 060
15½"	15.500	48.695	188.69	1.309	313 230	403 990
15¾"	15.750	49.480	194 83	1.352	323 410	417 130
16"	16.000	50.266	201.06	1.398	333 760	430 470
16¼"	16.250	51.051	207.39	1.440	344 270	444 030
16½"	16.500	51.836	213.82	1.485	354 950	457 800
16¾"	16.750	52.622	220.35	1.531	365 790	471 780
17"	17.000	53.407	226.98	1.578	376 790	485 960
17¼"	17.250	54.193	233.71	1.619	387 950	500 360

(continued)

Table E.3 US. *(continued)*

Drill size or diameter	Diameter, inches	Circumference, inches	Area, in.²	Area, ft²	Flow, cfh with Δp = 1"wc, K̄ = 1.0 air	natl gas
17½"	17.500	54.978	240.53	1.673	399 280	514 970
17¾"	17.750	55.763	247.45	1.719	410 770	529 790
18"	18.000	56.548	254.47	1.769	422 420	544 820
18¼"	18.250	57.334	261.59	1.816	434 230	560 060
18½"	18.500	58.120	268.80	1.869	446 210	575 510
18¾"	18.750	58.905	276.12	1.920	458 350	591 170
19"	19.000	59.690	283.53	1.969	470 660	607 040
19¼"	19.250	60.476	291.04	2.022	483 130	623 120
19½"	19.500	61.261	298.65	2.075	495 760	639 410
19¾"	19.750	62.047	306.35	2.125	508 550	655 910
20"	20.000	62.832	314.16	2.182	521 500	672 620
20¼"	20.250	63.617	322.06	2.237	534 620	689 540
20½"	20.500	64.100	000.00	2.292	547 910	706 670
20¾"	20.750	65.188	338.16	2.348	561 350	724 010
21"	21.000	65.974	346.36	0.405	574 000	741 560
21¼"	21.250	66.759	354.66	2.463	588 730	759 320
21½"	21.500	67.544	363.05	2.521	602 660	777 290
21¾"	21.750	68.330	371.54	2.580	616 760	795 470
22"	22.000	69.115	380.13	2.640	631 020	813 860
22¼"	22.250	69.901	388.82	2.700	645 440	832 470
22½"	22.500	70.686	397.61	2.761	660 030	851 280
22¾"	22.750	71.471	406.49	2.823	674 780	870 300
23"	23.000	72.257	415.48	2.885	689 690	889 530
23¼"	23.250	73.042	424.56	2.948	704 760	908 990
23½"	23.500	73.828	433.74	3.012	720 000	928 650
23¾"	23.750	74.613	443.01	3.076	735 400	948 490
24"	24.000	75.398	452.39	3.142	750 970	968 570
24¼"	24.250	76.184	461.86	3.207	766 690	988 850
24½"	24.500	76.969	471.44	3.274	782 580	1 009 300
24¾"	24.750	77.755	481.11	3.341	798 640	1 030 000
25"	25.000	78.540	490.87	3.409	814 850	1 051 000
25¼"	25.250	79.325	500.74	3.477	831 230	1 072 100
25½"	25.500	80.111	510.71	3.547	847 770	1 093 400
25¾"	25.750	80.896	520.77	3.616	864 480	1 115 000
26"	26.000	81.682	530.93	3.687	881 340	1 136 700
26¼"	26.250	82.467	541.19	3.758	898 370	1 158 700
26½"	26.500	83.252	551.55	3.830	915 570	1 180 900
26¾"	26.750	84.038	562.00	3.903	932 920	1 203 200
27"	27.000	84.823	572.56	3.976	950 440	1 225 800
27¼"	27.250	85.609	583.21	4.050	968 120	1 248 600
27½"	27.500	86.394	593.96	4.125	985 970	1 271 700
27¾"	27.750	87.179	604.81	4.200	1 004 000	1 294 900
28"	28.000	87.965	615.75	4.276	1 022 100	1 318 300
28¼"	28.250	88.750	626.80	4.353	1 040 500	1 342 000

(continued)

Table E.3 US. *(concluded)*

Drill size or diameter	Diameter, inches	Circumference, inches	Area, in.²	Area, ft²	Flow, cfh with Δp = 1"wc, K = 1.0	
					air	natl gas
28¹/₂"	28.500	89.536	637.94	4.430	1 059 000	1 365 900
28³/₄"	28.750	90.321	649.18	4.508	1 077 600	1 389 900
29"	29.000	91.106	660.52	4.587	1 096 500	1 414 200
29¹/₄"	29.250	91.892	671.96	4.666	1 115 400	1 438 700
29¹/₂"	29.500	92.677	683.49	4.746	1 134 600	1 463 400
29³/₄"	29.750	93.463	695.13	4.827	1 153 900	1 488 300
30"	30.000	94.248	706.86	4.909	1 173 400	1 513 400
31"	31.000	97.390	754.77	5.241	1 252 900	1 616 000
32"	32.000	100.53	804.25	5.585	1 335 100	1 721 900
33"	33.000	103.67	855.30	5.940	1 419 800	1 831 200
34"	34.000	106.81	907.92	6.305	1 507 100	1 943 900
35"	35.000	109.96	962.11	6.681	1 597 100	2 059 900
36"	36.000	113.10	1017.9	7.069	1 689 700	2 179 300
37"	37.000	116.24	1075.2	7.467	1 784 800	2 302 000
38"	38.000	119.38	1134.1	7.876	1 882 600	2 428 100
39"	39.000	122.52	1194.6	8.296	1 983 000	2 557 600
40"	40.000	125.66	1256.6	8.727	2 086 000	2 690 500
41"	41.000	128.81	1320.3	9.168	2 191 600	2 826 700
42"	42.000	131.95	1385.4	9.621	2 229 800	2 966 200
43"	43.000	135.09	1452.2	10.08	2 410 700	3 109 200
44"	44.000	138.23	1520.5	10.56	2 524 100	3 255 500
45"	45.000	141.37	1590.4	11.04	2 640 100	3 405 200
46"	46.000	144.51	1661.9	11.54	2 758 800	3 558 100
47"	47.000	147.66	1734.9	12.04	2 880 000	3 714 500
48"	48.000	150.80	1809.6	12.57	3 003 900	3 874 300
49"	49.000	153.94	1885.7	13.10	3 130 300	4 037 400
50"	50.000	157.08	1963.5	13.64	3 259 400	4 203 800

Table E.4 US.   Approximate hardness number equivalents for steel

Brinell (3000 kg)		Rockwell hardness number				Rockwell superficial hardness number			Shore scleroscope number	Vickers DPH number	Approximate tensile strength, psi
Diam mm	Num-ber*	A	B	C	D	15-N	30-N	45-N			
–	–	85.6	–	68.0	76.9	93.2	84.4	75.4	97	940	–
–	–	85.3	–	67.5	76.5	93.0	84.0	74.8	96	920	–
–	–	85.0	–	67.0	76.1	92.9	83.6	74.2	95	900	–
–	767	84.7	–	66.4	75.7	92.7	83.1	73.6	93	880	–
–	757	84.4	–	65.9	75.3	92.5	82.7	73.1	92	860	–
2.25	745	84.1	–	65.3	74.8	92.3	82.2	72.2	91	840	–
–	733	83.8	–	64.7	74.3	92.1	81.7	71.8	90	820	–
–	722	83.4	–	64.0	73.8	91.8	81.1	71.0	88	800	–
2.30	712	–	–	–	–	–	–	–	–	–	–
–	710	83.0	–	63.3	73.3	91.5	80.4	70.2	87	780	–
–	698	82.6	–	62.5	72.6	91.2	79.7	69.4	86	760	–
–	684	82.2	–	61.8	72.1	91.0	79.1	68.6	84	740	–
2.35	682	82.2	–	61.7	72.0	91.0	79.0	68.5	84	737	–
–	670	81.8	–	61.0	71.5	90.7	78.4	67.7	83	720	–
–	656	81.3	–	60.1	70.8	90.3	77.8	66.7	81	700	–
2.40	653	81.2	–	60.0	70.7	90.2	77.5	66.5	81	697	–
–	647	81.1	–	59.7	70.5	90.1	77.2	66.2	–	690	–
–	638	80.8	–	59.2	70.1	89.8	76.8	65.7	80	680	329 000
–	630	80.6	–	58.8	69.8	89.7	76.4	65.3	–	670	324 000
2.45	627	80.5	–	58.7	69.7	89.6	76.3	65.1	79	667	323 000
2.50	601	79.8	–	57.3	68.7	89.0	75.1	63.5	77	640	309 000
2.55	578	79.1	–	56.0	67.7	88.4	73.9	62.1	75	615	297 000
2.60	555	78.4	–	54.7	66.7	87.8	72.7	60.6	73	591	285 000
2.65	534	77.8	–	53.5	65.8	87.2	71.6	59.2	71	569	274 000
2.70	514	76.9	–	52.1	64.7	86.5	70.3	57.6	70	547	263 000
2.75	495	76.3	–	51.0	63.8	85.9	69.4	56.1	68	528	253 000
2.80	477	75.6	–	49.6	62.7	85.3	68.2	54.5	66	508	243 000
2.85	461	74.9	–	48.5	61.7	84.7	67.2	53.2	65	491	235 000
2.90	444	74.2	–	47.1	60.8	84.0	65.8	51.5	63	472	225 000
2.95	429	73.4	–	45.7	59.7	83.4	64.6	49.9	61	455	217 000
3.00	415	72.8	–	44.5	58.8	82.8	63.5	48.4	59	440	210 000
3.05	401	72.0	–	43.1	57.8	82.0	62.3	46.9	58	425	202 000
3.10	388	71.4	–	41.8	56.8	81.4	61.1	45.3	56	410	195 000
3.15	375	70.6	–	40.4	55.7	80.6	59.9	43.6	54	396	188 000
3.20	363	70.0	–	39.1	54.6	80.0	58.7	42.0	52	383	182 000
3.25	352	69.3	–	37.9	53.8	79.3	57.6	40.5	51	372	176 000
3.30	341	68.7	–	36.6	52.8	78.6	56.4	39.1	50	360	170 000
3.35	331	68.1	–	35.5	51.9	78.0	55.4	37.8	48	350	166 000
3.40	321	67.5	–	34.3	51.0	77.3	54.3	36.4	47	339	160 000
3.45	311	66.9	–	33.1	50.0	76.7	53.3	34.4	46	328	155 000

* Numbers above 429 are for a tungsten carbide ball. Numbers 81-429 are for a standard, Hultgren, or tungsten carbide ball.

(continued)

Table E.4 US. *(concluded)*

Brinell (3000 kg)		Rockwell hardness number				Rockwell superficial hardness number			Shore scleroscope number	Vickers DPH number	Approximate tensile strength, psi
Diam mm	Num-ber*	A	B	C	D	15-N	30-N	45-N			
3.50	302	66.3	–	32.1	49.3	76.1	52.2	33.8	45	319	150 000
3.55	293	65.7	–	30.9	48.3	75.5	51.2	32.4	43	309	145 000
3.60	285	65.3	–	29.9	47.6	75.0	50.3	31.2	–	301	141 000
3.65	277	64.6	–	28.8	46.7	74.4	49.3	29.9	41	292	137 000
3.70	269	64.1	–	27.6	45.9	73.7	48.3	28.5	40	284	133 000
3.75	262	63.6	–	26.6	45.0	73.1	47.3	27.3	39	276	129 000
3.80	255	63.0	–	25.4	44.2	72.5	46.2	26.0	38	269	126 000
3.85	248	62.5	–	24.2	43.2	71.7	45.1	24.5	37	261	122 000
3.90	241	61.8	100.0	22.8	42.0	70.9	43.9	22.8	36	253	118 000
3.95	235	61.4	99.0	21.7	41.4	70.3	42.9	21.5	35	247	115 000
4.00	229	60.8	98.2	20.5	40.5	69.7	41.9	20.1	34	241	111 000
4.05	223	–	97.3	–	–	–	–	–	–	234	–
4.10	217	–	96.4	–	–	–	–	–	33	228	105 000
4.15	212	–	95.5	–	–	–	–	–	–	222	102 000
4.20	207	–	94.6	–	–	–	–	–	32	218	100 000
4.25	201	–	93.8	–	–	–	–	–	31	212	98 000
4.30	197	–	92.8	–	–	–	–	–	30	207	95 000
4.35	192	–	91.9	–	–	–	–	–	29	202	93 000
4.40	187	–	90.7	–	–	–	–	–	–	196	90 000
4.45	183	–	90.0	–	–	–	–	–	28	192	89 000
4.50	179	–	89.0	–	–	–	–	–	27	188	87 000
4.55	174	–	87.8	–	–	–	–	–	–	182	85 000
4.60	170	–	86.8	–	–	–	–	–	26	178	83 000
4.65	167	–	86.0	–	–	–	–	–	–	175	81 000
4.70	163	–	85.0	–	–	–	–	–	25	171	79 000
4.80	156	–	82.9	–	–	–	–	–	–	163	76 000
4.90	149	–	80.8	–	–	–	–	–	23	156	73 000
5.00	143	–	78.7	–	–	–	–	–	22	150	71 000
5.10	137	–	76.4	–	–	–	–	–	21	143	67 000
5.20	131	–	74.0	–	–	–	–	–	–	137	65 000
5.30	126	–	72.0	–	–	–	–	–	20	132	63 000
5.40	121	–	69.8	–	–	–	–	–	19	127	60 000
5.50	116	–	67.6	–	–	–	–	–	18	122	58 000
5.60	111	–	65.7	–	–	–	–	–	15	117	56 000
5.75	105	–	62.3	–	–	–	–	–	–	110	–
–	95	–	56.2	–	–	–	–	–	–	100	–
–	90	–	52.0	–	–	–	–	–	–	95	–
–	86	–	48.0	–	–	–	–	–	–	90	–
6.45	81	–	41.0	–	–	–	–	–	–	85	–

* Numbers above 429 are for a tungsten carbide ball. Numbers 81-429 are for a standard, Hultgren, or tungsten carbide ball.

Table E.5.   Maximum temperature for scaling resistance of some carbon and stainless steels and heat resistant alloys

Temperatures are approximate limits for reasonably long service without destruction by scaling in an oxidizing environment. These limits do not apply to reducing atmospheres or atmospheres contaminated by sulfur compounds or other impurities that may accelerate scaling.

Design of parts for elevated temperature service should consider all engineering properties of an alloy; scaling resistance alone is not a guarantee of satisfactory performance.

Maximum temperature for scaling resistance in air, F *(C)*	Continuous exposure	Intermediate exposure
0.20% C Steel	800 *(427)*	1000 *(538)*
302, 304, 316, 321, and 347 Stainless	1650 *(899)*	1500 *(816)*
309 Stainless	2000 *(1093)*	1800 *(982)*
310 Stainless	2000 *(1093)*	1900 *(1038)*
410 Stainless	1300 *(704)*	1450 *(788)*
416 Stainless	1250 *(677)*	1400 *(760)*
430, 431 Stainless	1500 *(816)*	1600 *(871)*
446 Stainless	1950 *(1066)*	2050 *(1121)*
Inconel* 600	2050 *(1121)*	2150 *(1177)*

The left column is labeled "Alloy" (rotated text).

* Inconel is a registered trademark of International Nickel Co., Inc.

Table E.6. Full load current of electric motors[a]

Current, amperes	Direct current motors[c]		Alternating current, induction type, squirrel cage and wound rotor motors[b]									
			Single phase[d]		Two phase (4-wire)[e,f]				Three phase[d,f,g]			
	120 V	240 V	115 V	230 V	115 V	230 V	460 V	575 V	115 V	230 V	460 V	575 V
1/8	—	—	4.4	2.2	—	—	—	—	—	—	—	—
1/4	3.1	1.6	5.8	2.9	—	—	—	—	—	—	—	—
1/3	4.1	2.0	7.2	3.6	—	—	—	—	—	—	—	—
1/2	5.4	2.7	9.8	4.9	4	2	1	0.8	4	2	1	0.8
3/4	7.6	3.8	13.8	6.9	4.8	2.4	1.2	1.0	5.6	2.8	1.4	1.1
1	9.5	4.7	16	8	6.4	3.2	1.6	1.3	7.2	3.6	1.8	1.4
1 1/2	13.2	6.6	20	10	9	4.5	2.3	1.8	10.4	5.2	2.6	2.1
2	17	8.5	24	12	11.8	5.9	3	2.4	13.6	6.8	3.4	2.7
3	25	12.2	34	17	—	8.3	4.2	3.3	—	9.6	4.8	3.9
5	40	20	56	28	—	13.2	6.6	5.3	—	15.2	7.6	6.1
7 1/2	58	29	80	40	—	19	9	8	—	22	11	9
10	76	38	100	50	—	24	12	10	—	28	14	11
15	—	55	—	—	—	36	18	14	—	42	21	17
20	—	72	—	—	—	47	23	19	—	54	27	22
25	—	89	—	—	—	59	29	24	—	68	34	27
30	—	106	—	—	—	69	35	28	—	80	40	32
40	—	140	—	—	—	90	45	36	—	104	52	41
50	—	173	—	—	—	113	56	45	—	130	65	52
60	—	206	—	—	—	133	67	53	—	154	77	62
75	—	255	—	—	—	166	83	66	—	192	96	77
100	—	341	—	—	—	218	109	87	—	248	124	99
125	—	425	—	—	—	270	135	108	—	312	156	125
150	—	506	—	—	—	312	156	125	—	360	180	144
200	—	675	—	—	—	416	208	167	—	480	240	192

[a] Table compiled from the 1978 *National Electrical Code*, copyright 1977, reproduced by permission from the National Fire Protection Association, Boston, MA. Currents listed are for system voltage ranges of 110-120, 220-240, 440-480, 550-600 V.

[b] These values of full load current are for motors running at speeds usual for belted motors and motors with normal torque characteristics. Motors built for especially low speeds or high torques may require more running current, in which case the nameplate current rating should be used. Listed voltages are rated motor voltages.

[c] For d-c motors running at base speed; voltages are average d-c armature voltage ratings.

[d] For full load current of 208 and 200 V motors, full load current will be 1.10 and 1.15 times value given for 230 V respectively.

[e] The current in the common conductor of a 2-phase, 3-wire system will be 1.41 times the value given.

[f] For 90 or 80% power failure, multiply by 1.1 or 1.25 respectively.   [g] Induction type, squirrel-cage and wound motor.

Figure E.7.   Ringleman charts for estimating smoke densities. The charts below
are proportional reductions of standard charts issued by the U.S. Bureau of Mines. These charts
are used in the following manner: Make observations from a point between 100 and 1300 ft from
the smoke. The observer's line of sight should be perpendicular to the direction of smoke travel.
Place the below charts approximately 15 ft in front of the observer and as close as possible to his
line of sight. (Standard ASME or U.S. Bureau of Mines charts should be placed 50 ft from the
observer.) Open sky makes the best background for observations. Compare the smoke density
with the charts (which, at 15 ft, are shades of gray instead of individual lines) and classify the
smoke according to the Ringleman chart number. Ringleman Nos. 0 and 5 are 0% and 100%
black, respectively. Charts for these are solid white and black (not shown).

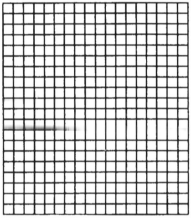

1. Equivalent to 20 percent black.

2. Equivalent to 40 percent black.

3. Equivalent to 60 percent black.

4. Equivalent to 80 percent black.

Table E.8. Percent volume full of horizontal cylindrical tanks

% of level	% full
100	100
95	98.1
90	94.8
85	90.6
80	85.8
75	80.4
70	74.8
65	68.8
60	62.6
55	56.4
50	50
45	43.6
40	37.4
35	31.2
30	25.2
25	19.6
20	14.2
15	9.4
10	5.2
5	1.9
0	0

For this S-curve relationship, the 0 to 50 portion is a mirror image of the 50 to 100 portion.

The table data apply to any size tank, in any consistent set of units of volume; and is accurate for flat-ended horizontal cylindrical storage tanks, and "sufficiently accurate" for hemispherical-ended tanks. The data may be only approximate for dished-ended tanks.

Source: "A Practical Economical Approach for Obtaining Percentage Volume Readings on Horizontal Tanks" by Hazem Huss and David Festa, Burns and Roe, Inc., Oradell, NJ; published in the May 24, 1984 issue of PLANT ENGINEERING.

Table E.9. Carbon monoxide (CO gas) warnings

CHARACTERISTICS OF CARBON MONOXIDE (CO GAS):

A gas.   Not visible.   Odorless.   Tasteless.

Approximately the same density as air.

HAZARDS – KNOW THE DANGERS OF CARBON MONOXIDE (CO GAS):

Toxic.   Can be fatal if inhaled.

Can explode or burn.

SYMPTOMS OF CARBON MONOXIDE POISONING:

Headache.   Throbbing sensation.   Sleepiness.

Nausea.   Vomiting.   Shortness of breath.

CORRECTIVE ACTIONS FOR VARIOUS CO GAS CONCENTRATION LEVELS:

CO concen-tration, ppm_v*	Life-threatening and health-threatening conditions	What action to take
50	OSHA permissible 8-hour exposure limit	Do not enter confined space. Determine source; take corrective action.
75	OSHA permissible 8-hour limit is 50 ppm	Danger. Ventilate. Avoid confined spaces. Eliminate CO sources.
200	Immediately dangerous to life and health	Evacuate the area, or wear self-contained breathing apparatus.
1500	IMMINENT DANGER to life and health	IMMEDIATELY EVACUATE the area unless wearing self-contained breathing apparatus.

* parts per million, by volume

Adapted from Tony Fennell's script of a mill bulletin board warning.

GLOSSARY
(including abbreviations, acronyms, Greek letters, symbols)

Definitions listed herewith are for the purpose of conveying a better under-standing of the meanings of the terms used in this handbook. They are not necessarily legal definitions. Some of these definitions are abstracted from the following references:

G.a   "ABBR Abbreviations for Scientific and Engineering Terms", Canadian Standards Association, Toronto, Ontario, Canada, 1983.

G.b   **ANSI Z2101-1973:** "Metric Practice Guide", American Society for Testing and Materials, Philadelphia, PA, 1973.

G.c   **Begell, W.:** "Glossary of Terms in Heat Transfer, Fluid Flow, and Related Topics (English, Russian, German, French, Japanese)".

G.d   **Cook, E. M. and DuMont, H. D.:** "Process Drying Practice", McGraw-Hill, Inc., New York, NY, 1991.

G.e   **Cowell, G. W.:** "Dictionary of Metalworking Terms", Advance Book Publishing Co., Cincinnati, OH, 1971.

G.f   **Cubberly, W. H. (ed.):** "Comprehensive Dictionary of Instrumentation and Control, Reference Guides", Instrument Society of America, Research Triangle Park, NC 27709, 1988.

G.g   **Dreyfuss, H.:** "Symbol Sourcebook", McGraw-Hill Book Co., New York, NY, 1974.

G.h   **Engineers Joint Council:** "Thesaurus of Engineering Terms", EJC, 345 E. 47 St., New York, NY; 1964. Hemisphere Publishing Corp., New York, NY, 1983.

G.i   **Factory Insurance Association:** "Recommended Good Practice for Combustion Safeguards on Single Burner Boiler-Furnaces", Industrial Risk Insurers, Chicago, IL, 1963.

G.j   **Freedman, A.:** "The Computer Glossary", 4th ed., American Management Association, Point Pleasant, PA 18950-0265, 1989.

G.k   **Gilpin, A.:** "Dictionary of Fuel Technology", Philosophical Library, Inc., New York, NY, 1969.

G.l   **Harbison-Walker:** "Modern Refractory Practice", Harbison-Walker Refractories Co., Pittsburgh, PA, 1961.

G.m **IEEE:** "Standard Dictionary of Electrical and Electronics Terms", Institute of Electrical and Electronics Engineers, Piscataway, NJ 08855-1331, 1993.

G.n   **Jay, F.:** "IEEE Standard Dictionary of Electrical and Electronic Terms", 1993.

G.o   **Lapedes, D. H. (ed.):** "Dictionary of Scientific and Technical Terms", McGraw-Hill Book Co., New York, NY 1956.

G.p   **Parker, S. P. (ed.):** "Dictionary of Electronics and Computer Technology", McGraw-Hill Book Co., New York, NY, 1984.

G.q   **Parker, S. P. (ed.):** "Dictionary of Mechanical and Design Engineering", McGraw-Hill Book Co., New York, NY, 1984.

G.r   **Parker, S. P. (ed.):** "McGraw-Hill Dictionary of Scientific and Technical Terms", 5th ed., McGraw-Hill Book Co., New York, NY, 1994.

G.s   **Parker, S. P. (ed.):** "McGraw-Hill Concise Encyclopedia of Science and Technology", McGraw-Hill, Book Co., New York, NY, 1984.

G.t   **Rose, A. & E.:** "The Condensed Chemical Dictionary", 6th ed., Reinhold Publishing Corporation, New York, NY, 1956.

G.u   **Sax, N. I. and Lewis, R. J.:** "Hazardous Chemicals Desk Reference", Van Norstrand Reinhold Co., Inc., New York, NY, 1987.

G.v   **Terrell, C. E. (ed.):** "AGA Gas Measurement Manual", American Gas Association, Arlington, VA, 1963.

G.w **U. S. Refractories Div.:** "Glossary of Terms", General Refractories Company, Pittsburgh, PA 15219.

G.x   **U. S. Environmental Protection Agency:** "Glossary of Environmental Terms and Acronym List", Office of Public Affairs (A-107), Washington, DC, 1988.

G.y   **Van Schoick, E. C. (ed.):** "Ceramic Glossary", American Ceramic Society, Columbus, OH, 1963.

G.z   **Zimmerman, O. T. and Lavine, I.:** "Conversion Factors and Tables", 3rd ed., Industrial Research Service, Inc., Dover, NH, 1961.

GLOSSARY for Volumes I and II
(including abbreviations, acronyms, Greek letters, and symbols)

Explanations listed below are for the purpose of conveying better understanding of the meanings of the terms used in this handbook. They are not necessarily legal definitions. Some of these definitions were abstracted from other references, listed at the end of this glossary, or from:

North American's Handbook Supplement Nos.
    113 (oil),
    196 (process control),
    236 (combustion, guiding),
    248 (NOx),
    253 (pollution control); and
"Incineration of Hazardous, Toxic, and Mixed Wastes", by Gill and Quiel, published by North American Mfg. Co.

A = **ampere**, unit of electric current.

Å = **Angstrom unit**, a measure of length, particularly electromagnetic wavelengths, = one ten-millionth of a millimetre.

absolute humidity = weight of water vapor per unit weight of dry air.

absolute pressure (abs press) = gauge pressure plus barometric pressure. Absolute pressure can be zero only in a perfect vacuum.

absolute temperature (K & R) = the temperature relative to absolute zero. Molecular motion stops at absolute zero, which is −273.16 Celsius or −459.69 Fahrenheit. Absolute temperature scales are Kelvin and Rankine. See Tables C.6 and C.7 for conversions.

absolute viscosity (abs visc) = by definition, the product of a fluid's kinematic viscosity times its density. Absolute viscosity is a measure of a fluid's tendency to resist flow, without regard to its density. Sometimes termed dynamic viscosity. Usually designated μ (mu) in poise, lb mass/sec ft, or Pascal · seconds. (By contrast, see kinematic viscosity.)

absorption — See sound absorption.

absorptivity = ability of a surface to absorb radiant energy, expressed as a decimal compared to the ability of a black body, absorptivity of which is 1.0.

a-c = **alternating current.**

ACerS = **American Ceramic Society; Columbus, OH.**

acf = actual cubic feet (at existing temperature and pressure).

ACGIH = **American Conference of Governmental Industrial Hygienists; Cincinnati, OH.**

acid rain = a condition resulting from complex atmospheric and chemical phenomena, often far from the original sources, in which emissions of sulfur and nitrogen compounds and other substances are deposited on the earth as rain, snow, or fog.

acoustic absorptivity = the ratio of sound absorbed by a surface relative to the incident sound.

acoustic power = see sound power.

acoustics = the study of sound — production, control, transmission, reception, and effects of sound and of hearing phenomena.

acoustic velocity = the speed of sound in a given medium.

ACS = American Ceramic Society; Washington, DC.

adi = austempered ductile iron.

adiabatic flame temperature = a theoretical flame temperature calculated for a condition with no heat loss. See flame temperature.

adjustable port valve = a special kind of rotary plug control valve, most commonly used for automatically controlling input to a furnace. One dimension of the rectangular port opening in the rotary plug is manually adjustable, permitting on-site optimizing of the valve's resistance relative to that of the entire pipeline in which it is installed.

adsorption = the adhering of molecules, atoms, or ions (solid, liquid, or gaseous) to surface (unlike absorption which is penetration within the bulk of a liquid or solid).

aeration = addition and mixing of air, % aeration compares actual aeration with the stoichiometrically correct amount, e.g. 60% primary aeration on a premix nozzle means that 60% of the stoichiometric air requirement is supplied through the mixer and nozzle and 40% from secondary air surrounding the nozzle.

afbc = atmospheric fluidized bed combustion.

a/f ratio = air/fuel ratio. For gaseous fuels, usually the ratio of volumes in the same units. For liquid and solid fuels, it may be expressed as a ratio of weights in the same units, but it is often given in mixed units such as ft³ air/pound or ft³ air/gallon, a/f ratio is the reciprocal of f/a ratio. See Table C.10.

AFS = American Foundry Society; Des Plains, IL.

AGA = American Gas Association; Arlington, VA.

agglomerating characteristics = the tendency of a coal to bind together into a larger mass when heated.

AIChE = American Institute of Chemical Engineers; New York, NY.

AIHA = American Industrial Hygiene Association; Fairfax, VA.

air-directed burner = a burner in which the flow or pressure energy of the air supply stream is used to control the aerodynamics of air-fuel mixing and flame formation, thus needing higher air pressure, but less fuel pressure than a Fuel Directed® burner.

air-flow proving switch = a device installed in an air stream which senses air flow or loss thereof and electrically transmits the resulting impulses to the flame supervising circuit.

air/fuel ratio = the proportion of air to fuel = the reciprocal of fuel/air ratio, both expressed in volumes for gaseous fuels, but more often in weights for liquid and solid fuels.

air-jet mixer (aspirator) = a mixer using the kinetic energy of a stream of air issuing from an orifice to entrain the gas required for combustion. In some cases this type of mixer may be designed to entrain some of the air for combustion as well as the gas.

air primary = a system of fuel/air ratio control in which the demand for heat adjusts the air flow to the combustion system, and the automatic ratio control then makes a corresponding adjustment in the fuel flow.

air-ramming = a method of forming refractory shapes, furnace hearths, or other furnace parts by means of pneumatic hammers, using a plastic ramming mix.

air register = a type of burner mounting that can admit secondary air to the combustion space through openings around the burner. Also used for primary air in windbox burners.

air-setting refractories = compositions of ground refractory materials that develop a strong bond upon drying. These refractories include mortars, plastic refractories, ramming mixes, and gunning mixes. They are available in both wet and dry condition, the latter requiring addition of water to develop the necessary consistency.

air shutter = an adjustable shutter on a burner air register by means of which the amount of air induced into the furnace through the register can be controlled.

AISE = Association of Iron and Steel Engineers; Pittsburgh, PA.

AISI = American Iron and Steel Institute; Washington, DC.

ait = autoignition temperature = t at which a substance ignites spontaneously from the heat of its environment. See also minimum ignition temperatures, Table 1.10, Volume I, and "Industrial Explosion and Protection," McGraw-Hill, 1980.

Al = aluminum.

aldehyde = a class of organic compounds containing the CHO radical. Examples: acetaldehyde (ethanal), benzaldehyde (almond oil), formaldehyde, furfural. A product of incomplete combustion (pic). Aldehydes have been known to cause eye irritation, headaches, and nausea near ovens when recirculating air quenches the burner flames.

alt = altitude.

alumina (Al_2O_3) = the oxide of aluminum having a melting point of 3720 F (2050 C). In combination with H_2O, alumina forms the minerals diaspore, bauxite, and gibbsite; in combination with SiO_2 and H_2O, alumina forms kaolinite and other clay minerals.

ambient noise = the total of all sounds associated with a surrounding area.

ambient temperature = the air temperature within a given enclosure or surroundings.

AMCA = Air Moving and Conditioning Assn.; Arlington Heights, IL.

amorphous = lacking crystaline structure or definite molecular arrangement; without definite external form.

amplitude = the magnitude of a wave's variation (such as sound pressure or electrical current).

anneal = to remove internal stress by first heating and then cooling slowly.

annular orifice = a flow-measuring device used in a pipe where the velocity profile may be non-symmetrical; consists of a circular 'target' plate centered in the pipe by supporting spider-like mounts; as opposed to a concentric orifice, which has a hole in the center.

ANSI = American National Standards Institute; New York, NY.

a-od = argon-oxygen decarburizing.

a-O mix = air-oxygen mixture as in oxygen-enriched air.

Apachi = a commercial gas mix, the major constituent of which is propylene, C_3H_6.

API = American Petroleum Institute; Washington, DC. A scale adopted by the American Petroleum Institute to indicate the specific gravity of a liquid. API gravity readings are higher for less dense liquids. Therefore the API gravity for a liquid rises as its temperature rises. See also gravity. Water has an API gravity of 10°; #2 fuel oil, about 35° API.

aromatics = unsaturated hydrocarbons typified by a benzene ring structure, such as benzene (C_6H_6), toluene ($C_6H_5CH_3$), xylene [$C_6H_4(CH_3)_2$]. Aromatics are chemically active, and relatively heavy, having a high carbon/hydrogen ratio. Cracked oils containing aromatics tend to smoke or form soot when burned.

arrangement factor, F_a = a decimal expression of the portion of theoretical radiation flux that is actually "seen" by a receiving surface as compared with the maximum that a surface of the same absorptivity could receive in the ideal positioning configuration. See Table 4.9 and formula 4/1 in Volume I.

artificial fuels = man-made fuels, including all manufactured and by-product fuels. Examples are water gas, blast furnace gas, and coke.

ASHRAE = American Society of Heating, Refrigeration, and Air Conditioning Engineers; Atlanta, GA.

ASME = American Society of Mechanical Engineers; New York, NY.

ASM Intl., was American Society for Metals; Materials Park, OH.

AST = above-ground storage tank (*vs.* UST).

ASTM = American Society for Testing and Materials; Philadelphia, PA.

atmosphere (atm) = refers to a mixture of gases (usually that within a furnace). Also a unit of pressure equal to 14.696 lb per sq inch or 760 mm of mercury. Another meaning = the mixture of gases within a furnace, e.g. reducing (rich) atmosphere, oxidizing (lean) atmosphere, prepared atmosphere.

atmospheric pressure (atm press) = the pressure exerted upon the earth's surface by the weight of the air and water vapor above it. Equal to 14.696 psia or 760 mm Hg at sea level and 45° latitude.

atmospheric (ratio) regulator = also called a zero governor. A diaphragm type regulator that maintains gas pressure at atmospheric or "zero" pressure.

atmospheric system = apparatus for air/gas proportioning and mixing, using energy of a jet of low pressure gas (<14"wc) to entrain part of the required combustion air from the atmosphere.

atom = the smallest part of an element that retains the properties of that element. Sometimes used as an abbreviation for atomizing, as in atom air or atom steam.

atomization = the process of breaking a liquid into a multitude of tiny droplets. See also:

centrifugal	mechanical
compressed air	oil pressure
low pressure air	steam

atomizing air = that part of the air supplied through a burner (usually about 10%) that is used to break the oil stream into tiny droplets. The atomizing air is also used for combustion after it has broken up the oil stream.

autogenous = self-generating, exothermic.

autoignition temperature = the lowest temperature required to initiate self-sustained combustion in the absence of a spark or flame. It varies considerably with the nature, size, and shape of the hot surface, and other factors. Some vapors can be ignited by surfaces at temperatures as low as 500 F. Reference 1.j lists autoignition temperatures and other properties of many gases, liquids, and solids.

automatic control = an arrangement by which a system reacts to a change or an imbalance in one of its variables and compensates by adjusting the other variables to restore the system to the desired balance. For example, a system for automatic control of air/fuel ratio wherein a change in the combustion air input results in a corresponding change in the fuel input.

automatic fuel shutoff valve = a valve for stopping the flow of fuel automatically when a dangerous situation develops. The valve is closed by a spring force of at least 5 pounds which is tripped by de-energizing an electric or pneumatic hold-open mechanism when any connected interlock senses a dangerous condition. See automatic reset and manual reset automatic fuel shutoff valves.

automatic reset fuel shutoff valve = an automatic fuel shutoff valve that automatically reopens as soon as a normal operating condition is restored.

available carbon = carbon not combined chemically with oxygen in any way, and therefore available for combustion.

available heat = the gross quantity of heat released within a combustion chamber minus both the dry flue gas loss and the moisture loss. It represents the quantity of heat remaining for useful purposes and to balance losses to walls, openings, conveyors.

available hydrogen = hydrogen not chemically combined with oxygen in any way, and therefore available for combustion.

avdp = avoirdupois, a system of weight measure.

background noise = the ambient noise level above which signals must be presented or noise sources measured.

BACT = best available control technology.

bagasse = the fibrous material remaining after the extraction of the juice from sugar cane. Used as a fuel.

bag wall = a refractory baffle in a kiln for the purpose of channeling the course of the flame and poc.

ball clay = a highly plastic, very fine-grained, refractory bond clay that has a wide range of vitrification and burns to a light color; often high in carbonaceous matter.

ball valve = similar to a plug valve except that the rotatable element is spherical.

bar = unit of pressure ~ one atmosphere — see units in Appendix.

barometer = an instrument for measuring atmospheric pressure, usually in inches or millimetres of mercury column.

barometric damper = a balanced air valve placed so as to admit air to the flue pipe in order to maintain a constant amount of draft through a furnace. A minimum draft has the advantage of reducing the heat loss through the flue.

barometric pressure = the atmospheric pressure at a specific place according to the current reading of barometer. Standard barometric pressure is 14.696 psia, 29.92 in. Hg, or 760 mm Hg.

barrel = the unit by which petroleum products are sold. 1 barrel (bbl) = 42 US gallons.

base pressure = a standard to which measurements of a volume of gas are referred. This and base temperature should be defined in any gas measurement contract.

base temperature = a standard to which measurements of a volume of gas are referred. The standard value in the United States is 60 F for natural gas (per AGA Measurement Manual).

basic refractories = refractories consisting essentially of magnesia, lime, chrome ore, or forsterite, or mixtures of these. (By contrast, acid refractories contain a substantial proportion of free silica.)

batch-type furnace = a furnace shut down periodically to remove one load and add a new charge, as opposed to a continuous type furnace. An "in-and-out" furnace, a "periodic" kiln.

°Baumé = a scale for expressing specific gravity of liquids, designed to provide a linear scale on hydrometers. Nearly the same as the modern API. (1°Bé ≈ 1°API.) Named for Antoine Baumé (1727-1804), a French chemist widely known for his improvements in technical processes, but best known for his invention of the hydrometer which bears his name.

bbl = barrel. See units of volume in Appendix. (Many sizes.)

°Bé = degrees Baumé.

beat = periodic vibration resulting from interference of two sound waves of different frequency.

beehive kiln = a periodic downdraft kiln with a circular hearth and dome-like roof (crown).

bellows = a metallic accordian-like cylinder or box that can be compressed or expanded mechanically or with fluid pressure (like a spring), and which will return to its normal shape when the pressure is released. Bellows are used in oil expansion chambers, in some high pressure fuel shutoff valves in place of a diaphragm, and in pressure sensitive instruments and regulators.

benzene (C_6H_6) = a highly combustible liquid derived from the distillation of petroleum. Its flash point (closed cup) is 12 F (−11 C). Also known as benzol.

Bernoulli theorem = conservation of energy in the steady flow of an incompressible, inviscid fluid.

beta ratio, β = d/D, where d = orifice diameter, and D = pipe ID.

bhp = **boiler horsepower** (see definition in this glossary). The abbreviation bhp is also used for brake horsepower, which has another meaning and another value.

bisque = fired, unglazed ceramic ware.

black body = a theoretical physical concept of a body that would absorb all radiant energy incident upon it, and would emit the maximum possible radiation at a given temperature. The emissivity and the absorptivity of a black body are each 1.0. A black body would not necessarily be black in color. Lampblack and platinum black most nearly approximate a black body.

black body radiation = the theoretical rate of radiation from a black body at a given temperature.

blast burner (pressure burner) = a burner delivering a combustible mixture under pressure, normally above 0.3"wc to the combustion zone.

blast furnace gas = a gas of low Btu content recovered from a blast furnace as a by-product and used as a fuel.

blast gate = a shutoff air valve – not to be used for flow control.

blast tip = a small metallic or ceramic burner nozzle so made that flames will not blow away from it, even with high mixture pressures.

bleeder = a device designed to provide an intentional leak, usually used to reduce pressure in an impulse line.

blending = usually the addition of #2 distillate oil to residual oils to meet a certain specification (viscosity or sulfur, for example) or to make the residual oil easier to handle.

block. See burner tile.

block valve = **blocking valve** = a redundant fuel shutoff valve for protection in case of failure of the primary fuel shutoff valve. Usually automatic or manual reset type.

blow-off = lifting of a flame because feed stream velocity exceeds flame velocity.

blue (water) gas = an artificial fuel made by passing steam over incandescent carbon (usually coke), forming a mixture of hydrogen and carbon monoxide. $C + H_2O \rightarrow H_2 + CO$, the water gas reaction, is hazardous because of its high CO content.

bluff body = a solid obstruction in a fluid stream, having a broad flattened front and providing a shelter for small scale turbulence and zones of low velocity; a stability assister.

bof = **basic oxygen furnace.**

bogie hearth furnace = car hearth furnace = car bottom furnace.

boiler horsepower (bhp) = the equivalent of the heat required to change 34.5 lb per hour of water at 212 F to steam at 212 F. It is equal to a boiler heat output of 33 475 Btu/hr or 8439 kcal/h = 9.81 kW.

booster heater = a heater used to raise the temperature of oil from that required for pumping to that required for atomization. The booster heater is usually located close to the atomizer.

boundary layer = the portion of a fluid flowing in the immediate vicinity of a surface. The boundary layer has a reduced flow due to the forces of adhesion and viscosity.

Bourdon tube = a metallic tube of elliptical cross section, shaped into an arc or spiral with one end attached to an indicating, recording, or controlling device. It is used in industry to measure pressures. A pressure within the tube tends to make it less elliptical and more nearly circular, straightening its arc or unwinding its spiral, thus actuating the attached device.

Boyle's Law = Mariotte's law = the product of the volume of a gas and its pressure is a constant at a fixed temperature.

branch circuit = a secondary circuit leading from the main oil circulating loop to a burner or group of burners; a booster circuit or loop.

breeching = a passageway leading from a furnace to its chimney.

BR gal = British (Imperial) gallon.

bring-up time = the time required to raise a cold furnace, and its charge if any, to operating temperature.

British thermal unit (Btu) = the quantity of energy required to heat one pound of water from 59 F to 60 F at standard barometric pressure, = 0.252 kcal = 0.000 293 kWh.

bsw = **bottom sediment and water.** Impurities and foreign materials found in fuel oils, Part 2, Vol. I.

Btu = **British thermal unit(s).**

bulk density = the weight of a unit volume of a substance. The volume includes the volume of pores in the material, and the volume between particles, so the bulk density of a substance is usually lower than the absolute density of the substance.

bunker B, bunker C = two designations for heavy fuel oils. Now becoming obsolete. Both now fit the #6 classification, but bunker B was somewhat lighter and less viscous.

bunker oil = a heavy fuel oil formed by stabilization of the residual oil remaining after the cracking of crude petroleum.

Bunsen-type burner = a gas burner consisting of a straight tube with a gas orifice at one end. Primary air is entrained through adjustable openings around the gas orifice. The gas-air mixture burns with a short intense flame as it emerges from the tube. Those that operate on low pressure gas are called atmospheric burners. Named for R. W. Bunsen (1811-1899), who proved that furnaces of his day wasted 50% to 80% of the fuel heating value through the exhaust gases.

buoyancy controlled flame = diffusion flame in which the aspiration of combustion air into the combustible gas is controlled by the buoyancy of hot gases.

burble = a separation or breakdown of laminar flow past a body; eddying or turbulent flow resulting from this.

burn = the degree to which desired physical and chemical changes have been developed in the firing of a ceramic material. A batch of ceramic materials subjected to the firing process. To heat ceramic materials so as to change their properties.

burner = a device that positions a flame in the desired location by delivering fuel and air to that location in such a manner that continuous ignition is accomplished. Some burners include atomizing, mixing, proportioning, piloting, and flame monitoring devices.

burner refractory = refractory block with a conical or cylindrical hole through its center. The block is mounted in such a manner that the flame fires through this hole. The brick helps to maintain ignition, and reduces the probability of flashback or blow-off. Also called a burner block, burner tile, combustion tile, combustion block, refractory tile, refractory block, or quarl.

burner tile, refractory, block, quarl = a part of a burner that serves as a primary combustion chamber, often helping to determine the flame position, character, and stability. Usually constructed of high temperature refractory ceramic material.

burning (firing) = the final heat treatment in a kiln to which ceramic materials are subjected in the process of manufacture, for the purpose of developing bond and other necessary physical and chemical properties.

butane (C_4H_{10}) = a gaseous hydrocarbon fuel of the paraffin series. Often a component of LPG or bottle gas. A by-product of refinery and gas well operations. Commercial butane is a mixture of easily liquefiable hydrocarbon gases (consisting principally of butane) which is sold as "butane."

butterfly valve = a damper or throttle valve in a pipe; consisting of a rotatable disk.

butylene (C_4H_8) = a colorless, gaseous, hydrocarbon, also known as butene, C_4H_8.

by-product coke oven gas = gas given off during the process of making coke from coal. It consists chiefly of hydrogen and methane.

C = Celsius (formerly centigrade), temperature level, e.g. water freezes at 0 C; coulomb (a unit of electrical charge), one ampere second; carbon.

c = specific heat, Btu/lb °F or cal/gm °C.

C1....C9 = methane, ethane....nonane in the octane series, CH_4 through C_9H_{20}.

CAAA = Clean Air Act Amendments (USA).

calcining = roasting, a process of heating mineral materials to a temperature below their melting point for the purpose of driving off moisture or other volatiles, reducing volume, oxidizing, or reducing.

cal/g = calories per gram.

Calorie (Cal) = large calorie = kilogram calorie = kilocalorie = kcal = 1000 cal = the energy required to heat one kilogram of water (at standard atmospheric pressure) through one degree Celsius = 3.968 Btu.

calorie (cal) = gram calorie = small calorie = the quantity of energy required to heat one gram of water from 4 C to 5 C. A kilogram calorie is one thousand times as large.

calorific value, cv = the amount of heat chemically released by combustion of a unit weight or volume of a fuel; heating value.

cap or crown = the arched roof of a furnace, especially a glass tank furnace.

carbon deposition = the deposition of amorphous carbon, resulting from the decomposition of carbon monoxide gas into carbon dioxide and carbon within a critical temperature range. When deposited within the pores of a refractory brick, the carbon may build up such pressure that it destroys the bond and causes the brick to "bloat" or disintegrate.

carbon dioxide, ultimate % = the percentage of carbon dioxide that appears in the dry flue gases when a fuel is burned with its chemically correct fuel/air ratio. The theoretical maximum %CO_2 possible.

carbon/hydrogen ratio = the weight of carbon in a hydrocarbon fuel divided by the weight of hydrogen. Aromatics, for example, have higher carbon/hydrogen ratios than paraffins. Also see carbon residue and hydrocarbon.

carbon monoxide, CO = a product of incomplete combustion (pic); a colorless odorless gas harmful to humans if inhaled at a concentration >400 ppm for more than one hour.

carbon residue = the amount of carbonaceous material left from a sample of oil heated or distilled in the absence of air; an indication of carbon/hydrogen ratio; of most importance in domestic applications using #1 or #2 oil. Industrial burner systems are generally capable of handling oils with any normal carbon/hydrogen ratio, although sometimes special precautions must be taken to minimize formation of soot. Carbon residue is measured by: (1) the Conradson test with an open flame heating the distillation apparatus; or (2) the Ramsbottom test with the heat applied indirectly through a molten bath. The two methods do not agree exactly, but a conversion table is available.

car bottom furnace = car hearth furnace = bogie hearth furnace. A batch-type furnace in which the load is placed on piers on a refractory-covered car which can be loaded outside the furnace and rolled into the furnace (to constitute its floor), a door being closed behind it. This type of furnace is very versatile in types of loads it can handle and it saves time because the car and load can be rolled out for quicker cooling. Some "shuttle" types have 2 cars with doors at both ends.

carbureted blue (water) gas = water gas with gaseous hydrocarbons added, originally for the purpose of increasing the flame's luminosity, later to increase the Btu content.

CAS number = Chemical Abstract Service's assigned number used to identify a material.

castable refractory = a mixture of a heat-resistant aggregate and a heat-resistant hydraulic cement; for use, it is mixed with water and rammed or poured into place. Hydraulic setting refractory.

catalyst = a substance that can alter the rate of a chemical reaction, without itself entering into the reaction products, or undergoing a chemical change.

cc = **cubic centimetre** = 1 millilitre.

Celsius (formerly centigrade) = temperature scale. Named for Andres Celsius, a Swedish astronomer who devised the centigrade scale in 1742.

cem = continuous emissions monitoring.

CEMA = **Canadian Electrical Manufacturers Association.**

CEMS = continuous emission monitoring system.

centi, abbreviated c = prefix for submultiple 10^{-2}.

centipoise = a unit for measurement of absolute viscosity = one hundredth of a poise. The absolute viscosity of water at 20 C is approximately one centipoise.

centistoke = a unit of measurement of kinematic viscosity = one hundredth of a stoke. The kinematic viscosity in centistokes times the specific gravity equals the absolute viscosity in centipoise.

centrifugal atomizing oil burner = a burner in which oil is thrown by centrifugal force from a rotating cup into an air stream, causing the oil to break into a cone of spray.

centrifugal blower = a mechanical device for moving air by using the centrifugal force from a rotating fan, squirrel cage, or paddle wheel, to increase the air pressure in a collecting scroll. "Fan" usually implies a lower pressure than does "blower" (4 to 44 ounces per square inch, or 200 to 1900 mm of water column).

ceramics = "Products made of inorganic materials by first shaping them and later hardening them by fire." –F. Singer. Originally, only ware formed from clay and hardened by the action of heat, or the art of making such ware. Now understood to include all refractory materials, cement, lime, plaster, pottery, glass, enamels, glazes, abrasives, electrical insulating products, and thermal insulating products made from inorganic nonmetallic mineral substances.

cermet = a product consisting of a mixture of ceramic material and finely divided metal; also referred to as ceramals.

cf or cu ft = cubic foot.

CFC = ChloroFluoroCarbons.

cfd = computational fluid dynamics.

cfh = **cubic feet per hour** (ft³/hr).

CFR = **Code of Federal Regulations.**

cfs = **cubic feet per second** (ft³/sec).

CGA = **Canadian Gas Associaton; Arlington, VA.**

cgs = **centimetre-gram-second**, system of units (old metric).

CH_4 = methane, first in the paraffin series of hydrocarbons: C_2H_6 = ethane; C_3H_8 = propane; C_4H_{10} = butane; C_5H_{12} = pentane; C_6H_{14} = hexane; C_7H_{16} = heptane; C_8H_{18} = octane; C_9H_{20} = nonane; etc.

C_2H_2 = acetylene.

Charles' Law = Gay-Lussac law = for a constant mass of gas the volume varies directly with its absolute temperature if the pressure is constant; and absolute pressure varies directly with absolute temperature if the volume is constant.

checkers, checkerworks = refractory in furnace regenerators to recover heat from outgoing hot gases, and later to transmit the heat to cold air or gas entering the furnace; so-called because the brick are arranged in checkerboard patterns, with alternating brick units and open spaces.

check valve = a valve to prevent reverse flow, usually closed by reverse movement of the fluid.

Chemex = a commercial gas mix, the major constituent of which is propane, C_3H_8.

chr = condensation heat recovery.

circular kiln = a tunnel kiln in the shape of a circle, or doughnut.

circular mil = a unit for measuring the cross sectional area of wires; the area of a wire having a diameter of one mil, or one thousandth of an inch (0.001").

circulating loop = the main loop in which oil is circulated from the oil storage tanks to the branch circuits and then back to the storage tank.

closed burner = a sealed-in burner which in most cases supplies all the air for combustion through the burner itself.

closed circuit oil system = a system in which oil may be pumped completely through the circulating loop and back into the oil storage tank.

closed impeller = a blower impeller with cover plate discs attached to the sides of the blades to minimize short-circuiting. (An open impeller consists of a series of rotating blades or vanes similar to an old-fashioned paddle wheel. The blades rotate between the stationary walls of the blower housing, these walls channeling the air so that most of it flows out through the tips of the blades. Some air short circuits back to the impeller inlet.)

cm = centimetre(s), a unit of length in the cgs system of units, not recommended in the SI system; cm^2 = square centimetre(s); cm^3 = cubic centimetre(s) or cc; = 0.01 metre.

CO = carbon monoxide. ["C.O. gas" is occasionally used for coke oven gas, but it is suggested this be spelled out to avoid confusion with carbon monoxide gas.] CO_2 = carbon dioxide.

coal tar = a viscous mixture of organic compounds; a by-product of coke production. Limited fuel use.

Coanda effect = the tendency of a fluid issuing from a jet to follow the wall contour even where the wall's curvature is away from the axis of the jet; a characteristic utilized in the operation of fluidic elements; after a French engineer and inventor, born in 1885. A type E burner (Figure 6.2) is sometimes termed a Coanda burner.

co-combustion = capability of burning more than one fuel--requires Dual-Fuel™, multi-fuel, or combination burners.

co-current flow = parallel flow, opposite of counterflow.

coefficient of discharge = a factor used in figuring flow through an orifice. It takes into account the facts that a fluid flowing through an orifice will contract to a cross sectional area smaller than that of the orifice, and that there is some dissipation of energy due to turbulence.

co-firing = using a small portion of natural gas along with other fossil or waste fuels in utility, industrial, and waste-to-energy boilers to mitigate adverse environmental emissions from combustion (NOx, SOx, particulates, dioxins), and to improve operational flexibility and performance.

coke (general) = the solid product, principally carbon, resulting from the destructive distillation of coal or other carbonaceous materials in an oven or closed chamber. In gas and oil combustion, the carbonaceous material formed due to abnormal circumstances.

coke oven gas = gas saved for use as fuel when coke is made from coal in by-product ovens. Chiefly hydrogen and methane.

combination burner = a burner capable of burning either gas or oil. (Dual-Fuel™ is the North American Mfg. Co. trademark for combination burners.)

combined carbon (or hydrogen) = carbon (or hydrogen) chemically combined with oxygen and therefore unavailable for combustion.

combustibles = materials that can be burned, often including pic.

combustion = burning, or rapid oxidation.

combustion air = main air. All of the air supplied through a burner other than that used for atomization. May or may not include air induced through the burner register by a negative pressure in the combustion chamber.

combustion intensity = volumetric rate of combustion. The ratio of the fuel energy input to the flame volume.

combustion noise = flame noise, and see Figure 12.20.

combustion products = matter resulting from combustion, such as flue gases, water vapor, and ash.

combustion roar = flame noise. The driving force that sets resonant cavities (such as burner tiles) into resonance, further amplifying the original noise, e.g. on Figure 12.20 at 1000 Hz, a combustion roar of almost 80 dB resonates in a burner tile to produce almost 90 dB.

combustion safeguard = flame supervising system.

comeback = the time required for a batch-type furnace to return to temperature after the introduction of a load of ware.

composite walls = walls made up of a series of materials of various qualities. Used in heating chambers to resist temperature, abrasion, and heat loss.

compound = a distinct substance formed by the chemical combination of two or more elements in definite proportions.

compressed air = air generally at a pressure of 5 or more pounds per square inch above atmospheric pressure (>about 0.35 kg/cm² gauge).

compressed air atomizing oil burner = a burner in which compressed air is used to break the oil stream into a spray of tiny droplets.

compressibility effect = the change in density of gas or air under conditions of compressible flow.

compressible flow = flow of high pressure gas or air which undergoes a pressure drop sufficient to result in a significant reduction of its density.

condensible hydrocarbon content = the fraction of a gas that may condense as the gas is cooled. Natural gasoline is a condensible hydrocarbon present in raw natural gas.

conductance. See thermal conductance.

conduction = the transfer of heat through a material by passing it from molecule to molecule.

conductivity = ability to transfer heat or electricity. See k or thermal conductivity.

constant pilot. See pilot.

continuous furnace = a furnace operated on an uninterrupted cycle, in which the charge is being constantly added to, moved through, and removed from the furnace; as opposed to a batch-type furnace.

continuous pilot. See pilot.

controller = a device that detects a change in a process variable, and then automatically uses an external source of power to amplify the detected signal and to energize a mechanism that will correct the deviation in the process variable until it returns to a preset value. (Not the same as regulator.)

control zone = that section of a furnace within which temperature is controlled by one temperature sensing device and one set of valves or regulators (one each for air and each fuel).

convection = transfer of heat by moving masses of matter. Convection currents are set up in a fluid by mechanical agitation (forced convection) or because of differences in density at different temperatures (natural convection).

convection burner = a burner designed to produce a flue gas stream with considerable velocity, enabling convection heat transfer.

copper strip corrosion = an oil test standardized by ASTM as an indication of sulfur in the oil. Generally applied only to light distillates and gasoline.

corbel = a supporting projection of the face of a wall; an arrangement of brick in a wall in which each course projects beyond the one immediately below it to form a support, shelf, or baffle.

corebuster = device inserted inside a heat transfer tube to turbulate the flow and increase velocity.

counterflow. See Figure 9.11.

course = a horizontal layer or row of brick in a structure.

cp = combustion products--same as fg unless there is a recuperator or other heat recovery device; centipoise, unit of viscosity. (See Volume 1, Part 2.)

c$_p$ = specific heat at constant pressure, often simply written c and assumed to be at constant pressure unless otherwise specified.

Cr = chromium.

cracking = the process of breaking hydrocarbon molecules so that they recombine into both lighter and heavier molecules. Thermal cracking involves the use of high temperatures in the absence of air. Catalytic cracking uses lower temperatures and pressures in the presence of a catalyst. Catalytically cracked distillates are more stable than straight run or thermally cracked distillates, and they have a higher carbon/hydrogen ratio.

creosote = a light fuel produced by the distillation of coal tar. Limited fuel use.

criteria pollutants = CO (carbon monoxide), Pb (lead), NO_2 (nitrogen dioxide), O_3 (ozone), particulates, SO_2 (sulfur dioxide).

critical flow (sonic flow) = fluid flowing through an orifice at a velocity equal to the velocity of sound in the fluid. Under such conditions, the rate of flow may be increased (supersonic) by an increase in upstream pressure, but it will not be affected by a decrease in downstream pressure. Also, the (much lower) rate of flow at which the flow changes from the laminar to the turbulent form.

critical Reynolds number = the point where there is a radical change in the character of the flow. Below this number the fluid flow is laminar. Above this number, turbulent flow (irregular eddying and mixing motion) occurs.

cross-connected = a term used to describe two pipes or systems of flow connected to each other to provide an equalization or interplay of pressures, or to provide an impulse to a regulator for this purpose.

cross-connection = a low pressure pneumatic sensor signal line, usually conveying a burner air input pressure signal to an air/fuel ratio control regulator. Usually 3/8" or larger tubing tapped into the side of an 8d (minimum) straight air line. Also called an "impulse line".

cross-fired furnace = a furnace with fuel supplied from the side.

crossover network. See lead-lag control.

crown = a furnace roof, especially one which is dome-shaped; the highest point of an arch.

crown drift = kiln gases flowing from the firing zone toward the entrance of a tunnel kiln (counter-current to product travel). Gases flow due to pressure differential between firing zone and entrance. Most of the volume flow occurs between the top of the load and the crown.

crude oil = petroleum as it comes out of the ground. Crudes vary considerably in make-up, quality and appearance--some thin and light, others heavy and viscous. Crudes are sometimes used as fuel, after suitable cleaning and preheating.

CSA = Canadian Standards Association; Toronto, Ontario, Canada.

Cu = copper.

cu ft = cubic foot (or feet); ft^3 is preferred.

cu in. = cubic inch (or inches); $in.^3$ is preferred.

cupola = a form of shaft furnace for making cast iron.

C_v = coefficient of valve capacity = gallons of water per minute at 1 psi drop.

cv = calorific value = heating value = hv.

D = impeller diameter, pipe diameter, diameter in feet; total fluid temperature drop, degrees.

d = orifice diameter, diameter in inches.

da = dry air.

damping (of sound) = dissipation of vibration within a vibrating body to make it an efficient acoustic radiator.

Darcy friction factor = Weisbach friction factor = a factor used in determining resistance to flow of fluids in pipe. See formulas 5/17a and b, Figure 5.13a, Table 5.13b. The Darcy friction factor = 4 × the Fanning friction factor, and Darcy friction factor = 8 × the Stanton (Blasius) friction factor.

dars = data acquisition and reporting system.

DB = dry bulb temperature.

dB(A) = decibels, A-weighted scale. (Also dBA.) A unit of sound intensity, or sound pressure level taking into account a distribution of frequencies similar to the characteristics of the human ear.

dc = dense castable.

d-c = direct current.

dcp = dry combustion products--same as dfg unless there is a recuperator or other heat recovery device.

DCS = distributed control system.

dead end of a line = the end of a pipe line that does not lead back to an oil storage tank, so that the oil in that end of the line cannot be recirculated.

dead end system = an oil system that does not contain a return line to an oil storage tank; therefore the oil cannot be pumped around in a closed circuit.

dead weight relief valve = a valve in which the unrelieved weight of the plug is the force which tends to keep the valve closed. The valve opens when the pressure increases sufficiently to lift the plug against the force of gravity.

deca, abbreviated da = prefix for multiple 10.

deci, abbreviated d = prefix for submultiple 10^{-1}.

decibel = **dB** = a measure of sound intensity, a unit of sound pressure level = the ratio of a measured sound pressure to that at an arbitrary base. Also a unit of sound power level. See equations 12/3 through 12/6.

def = definition; deficiency.

deficiency of air = a supply of air which is inadequate for complete combustion of a fuel. This is the same as an excess of fuel.

deflagration = a chemical reaction accompanied by vigorous evolution of heat, such as flame or burning particles.

deg = degree (or degrees), °.

degree-day method = a procedure for estimating fuel requirements for heating buildings by adding the degrees below 65 F of the mean temperatures for each day. For example, if the mean temperatures for M-T-W were 50, 45, 40, the degree-days would total $(65 - 50) + (65 - 45) + (65 - 40) = 60$. If the degree-days for Th-F-S totaled 120, it would take twice as much fuel for Th-F-S.

delayed mixing = a process in which the fuel and air leave the burner nozzle unmixed, and thereafter mix relatively slowly, largely through diffusion. This results in a long, luminous flame, called a diffusion flame, luminous flame, or long flame. (Type F, Figure 6.2.)

delta p = pressure difference or pressure drop.

delta t = temperature difference or temperature drop.

de-NOx. See thermal de-NOx.

density = the weight of a unit volume of a substance, usually designated ρ (rho), in lb/ft^3. Also called specific weight.

detonation = an exothermic chemical reaction that propagates with such rapidity that the rate of advance of the reaction zone into the unreacted material exceeds the velocity of sound (Mach number >1.0) in the unreacted material; that is, the advancing reaction zone is preceded by a shock wave.

devitrification = the change from a glassy to a crystalline condition.

dew point = the highest temperature at which vapor condenses from a gas-vapor mixture that is being cooled.

dfg = dry flue gas. This abbreviation is peculiar to the combustion industry and may not be widely recognized in other fields.

dgb = dry gas basis.

dia = diam = diameter.

diaphragm = the thin, flexible material separating the various chambers in pressure sensing devices such as pressure regulators, pressure switches, gauges.

diaphragm burner = a burner which utilizes a porous refractory diaphragm at the port so that the combustion takes place over the entire area of this refractory diaphragm.

diatomaceous earth = a hydrous form of silica which is soft, light in weight, and consists mainly of microscopic shells of diatoms or other marine organisms. Widely used for furnace insulation.

diatomic molecule = a molecule having two atoms--such as N_2, O_2, and H_2.

diesel fuel = a distillate fuel oil similar to #2 fuel oil.

diffusion flame = a long luminous flame created by the slow diffusion mixing (delayed mixing) of parallel fuel and air streams in laminar flow; or, in a broad sense, any flame in which combustion follows from the gradual mixing of air and combustible gas after these have been introduced separately into the combustion region.

diffusivity = a measure of the rate with which heat diffuses through a material, evaluated as $k/c\rho$, conductivity divided by volume specific heat.

DIN = Deutsches Institut for Normung (German standards organization).

direct-fired heater = a heating device in which direct radiation and convection contact the load-- no muffle or radiant tube separating the poc from the product being heated.

direct spark ignition = direct electric ignition, use of an electric arc, as from a spark plug or other igniter, to light the main flame of a burner without the use of a gas or oil pilot as an intermediate step. See pilot.

discharge coefficient = a factor used in figuring flow through an orifice. It takes into account the facts that a fluid flowing through an orifice will contract to a cross sectional area smaller than that of the orifice, and that there is some dissipation of energy due to turbulence.

dissociation = the breaking up of combustion products into combustibles and oxygen (or compounds containing less oxygen), accompanied by absorption of heat. This usually occurs at high temperatures, and is one of the factors limiting the maximum temperature of a flame.

distillate = an oil obtained by condensation of hydrocarbons that have been vaporized by heating, usually in a fractionation or distillation column.

distillate oil = oil separated from crude oil by fractional distillation.

distillation = the process by which an oil (crude) is vaporized by the application of heat, and in which the products are selectively condensed so as to result in oils or fuels with desirable properties. Straight run distillation of crude oil results in the collection of gasoline, naphtha, kerosene, and #2 distillate oil (in order of increasing gravity or molecular weight). The residuals, reduced crude, or #5 and #6 fuel oils, or residuum, are collected from the bottom of the column.

distillation temperature = distillate fuel oils are specified, partially, by means of distillation temperatures; intitial boiling point, 10% point, 50% point, 90% point, and end point. These figures are obtained within a standardized test, with a closely controlled rate of heating. Used in fuel oil specification.

dm = **decimetre** = $^1/_{10}$ of a metre.

dm³ = cubic decimetre--see L.

doc = dilute oxygen combustion.

DOE = (US) Department of Energy.

downdraft kiln = an enclosed periodic kiln, round or rectangular. Hot combustion gases collect under the crown and then pass down through the ware to flues in the hearth.

DP = dew point temperature.

draft = a difference of pressure that causes a flow of air or gases through a furnace or chimney.

DRE = destruction and removal efficiency.

dry basis = flue gas analysis by an analyzer that scrubs or dries the sample before analysis; or an analysis in which all wet basis readings (except water vapor) have been divided by (1 − summation of all dry gas percentages/100%).

dry bulb temperature, DB = temperature of air. (By contrast, see wet bulb temperature.)

dry flue gas (dfg) = gaseous products of combustion exclusive of water vapor. Separation of the vapor from the flue gas (a practical impossibility) is a theoretical concept used in combustion calculations.

DSC = distributed control system.

dscf = dry standard cubic feet.

Dual-Fuel™ Burner = the North American Mfg. Co. trademark for combination gas-oil burners designed for rapid and convenient conversion from one fuel to the other by simply opening and closing valves.

Dutch oven = a combustion chamber built outside and connected with a furnace.

dynamic pressure = see velocity pressure.

e = **emissivity or emittance.** See definition of those terms. Also, a measure of the degree of oxygen enrichment, varying from 20.9% to 100%.

EAF = electric arc furnace.

EASA = Electrical Apparatus Service Association, a standards group.

E$_d$ = equivalent pipe length (of a fitting) expressed in diameters.

EDR = the rate of heat transfer (by both radiation and convection) from a radiator or convector. The equivalent direct radiation is expressed in terms of the number of square feet of surface of an imaginary standard radiator that would be required to transfer heat at the same rate as does the unit in question. One square foot of EDR gives off 240 Btu/hr for steam heating units, or 150 Btu/hr for hot water heating.

effective area of furnace openings = the area of an opening in an infinitely thin furnace wall that would permit a radiation loss equal to that occuring through an actual opening in a wall of finite thickness. The effective area is always less than the actual area because some radiation always strikes the sides of the opening and is reflected back into the furnace.

effective chimney height = the height above that necessary to overcome the pressure drop caused by the friction of gas flow.

efficiency = the percentage of gross Btu input that is realized as useful Btu output of a furnace.

EGR = FGR = exhaust gas recirculation = flue gas recirculation.

EJC = Engineers Joint Council; New York, NY.

element = one of the 103 basic substances of which all matter (compounds and mixtures) is composed.

emissivity = a measure of ability of a material to radiate energy = the ratio (expressed as a decimal fraction) of the radiating ability of a given material to that of a black body. (A "black body" emits radiation at the maximum possible rate at any given temperature, and has an emissivity of 1.0.) Assumed to be total hemispheric emissivity (all wavelengths, all directions) unless otherwise specified. By contrast, see emittance.

emittance = the ability of a surface to emit or radiate energy, as compared with that of a "black body", which emits radiation at the maximum possible rate at any given temperature, and which has an emittance of 1.0. (Emissivity denotes a property of the bulk material independent of geometry or surface condition, whereas emittance refers to an actual piece of material.)

emittance factor, F_e = the combined effect of the emittances of two surfaces, their areas, and relative positions. See Table 4.9 and formula 4/1 in Volume I.

enclosed combustion burner = a burner that confines the combustion in a small chamber or miniature furnace and only the high temperature, completely combusted gases, in the form of high velocity jets or streams, are used for heating.

endothermic reaction = a chemical reaction that absorbs heat.

end point = upper temperature limit in the distillation range of crude oil. IBP is the initial boiling point and EP is the end point.

Engler degrees = a scale of kinematic viscosity. See Volume I, Part 2.

enthalpy = total heat content, expressed in Btu per pound or kcal/kg, above that at an arbitrary set of conditions chosen as the base or zero point.

EPA = Environmental Protection Agency; Washington, DC (U.S. unless otherwise stated).

EPIC® = North American Mfg. Company's trademark for Electronic Pressure Indicating Control for furnace pressure.

EPRI = Electric Power Research Institute; Palo Alto, CA.

equilibrium = as applied to a furnace, the condition that exists when its walls have absorbed all the heat they can hold at a specific furnace temperature, so that any further flow of heat to the walls results in an equal amount of heat being transferred to the outside.

equiv. = equivalent.

equivalence ratio, φ = a means of expressing air/fuel ratio = the actual amount of fuel expressed as a decimal ratio of the stoichiometrically correct amount of fuel. (See Table C.10 in the Appendix.)

equivalent diameter = the diameter of a circle having an area equal to the area of the shape being considered = 4 × hydraulic radius.

equivalent inches of firebrick = the thickness of firebrick having the same insulating value as the material being considered.

equivalent length = the length of straight pipe that would produce the same pressure drop as a fitting or valve of the same pipe size.

equivalent pure oxygen = a measure of the dollar value of various commercial oxygen purities. See Example 13-1.

equivalent thickness = for refractory walls, the term refers to the thickness of firebrick wall that has the same insulating capability as a wall of other refractory materials. See Figures and Tables 4.5 in Volume I. For pipe insulation, this term refers to an imaginary pipe thickness that simplifies pipe heat loss calculations by relating all calculations to the outside surface area as in formula 4/6d.

ethane (C_2H_6) = hydrocarbon gas used as a fuel, frequently a component of natural gas.

évasé = a diffuser or diverging section of a duct, fan or stack. The diverging tail section of a venturi. A flow passage in which kinetic energy is converted to static pressure. (Pronounced aye-vaaz-aye.)

excess air, abbreviated XSAir = the air remaining after a fuel has been completely burned, or that air supplied in addition to the quantity required for stoichiometric combustion. A lean air/fuel ratio. Equivalence ratio less than 1.0. (See Table C.10 in the Appendix.)

excess oxygen, like excess air, is an indication of how lean or how oxidizing the combustion reaction is. For most fuels, the % excess oxygen in the flue gas is about 1/3 of the % excess air (up to about 3% oxygen or 15% excess air, above which point the ratio is progressively less than 1/3).

excess pure oxygen = a measure of degree of oxygen enrichment above pure air. For example, with 25% oxygen concentration instead of 20.9, the "excess pure oxygen" would be 25 − 20.9 = 4.1%. See Part 13.

exit temperature = the temperature of combustion gases as they leave a furnace.

exothermic reaction = a chemical reaction that liberates heat, such as the burning of a fuel.

extrusion = a process in which plastic material is forced through a die by the application of pressure.

F or °F = **Fahrenheit** (temperature level, e.g. water freezes at 32 F). This is in contrast to ° or °F which indicates temperature change or difference, e.g. the drop across the wall is 355°F, or the air temperature rose 60°F between dawn and noon.

f = friction factor.

#f = **pounds force**, as opposed to the less frequently used pounds mass.

F_a = arrangement factor. See that definition.

f/a ratio = fuel/air ratio. For gaseous fuels, this is usually the ratio of volumes in the same units. For liquid and solid fuels it may be the ratio of weights in the same units. f/a is the reciprocal of a/f ratio. (See Table C.10 in the Appendix.)

fan mixer = an air blower in which fuel gas is admitted to the inlet to be mixed with air.

Fanning friction factor = a factor used in determining resistance to flow of fluids in a pipe = 1/4 of the Darcy or Weisbach friction factor (formulas 5/17a and b, Figure 5.13a, Table 5.17b in Volume I).

fb = **firebrick.**

fbc = fluidized bed combustion (or fluid-bed combustion).

f_e = emittance factor. See that term.

feedback = a sensor message, in an automatic control system, that measures a result of the control action and re-corrects it as necessary.

feedforward = a sensor message, in an automatic control system, that anticipates the need for a correction.

femto, abbreviated f = prefix for submultiple 10^{-15}.

FERC = **Federal (US) Energy Regulation Commission.**

fg = **flue gas**--same as combustion products unless there is a recuperator or other heat recovery device.

FGD = flue gas desulfurization.

FGR = **EGR** = flue gas recirculation = exhaust gas recirculation.

FIA = **Forging Industry Association; Cleveland, OH.**

FID = flame ionization detector.

film coefficient, h_c = convection heat transfer coefficient = the reciprocal of the boundary layer film resistance = the rate of heat flow per unit area (heat flux) for each degree of temperature difference between the surface and the bulk of the fluid stream.

filter = a porous mass used to remove particles suspended in a fluid. An oil filter is actually a wire mesh strainer, but is frequently called a filter.

firebox = a combustion chamber. If external to the furnace or kiln, also called a Dutch oven or doghouse.

fireclay brick = a refractory brick manufactured substantially or entirely from fireclay.

fire point = the minimum oil temperature at which a flame is sustained for at least 5 seconds. By contrast, see flash point.

firing (burning of ceramic materials) = the final heating process to which ceramic shapes are subjected for the purpose of developing bond and other necessary physical and chemical properties.

firing rate = the rate at which air, fuel, or a fuel-air mixture is supplied to a burner, or furnace. It may be expressed in volume, weight, or heat units supplied per unit time.

firing zone = that portion of a furnace, usually a continuous furnace, through which the load passes and which is at or near the maximum process temperature.

flame = a visible shape within which combustion occurs, flame envelope.

flame blow-off = the phenomenon that occurs when a flame moves away from a burner. This often results in the flame being extinguished. A flame blows off when the fuel-air mixture leaves the burner at a velocity greater than the velocity with which the flame front progresses into the mixture.

flame character = the nature of a flame, e.g. length, size, shape, color, luminosity, velocity--usually determined by the design of the flame holder and refractory quarl, and by pressures, velocities, and directions of fuel jets and air jets.

flame front = the plane along which combustion starts, or the root of a flame.

flame holder = burner nozzle, a part of a burner that positions the flame, determines the character of the flame (length, shape, luminosity, velocity), and provides flame stability.

flame monitoring device = flame "sensor", flame "scanner", formerly flame safety device. A device for flame surveillance--ultraviolet detector, flame rod, flicker detector, infrared detector, photocell, thermopile, bimetal warp switch.

flame noise – see combustion roar and superturbulent combustion noise. See Part 12.

flame retaining nozzle = any burner nozzle with built-in features to hold the flame at high mixture pressures.

flame retention burner (flame retaining nozzle) = a burner whose nozzle is surrounded with small ports that act as pilots to relight the main flame if it blows off. The velocity through the small ports is less, so the flame almost never blows off. Also called a stick-tight nozzle.

flame speed = the rate at which a flame propagates through a combustible mixture. See flame velocity.

flame supervising system (formerly combustion safeguard) = a safety control responsive directly to flame properties, except where supplemented by the words "gas analyzer type". It senses the presence of flame and causes fuel to be shut off in the event of flame failure. See "Supervising Controls" in Part 7.

flame temperature. Theoretical flame temperature is calculated in the same way as hot mix temperature, but usually for stoichiometric air/fuel ratio. It may or may not be corrected for dissociation. See also hot mix temperature.

flame velocity = the speed at which a flame progresses into a mixture relative to the speed of the mixture. Also called flame speed, ignition velocity, rate of flame propagation. The latter sometimes refers to flame front movement in a tube whereas the other forms usually refer to measurements in quiescent mixtures or in perfectly streamlined (laminar) flames. The turbulence encountered in the tube measurements usually results in velocities about twice as great as by the other methods.

Flamex = a commercial gas mix, the major constituent of which is propane, C_3H_8.

flammability limits = the maximum and minimum percentages of a fuel in a fuel-air mixture which will burn. Sometimes called limits of inflammability.

flange taps = pressure taps that are 1 in. upstream and downstream from a thin metering orifice. Standard flanges can be purchased with these taps already drilled.

flashback = the phenomenon that occurs when a flame front moves back through a burner nozzle (and possibly back to the mixing point). Flashback occurs because the flame velocity exceeds the fuel-air mixture velocity through the burner nozzle.

flash heat = refers to an application of heat to an object in a very short time period, utilizing direct flame impingement or a high thermal head.

flashing (brick) = firing a kiln under reducing conditions to obtain certain desired colors. May be effected by adding maganese, salt, or zinc.

flash point = the temperature at which enough of an oil is vaporized to produce a flash of burning oil vapor when ignited by an external flame. Flash point and fire point are obtained (1) by the Pensky-Martens; or tag, closed cup, and (2) by the Cleveland, open cup. By contrast, see fire point.

flat arch = a furnace roof in which both outer and inner surfaces are horizontal planes. Formed of special tapered brick and held in place by their keying action. A jack arch.

Flat Flame™ = North American Mfg. Company's trademark for a Type E flame that spreads radically across the adjacent furnace wall. In scientific literature, a flat flame is one formed by a slot-like nozzle.

flow coefficient = a correction factor used for figuring volume flow rate of a fluid through an orifice ($\overline{K}$ in formula 5/37). This factor includes the effects of contraction and turbulence loss (covered by the coefficient of discharge), plus the compressibility effect, and the effect of an upstream velocity other than zero. Since the latter two effects are negligible in many instances, the flow coefficient is often equal to the coefficient of discharge.

flow noise = that portion of burner noise produced exclusively from the flow of fuel and air through the burner, irrespective of combustion.

flue gas = all gases, combustion gas, products of combustion (poc) that leave a furnace by way of a flue, including gaseous products of combustion, water vapor, excess oxygen and nitrogen. Exit gases from recuperators, waste heat boilers, regenerators, and other heat recovery devices are termed "waste gases" or "stack gases".

flue gas analysis = a statement of the quantities of the various compounds of a sample of flue gas, usually expressed in percentages by volume.

flue gas loss = also called stack loss. The sensible heat carried away by the dry flue gas plus the sensible and latent heat carried away by the water vapor in the flue gas.

fluid = any liquid, vapor, or gas or their mixtures. The terms fluid bed and fluidized bed refer to solid particles activated into a fluid-like condition by forcing a gas up through them.

fluid bed = fluidized bed.

flux = rate of flow of heat or fluid across a unit area. See also fluxing.

fluxing = addition of a substance (flux) to promote fusing of minerals, metals, glass, enamels, and refractories, or to prevent oxide formation.

FM = Factory Mutual Insurance Association; Norwood, MA.

foot valve = a check valve at the bottom end of a suction line in a tank to prevent the oil system from draining back into the tank when the pump is not operating.

forced convection = convection heat transfer by artificial fluid agitation.

forced draft = the difference in pressures that blows air into a furnace, usually produced by a fan located in the inlet air passage to the furnace.

fph, fpm, fps = **feet per hour, feet per minute, feet per second**; American or English foot-pound-second system units for velocity.

fractional distillation = the process of heating, evaporating, and condensing crude oils into various grades (or fractions) of oil. This is accomplished in a fractionating tower with the lightest materials (those with the lowest boiling points) such as gasoline and naphtha condensing at the top, and the heaviest materials (those with the highest boiling points) such as the heavy gas oils condensing at the bottom. The fraction that does not evaporate, the residual oil, is also collected in this process.

free field = an area in which the boundaries have no effect on noise measurements.

frequency (of sound) = the speed with which its cyclical pressure variation occurs, dp/dt (the differential of pressure relative to time). Frequency = the reciprocal of periodic time. (See also low-frequency sounds, and high-frequency sounds.)

friable = easily reduced to a granular or powdery condition.

friction factor = a factor used in calculating loss of pressure due to friction of a fluid flowing through a pipe. See Darcy and Fanning.

FRP = fiberglass-reinforced plastic.

ft = foot (or feet); **ft²** = square foot (or feet); **ft³** = cubic foot (or feet).

fuel = any substance used for combustion as a heat source.

fuel/air ratio = the ratio of the fuel supply flow rate to the air supply flow rate when both rates are measured in the same units under the same conditions; the reciprocal of air/fuel ratio. (These terms are often used interchangeably in qualitative discussions.)

Fuel-Directed® burner = the North American Mfg. Co. trademark for a burner in which the flow or pressure energy of the fuel supply stream is used to control the aerodynamics of air-fuel mixing and flame formation, thus needing higher fuel pressure, but less air pressure than an air-directed burner.

fuel-lean = ϕ = <1.0, less fuel than required for stoichiometric combustion.

fuel NO = nitric oxide from N atoms in the fuel in forms other than N_2, i.e. from fixed nitrogen.

fuel oil = a petroleum product used as a fuel. Common fuel oils are classified as:
  #1 = distillate oil for vaporizing type burners.
  #2 = distillate oil for general purpose use, and for burners not requiring #1.
  #4 = blended oil intended for use without preheating.
  #5 = blended residual oil for use with preheating facilities. Usual preheat temperatures are 120 to 220 F.
  #6 = residual oil, for use in burners with preheaters permitting a high viscosity fuel. Usual preheat temperatures are 180 to 260 F.

fuel primary = a system of air/fuel ratio control in which the demand for heat adjusts the fuel flow to the combustion system, and the automatic ratio control then makes a corresponding adjustment in the air flow.

fuel-rich = φ = >1.0, more fuel than required for stoichiometric combustion.

fuel shutoff valve = any valve for stopping the flow of fuel, but this term is frequently used to mean an automatic fuel shutoff valve, which is spring loaded and tripped by de-energizing an electric or pneumatic hold-open mechanism when any connected interlock senses a dangerous condition. Automatic reset fuel shutoff valves reopen as soon as a normal operating condition is restored. Manual reset automatic fuel shutoff valves must be reopened by hand after the dangerous condition is rectified and the hold-open mechanism re-energized.

fuel train = the fuel handling system between the source of fuel and the burner. It may include regulators, shutoff valves, pressure switches, flow meters, control valves.

fully-metered flow control = use of volumetric flow measuring devices to control air/fuel ratio, as opposed to the simpler pressure-balanced a/f ratio control or linked valves (area control).

fundamental frequency = the lowest frequency at which a column of gas or a body vibrates. Also called the first harmonic.

furnace = an enclosed space in which heat is intentionally released by combustion, electrical devices, or nuclear reaction.

furnace pressure = the gauge pressure that exists within a furnace combustion chamber. The furnace pressure is said to be positive if greater than atmospheric pressure, negative if less than atmospheric pressure, and neutral if equal to atmospheric pressure.

Furol, Saybolt = a scale used for measuring the viscosity of heavy oils. The instrument has a larger orifice and is used at a higher temperature than the Saybolt Universal instrument used for lighter oils.

fusion = a state of fluidity or flowing, in consequence of heat; the softening of a solid body, either through heat alone or through heat and the action of a flux, to such a degree that it will no longer support its own weight, but will slump or flow. Also the union or blending of materials, such as metals, upon melting, with the formation of alloys.

fusion point = the temperature at which the solid and liquid states of a substance can exist together in equilibrium; also designated the melting point or freezing point of a substance.

G = gas or density of a gas relative to that of standard air. See SGA and SGW; giga--see below.

g = **gram(s)** = 0.00220 pound avdp; acceleration of gravity = 32.2 ft/sec² or 0.908 m/s².

gal = gallon (US gal unless specified imperial).

GAMA = Gas Appliance Manufacturers Association.

gas = in broad general terms, any substance that is not a liquid or a solid--air, for example. A perfect gas is one that conforms to Charles' Law, and the characteristic equation of a perfect gas, $pV = mRT$. (By contrast, see vapor.) See also the following types of gases: blast furnace, blue (water), butane, carbureted water, coke oven, high pressure, liquefied natural, liquefied petroleum, low pressure, natural, producer, propane, synthetic, water.

gas booster = a gastight blower used to increase gas pressure.

gas gravity = the ratio of the weight of a given volume of a gas to the weight of an equal volume of air (0.0765 lb/ft³). Sometimes called specific gravity.

gas-jet mixer (inspirator) = a mixer using the kinetic energy of a jet of gas issuing from an orifice to entrain all or part of the air required for combustion.

gas mixer = any device for mixing gas and air, such as a fan, aspirator, or inspirator.

gas oil = a heavy distillate oil; a product of straight run distillation, slightly heavier than #2 fuel oil.

gas-oil burner = a burner designed to burn gas and oil simultaneously.

gasoline = a fuel for internal combustion engines. A distillate lighter than kerosene. May be straight run, cracked, blended. Molecular weight is similar to octane, C_8H_{18}.

gas train = the fuel train for gas.

gauge pressure = the difference between atmospheric pressure and the pressure the gauge is measuring. Also written gage pressure.

GHV = **gross heating value** = higher heating value--see those definitions.

giga, abbreviated G = prefix for multiple 10^9.

gigajoule = **GJ** = 948 000 Btu.

glass = an inorganic product of fusion that has cooled to a rigid condition without crystalizing.

glost fire = the process of kiln-firing bisque ware to which glaze has been applied.

gph, gpm, gps = gallons per hour, gallons per minute, gallons per second.

gr = grain(s) avoirdupois = 1/7000th of a pound.

grades of fuel oil = voluntary commercial standards recommended by the ASTM for different classifications of fuel oils, based on several characteristics among which the most important are specific gravity and viscosity. See listing under "fuel oil".

gravimetric analysis = an analysis based upon the weights of the compound parts.

gravity, specific = a measure of the density of a liquid relative to that of water (62.43 lb/ft³). By contrast, see gas gravity.

gravity, standard = g, the standard accepted value for the force of the earth's gravity. A gravitational force that will produce an acceleration equal to 32.17 feet per second, or 9.80 metres/sec². The actual force of gravity varies slightly with altitude and latitude. The standard was arbitrarily established as that at sea level and 45 degrees latitude.

greenhouse gases = principally carbon dioxide, methane, CFCs, nitrous oxide, and ozone, all of which are opaque to reradiation of infrared heat from earth to space. (They are transparent to the wavelength of solar radiation--hence the thermal build-up as in a greenhouse.)

GRI = **Gas Research Institute; Chicago, IL.**

gross heating value = higher heating value = the total heat obtained from combustion of a specified amount of fuel and its stoichiometrically correct amount of air, both being at 60 F when combustion starts, and the combustion products being cooled to 60 F before the heat release is measured. Also called higher heating value. By contrast, see net or lower heating value. Calorific value. Heat of combustion.

grout = a suspension of mortar material in water, of such consistency that when it is poured upon horizontal courses of brick masonry, it will flow into vertical open joints.

H, H$_2$ = atomic hydrogen, molecular hydrogen.

h = enthalpy = heat content, usually in Btu/lb; or h = heat transfer coefficient--a film coefficient for convection or a surface coefficient for radiation = rate of heat flow per unit area (heat flux) for each degree of temperature difference between the surface and the heat receiver (fluid stream in the case of convection or another surface in the case of radiation); or h = hour(s).

hack = a stack of bricks in a kiln or on a kiln car.

harmonic = a multiple of the fundamental sound frequency.

h$_c$ = heat transfer coefficient for convection; also called film coefficient, $h_{c_o} = h_{o_c}$ = the coefficient on the outside surface; $h_{c_i} = h_{i_c}$ = the coefficient on the inside surface.

HCl = hydrochloric acid.

HCN = hydrogen cyanide = hydrocyanic acid = an intermediate compound formed in the process of chemical NOx formation.

head = the pressure difference that causes the flow of a fluid in a system. When applied to liquids it is usually measured in height of liquid column. See also velocity head.

header = in piping, manifold or supply pipe to which a number of branch pipes are connected. In brickwork, a brick laid flat with its longest dimension perpendicular to the wall face.

heat content = the sum total of latent and sensible heat stored in a substance minus that contained at an arbitrary set of conditions chosen as the base or zero point. It is usually designated h, in Btu per pound, but may also be expressed in such units as Btu per gallon and Btu per cubic foot if the pressure and temperature are specified.

heat exchanger = any device for transferring heat from one fluid to another without allowing the fluids to mix.

heat flux = q = Q/A = the rate of flow of heat through a unit area, in Btu/hr ft^2.

heat of combustion = the heat released by combustion of a unit quantity of a fuel, measured in calories, joules, or Btu. Heating value. Calorific value.

heat of fusion = the heat given off by a unit weight of a liquid freezing to a solid or gained by a solid melting to a liquid, without a change in temperature.

heat of vaporization = the heat given off by a unit weight of vapor condensing in a liquid or gained by a liquid evaporating to a vapor, without a change in temperature.

heat receiver = heat sink, load.

heat transfer = flow of heat by conduction, convection, or radiation. This term is often used to mean heat transfer rate.

heavy oil = a term denoting residual oil, as opposed to light or distillate oil. Fuel oils #4, 5, and 6 and "Bunker C" are "heavy oils".

hecto, abbreviated h = prefix for multiple 10^2.

heptane = C_7H_{16}.

hertz = **Hz** = cycles per second.

hexane = C_6H_{14}.

h_f = enthalpy of liquid at saturation temperature.

h_{fg} = change of heat content (or enthalpy) during vaporization of 1 lb of liquid; latent heat of vaporization.

H-fuel = a by-product liquid fuel, containing heptane, similar to naphtha, sold by Sun Oil Co. C/H ratio = 5.3. Gravity relative to water = 0.70. Higher heating value = 119 000 Btu/gal. Distillation: 5% at 147 F, 50% at 185 F, 90% at 225 F, EP = 290; recovery = 98%.

Hg = mercury; "Hg or mmHg measure height on a manometer to indicate pressure.

"Hg = inches of mercury column, a unit used in measuring pressure. One inch of mercury column equals a pressure of 0.491 psi. See Table C.5 in the Appendix.

"Hg absolute = total pressure, including atmospheric, measured in inches of mercury column. One inch of mercury column equals a pressure of 0.491 psi. Standard barometric pressure is 29.92 inches of mercury column.

"Hg vacuum = a scale for measuring pressures less than atmospheric. A reading of zero indicates atmospheric pressure, while 29.92 would indicate no pressure, or a perfect vacuum.

h_g = heat content (or enthalpy) of 1 lb of saturated vapor, i.e. vapor at the boiling point. Includes heat content of the liquid plus latent heat of vaporization.

HHV or hhv = higher heating value, gross heating value. (See below.)

h_i = heat transfer coefficient on the inside = $h_{i_c} + h_{i_r}$, where h_{i_c} = convection or film coefficient on the inside, and h_{i_r} = radiation or surface coefficient on the inside.

high-alumina refractories = alumina-silica refractories containing 45% or more alumina. The materials used in their production include diaspore, bauxite, gibbsite, kyanite, sillimanite, andalusite, and fused alumina (artificial corundum).

high-duty fireclay brick = fireclay brick which have a PCE not lower than cone 31½ or above 32½-33.

higher heating value = gross heating value = the total heat obtained from combustion of a specified amount of fuel and its stoichiometrically correct amount of air, both being at 60 F when combustion starts, and the combustion products being cooled to 60 F before the heat release is measured. By contrast, see net or lower heating value. Calorific value. Heat of combustion.

high fire = a relative term meaning that the input rate to a burner or combustion chamber is at or near its maximum.

high frequency sound = high-pitched (soprano) sound as from the right end of a keyboard instrument. (See also frequency.)

high pressure air system for gas-air proportioning and mixing (air pressure = 5 psig or higher) = a system using the momentum of a jet of high pressure air to entrain gas, or air and gas, to produce a combustible mixture.

high pressure gas = gas at pressures greater than 2 psi.

high pressure switch = a device to monitor liquid, steam, or gas pressure and arranged to shut down the burner at a preset high pressure (normally closed).

HiRam®, HiVam™ = North American Mfg. Company's trademarks for large high velocity (high momentum) burners.

hl = heat loss, Btu/ft² hr.

H₂O = water.

"H₂O = "wc = inches of water gauge = inches of water column = a measure of pressure. 1.732"wc = 1 osi. See Table C.6 in the Appendix.

h_0 = heat transfer coefficient on the outside = $h_{oc} + h_{or}$, where h_{oc} = convection or film coefficient on the outside, and h_{or} = radiation or surface coefficient on the outside.

horsepower (hp) = a unit of power equal to 550 foot pounds per second, or 33 000 foot pounds per minute. In contrast, see boiler horsepower.

hot mix temperature = a theoretical average temperature of all products of combustion of a fuel with air, including excess air assuming a plug flow reaction with no heat loss. Sometimes called flame temperature, but flame temperature more often implies an actual or measured temperature. See also flame temperature.

hp = horsepower.

HPG = a commercial gas mix, the major constituent of which is propylene, C_3H_6.

hr = hour(s). See h.

h_r = heat transfer coefficient for radiation; also called the surface coefficient. See formula 4/2, Volume I. $h_{ro} = h_{or}$ = the coefficient on the outside; $h_{ri} = h_{ir}$ = the coefficient on the inside.

hrt = horizontal return tubular boilers.

H₂S = hydrogen sulfide.

hs = heat storage, Btu/ft².

ht = heat.

hthw = high temperature hot water.

HTV = hi temp/hi vel drying (paint finishing).

humidity--see absolute humidity and relative humidity.

hydrate (verb) = to combine chemically with water.

hydraulic radius = the cross sectional area of a flow path divided by its perimeter.

hydrocarbon = any of a number of compounds composed of carbon and hydrogen atoms. The four basic classes of hydrocarbon compounds in petroleum are paraffins, olefins, aromatics, and naphthenics.

Hz = hertz, unit of frequency = 1 cycle/second. (See above.)

I = inlet fluid temperature, F.

i₁, i₂ = temperatures at 1st and 2nd interfaces from the hot face.

id, ID = inside diameter, inside dimension; induced draft.

ideal combustion = perfect combustion = stoichiometric combustion = 'on ratio' = combustion occurring at stoichiometric air/fuel ratio. See stoichiometric ratio.

ifb = insulating firebrick.

ignition = the act of starting combustion.

ignition pilot. See pilot.

ignition temperature = the lowest temperature at which a fuel-air mixture can proceed as flame with an oxidation rate that releases heat faster than heat is lost to the surroundings. See Table 1.10, Volume I.

ignition velocity = flame velocity--see that definition.

IGT = Institute of Gas Technology; Chicago, IL.

IHEA = Industrial Heating Equipment Association.

impact tube (Pitot tube) = a tube with an open end immersed in a stream, and pointed upstream. The other end of the tube is connected to a pressure gauge, which then indicates the total pressure due to the "ram effect" of the flow.

impeller = a blower impeller consisting of a series of rotating blades or vanes, similar to an old-fashioned paddle wheel, the purpose of which is to impart velocity to air or a gas.

impulse line = a small diameter pipe or tube used to convey pressure from a piping system to a diaphragm or bellows-operated mechanism. (See cross-connected.)

in. = " = inch(s); **in.²** = square inch(es); **in.³** = cubic inch(es). 1" = 0.02540 m.

inches of mercury column = "Hg = a unit used in measuring pressures. One inch of mercury column equals a pressure of 0.491 psi. See Table C.5 in the Appendix.

inches of water column = "wc = a unit used in measuring pressures. One inch of water column equals a pressure of 0.578 osi. See Table C.5 in the Appendix.

incomplete combustion = combustion in which fuel is only partially burned, and is capable of being further burned under proper conditions. An example is the case of carbon burning to form CO. With more air, it would burn to CO_2.

indirect-fired heater or dryer = a heating device in which the poc do not contact the load being heated--separated by a muffle or radiant tube.

induced air = air that flows into a furnace through openings because the furnace pressure is less than atmospheric pressure. Also air brought into a furnace by entrainment in a high velocity stream.

induced draft = gas flow caused by a furnace exit pressure less than the furnace pressure. It may be produced by natural or artificial means.

induced draft fan = a fan or blower that produces a negative pressure in the combustion chamber either by taking its suction from the combustion chamber or by induction means.

industrial heating = the direct application of heat to an industrial process, such as in the metallurigical and ceramic industries; as opposed to domestic, space, process, and power heating.

inerts = noncombustible substances in a fuel.

infrared burner = a radiation type burner. This term is most commonly applied to ceramic grid or alloy mesh gas burners for low temperature ovens or people-heating by direct radiation instead of "space" heating, but Figures 6.13, 6.14, and 6.28 show industrial type infrared burners.

input controller = firing rate control = a system that automatically determines when more heat is needed and automatically opens a valve or valves to meet that need. Usually temperature-actuated, or pressure-actuated as in the case of a steam boiler.

input rate = the quantity of heat, fuel, or air supplied per unit time, measured in volume, weight, or heat units.

inspirator = inspirator mixer = gas-jet mixer, a mixer using the kinetic energy of a jet of gas issuing from an orifice to entrain all or part of the air required for combustion.

insulating firebrick = ifb = a refractory brick characterized by low thermal conductivity and low heat capacity.

insulation = a material that is a relatively poor transmitter of heat. It is usually used to reduce heat loss from a given space.

interface = a surface regarded as the common boundary between two solids or fluids, as in composite furnace walls, or between air and fuel streams from a burner.

interlock = an electrical, pneumatic, or mechanical connection between elements of a control system that verifies conditions satisfactory to a proper operating sequence; and which commands a shutdown of the system when a dangerous or unwanted condition develops. Examples: excess furnace temperature limit, low gas pressure limit.

intermittent pilot. See pilot.

interrupted ignition = same timing and function as "interrupted pilot", but may be direct electric ignition or a pilot flame. See pilot.

interrupted pilot. See pilot.

inviscid fluid = ideal or non-viscous fluid; flows without energy dissipation; supports no shear stress.

i/o = input/output.

IR = infrared.

IRI = Industrial Risk Insurers; Hartford, CT.

ISA = Instrument Society of America; Research Triangle Park, NC.

ISO = International Standards Organization.

iso = a prefix before a chemical compound indicating an isomer atomic arrangement.

iso-C_4H_{10} = isomer of butane, iso-butane.

J = joule(s) = unit of energy, work, or heat = 1 N·m = 0.000 948 Btu = 0.239 cal.

jamb = a vertical structural member forming the side of an opening in a furnace wall.

JIC = Joint Industrial Council, an electrical standards organization.

K or °K = Kelvin (absolute Celsius). See unit conversions in Appendix.

$\overline{K}$ = flow coefficient. See Table 5.21, Volume I.

k = thermal conductivity of insulation, Btu in./ft² hr °F; k = (prefix, pronounced kilo) = thousands, × 10³; k = resistance coefficient (see Table 5.21, Volume I).

kaolin = a white-burning clay having kaolinite as its chief constituent. Specific gravity = 2.4 to 2.6. The PCE of most commercial kaolins ranges from cone 33 to cone 35.

kcal = kilogram-calorie. See calorie.

k-factor = the thermal conductivity of a material.

kerosene = a light liquid petroleum fuel, mostly nonane, C_9H_{20}. A constituent in #1 and #2 fuel oils and jet propulsion fuels.

ketone = a class of liquid organic compounds in which the carbonyl group, C double bonded to O, is attached to two alkyl groups. Used primarily as solvents. Examples: acetone, MEK.

key = in furnace construction, the uppermost or the closing brick of a curved arch.

kg = kilogram (force or mass); 1 kg = 2.205 lb avdp.

kg/m³ = kilograms per cubic metre, a measure of density. 1 kg/m³ = 0.062 43 lb/ft³. (See Table C.6 in the Appendix.)

kiln = a heated chamber used for the burning, hardening, and/or drying stage of an industrial (usually ceramic) process.

kilo, abbreviated k = prefix for multiple 10³.

kilocalorie (kcal) = 1000 calories = the energy required to raise one kilogram of water one degree Celsius. (See Table C.6 in the Appendix.)

kilohertz = kHz = 1000 Hz.

kilowatt-hour (kWh) = a unit of work or heat, equivalent to that resulting from the use of electricity at the rate of one kilowatt for one hour. It is equal to 3412 Btu.

kinematic viscosity (kin visc) = the relative tendency of fluids to resist flow; equal to the absolute viscosity divided by the density. Usually designated ν (nu) in stokes, ft²/sec, SSU, SSF. See Table 2.6 and Figure 2.7, Volume I.

kip – thousand pound.

kk = million.

km = kilometre. 1 km = 3281 ft.

knot = 1 nautical mile per hour = 1.15 statute miles per hour.

kPa = **kiloPascal** = 0.145 psi pressure.

kp(T) = equilibrium constant at constant pressure = a measure of concentration of molecules in an end product mixture, relative to the concentration; and of molecules in the starting mixture, as a function of temperature.

kW = **kilowatt.** 1 kW = 3412 Btu/hr.

kWh = **kilowatt-hour(s).**

L = **litre** = dm^3. Also length or thickness of a conduction path, also designated X in some textbooks. See formulas 4/3 and 4/6, Volume I; = length of pipe, ft, corresponding to drop D.

ladle = a refractory-lined vessel used for the temporary storage or transfer of molten metals.

laminar flow = streamline flow = viscous flow = the flow of a viscous fluid in which the particles of fluid move in straight lines parallel to the direction of flow.

large port burner = a burner characterized by a single large discharge opening or nozzle.

latent heat = heat absorbed or given off by a substance without changing its temperature, as when melting, solidifying, evaporating, condensing, changing crystalline structure.

latent heat of fusion = heat given off by a unit weight of a liquid freezing to a solid or gained by a solid melting to a liquid, without a change in temperature.

latent heat of vaporization = heat given off by a unit weight of a vapor condensing to a liquid or gained by a liquid evaporating to a vapor, without a change in temperature.

lb = pound(s) (avdp unless otherwise stated); unit of force or mass.

lb/ft^3 = pounds per cubic foot = a measure of density. 1 lb/ft^3 = 16.02 kg/m^3. (See Table C.6 in the Appendix.)

LCD = liquid crystal display.

L-D process = a process for making steel by blowing oxygen upon or through molten pig iron, whereby most of the carbon and impurities are removed by oxidation.

lea = low excess air.

lead-lag control = control scheme to prevent going rich when changing firing rate--via crossover network or stepwise limits.

lean mixture = a mixture of fuel and air in a premix burner system in which an excess of air is supplied in relation to the amount needed for complete combustion.

lean ratio = a proportion of fuel to air in which an excess of air is supplied in relation to the amount needed for complete combustion of the fuel.

led = light-emitting diode. A solid state diode that emits light when forward biased.

LEL = lower explosive limit.

lhv = lower heating value = net heating value.

lifting = blow-off of a flame because fuel stream velocity exceeds flame velocity.

light fuel oil = a term designating a distillate fuel oil, generally grade #1 or #2.

light-up = the entire procedure of igniting a burner or system of burners.

limit control = interlock.

limiting orifice = a flow restriction, often a very sensitive adjustable valve as a V-port, for restricting the flow of a fluid. Specifically for initial setting of air/fuel ratio by limiting the fuel flow.

limiting orifice valve – a fuel flow control device, usually manually adjustable, for setting fuel/air ratio.

limits of flammability = see flammability limits, explosive limit, low limit.

line burner = a burner whose flame is a continuous "line" from one end to the other.

lintel = a horizontal supporting member spanning a wall opening.

liquefied petroleum gas (LPG) = "bottle gas" = propane and/or butane (often with small amounts of propylene and butylene) sold as a liquid in pressurized containers. Usually a by-product of natural gas or gasoline processing.

litre, formerly liter (L) = a measure of volume in the old cgs metric system, not recommended in the new SI metric system = 1000 cc = 0.001 m^3 = 0.2642 US gallons. See Table C.6 in the Appendix.

lmtd = logarithmic mean temperature difference.

ln, log_e = natural logarithm, logarithm of the base e, where e = 2.718. Example: "ln 2 = 0.69" means the power to which e must be raised to produce 2 is 0.69, or $e^{0.69}$ = 2.

LNG = liquefied natural gas.

LNI = low NOx injection system using nozzles for feeding air, oxygen, or fuel around the periphery of a flame to reduce NOx emissions, as in fuel-staging or air-staging. (Please note that the nozzles do **not** inject NOx.)

load, boiler = the quantity of steam a boiler must produce, usually measured in lb per hr.

load coupled combustor = a combustion chamber in which the combustion zone and the load share a common wall and are close together so that there will be appreciable heat transfer from the combustion zone.

log = log_{10} = log to the base 10 = common logarithm.

logarithmic mean temperature difference (lmtd) = a term used in evaluating heat exchanger performance; defined in formula 4/8, Volume I.

log_e, ln = logarithm to the base e, where e = 2.718; natural logarithm.

low Btu gas = a fuel gas of low calorific value, such as producer gas.

low CV gas = low Btu gas.

lower heating value (lhv) = net heating value (nhv) = net calorific value. The gross heating value minus the latent heat of vaporization of the water vapor formed by the combustion of the hydrogen in the fuel. For a fuel with no hydrogen, net and gross heating values are the same.

low fire = a relative term meaning that the input rate to a burner or combustion chamber is at or near the minimum.

low-fire start = the firing of a burner with fuel controls in a low-fire position to provide safer operating conditions during light-off.

low frequency sounds = the low-pitched (bass) sounds, as from the left end of a piano keyboard. (See also frequency.)

low pressure air = air generally at a pressure less than two pounds per square inch gauge pressure. Produced by a centrifugal blower or a turbo blower.

low pressure air atomizing oil burner = a type of burner that uses a relatively large amount of air at a pressure between 1 and 2 psi to atomize the oil.

low pressure air system for gas/air proportioning and mixing (air pressure up to 5 psig) = a system using the momentum of a jet of low pressure air to entrain gas to produce a combustible mixture.

low pressure gas = gas at a pressure less than 2 psi.

low pressure switch = a device to monitor liquid, steam, or gas pressure and arranged to shut down the burner if the pressure falls below a preset low pressure limit (normally open).

L_p = SPL = sound pressure level, dB.

l-p = low pressure.

LPG = liquefied petroleum gas, usually propane and/or butane. Bottled gas. From natural gas liquids or a refinery by-product.

LSI = large scale integration.

luminous flame burner = a long flame burner. A burner that purposely has poor mixing of fuel and air so that the fuel takes more time to burn, thus producing a long, luminous flame.

L_w = PWL = sound power level, dB.

M (prefix, pronounced mega) = millions, $\times 10^6$ (formerly, thousands as per the Roman numeral M).

m = metre; (as a prefix, pronounced milli) = thousandths, $\times 10^{-3}$; measure of length; m^3 = cubic metre.

#m = pounds mass, as opposed to the more common meaning of pounds, which is force.

mA = milliampere.

main air = combustion air = the air supplied through a burner, but not including that air used for atomizing.

main circulating loop = the section of an oil handling system that delivers the oil from storage to the branch circuits and returns the unused oil to the storage tank.

makeup air = air supply to a building to make up for exhausted air.

makeup water = water added to a boiler, tank, or some other container to replace water which has been vaporized, thus maintaining the proper water level.

manifold = header = a supply pipe from which a number of branch pipes are fed. (See also plenum.)

manometer = a U-shaped tube, with liquid in the bottom of the U, used for measuring gauge pressure or pressure differences of fluids. The U-tube is partially filled with a liquid of density different from that of the fluid being measured. When different pressure lines are connected to the two ends of the U-tube, the liquid level rises in the low pressure side, and falls correspondingly in the high pressure side. The difference in height of the two liquid columns is proportional to the difference in pressure, and is measured in inches or millimetres of liquid column.

manual ignition = the lighting of a burner by use of a manually applied portable gas torch, or gas pilot. (The terms manual and manually do not necessarily imply that the pilot is hand-held--but that it is not automatically ignited.)

manual reset automatic fuel shutoff valve = manual reset valve = M-R valve, a fuel shutoff valve that automatically closes by spring action when its hold-open mechanism is electrically or pneumatically tripped by any connected interlock sensing a dangerous condition. It must be reopened by hand after the dangerous condition is rectified and the hold-open mechanism re-energized.

manufactured gas = any gas made artificially on a large scale for use as fuel.

Mapp gas = stabilized methylacetylene propadiene. A fuel patented by Dow Chemical Co., originally as a substitute for acetylene. See Table 2.12, Volume I.

mass flow air/fuel ratio control = fully metered air/fuel ratio control with temperature compensation to correct volumetric metering to weight flow metering.

matter = anything that has mass or weight, and occupies space.

max = maximum.

mean stack temperature = the average temperature of flue gases in a chimney, sometimes approximated by measuring the temperature of the gases at a point midway between the breeching inlet and the top of the chimney.

mechanical draft = a pressure differential produced by machinery, as a fan or blower.

mechanical mixer = a device that uses mechanical means to mix gas and air, and compress the mixture to a pressure suitable for delivery to its point of use. These utilize either a centrifugal fan or mechanical compressor with a proportioning device on its intake.

mechanical pressure atomizing burner = a burner in which oil under pressure is permitted to expand through a small orifice, causing the oil to break into a spray of fine droplets. (The same principle as in a garden hose nozzle.)

mega, abbreviated M = prefix for the multiple 10^6.

melting point = fusion point, the temperature at which crystalline (solid) and liquid phases having the same composition coexist in equilibrium. Few refractory materials have true melting points. Most melt progressively over a relatively wide range of temperatures.

mesh = mesh size, a size of screen or of particles passed by a screen in terms of the number of openings per linear inch.

methane (CH_4) = a gaseous hydrocarbon fuel. It is a principal component of natural gas, marsh gas, and sewage gas.

methanol = methyl alcohol = wood alcohol = CH_3OH, a poisonous, colorless, volatile, water-soluble liquid produced by distillation of wood or incomplete oxidation of natural gas. Its liquid form has a specific gravity of 0.793; its vapor form, a gas gravity of 1.1052. Reid vapor pressure is 4.5 psi. Viscosity at 20 C is 0.0593 centipoise, 0.748 centistokes. Heating values are 64 000 gross Btu/gal, 56 000 net Btu/gal. Open cup flash point is 52 to 60 F. Coefficient of cubical expansion is 0.0012/°C. Corrosive to aluminum and lead. See Tables 1.8, 1.10, 2.1, Volume I and Table 8.9, Volume II.

metre (m), formerly meter = SI unit of length. See Table C.6 in the Appendix.

mfg = manufactured.

Mg = magnesium.

m^3/hr = cubic metres per hour = a measure of volume flow rate. 1 m^3/hr = 0.5887 cfm = 4.403 US gpm. See Table C.6 in the Appendix.

MHF = multiple hearth furnace.

micro, abbreviated μ (mu) = one millionth = prefix for submultiple 10^{-6}.

microbar = one millionth of a bar (unit of pressure).

micron (μ) = one thousandth of a millimetre (0.001 mm) = one millionth of a metre.

milli, abbreviated m = prefix for submultiple 10^{-3}.

millilitre = 0.001 L = 1 cc. See litre. (See Table C.6 in the Appendix.)

min = minimum.

mineral = a natural inorganic substance sometimes of variable chemical composition and physical characteristics. Most minerals have a definite crystalline structure; a few are amorphous.

mineral matter = inorganic elements or compounds found in a natural state. As far as combustion is concerned, mineral matter, water, inert gases, and oxygen comprise the non-combustible parts of fuels. Thus all ash is mineral matter.

minimum firing rate = the lowest input rate for a burner or a process.

minimum ignition temperature = the lowest temperature at which combustion of a given fuel can start. (Sometimes shortened to "ignition temperature".)

mixer, gas = a device used to mix gas and air before delivery to a burner; an aspirator, an inspirator, or a fan mixer.

mixture, lean - an air-fuel mixture containing too little fuel or too much air for perfect combustion.

mixture, rich = an air-fuel mixture containing too much fuel or too little air for perfect combustion.

MJ = megajoule = 1 000 000 joules = 947 Btu = 239 kcal.

mm = millimetre = one thousandth of a metre.

mm Hg = millimetres of mercury column, a measure of pressure. 1mm Hg = 0.019 33 psi. (See Table C.5 in the Appendix.)

mm wc = mm H_2O = millimetres of water column, a measure of pressure. 25.4 mm wc = 1"wc. (See Table C.5 in the Appendix.)

Mn = manganese.

modulating control = proportional control, but sometimes used to refer to any system of automatic control that provides an infinite number of control positions, as opposed to systems with a finite number of positions such as two-position control.

modulus of elasticity = a measure of the elasticity of a material; the ratio of stress (force) to strain (deformation) within the elastic limit.

modulus of rupture = a measure of the transverse of "cross-breaking" strength of a solid body.

mol = mole = pound mol, the molecular weight of a substance expressed in pounds, e.g. 32 pounds of O_2 constitute 1 pound mol of oxygen (or 32 grams of O_2 constitute 1 gram mol). For perfect gases, a pound mol occupies 379 ft³ at stp; a gram mol, 22.414 litres at 32 F and 760 mm Hg.

mol fraction = mol percent, or volumetric analysis for a gas.

molecular weight = the sum of the atomic weights of the atoms forming one molecule of a substance.

molecule = the smallest part of a compound that retains the properties of that compound.

momentum = mass times velocity; or (often with gases) density times velocity.

monolithic lining = a furnace lining without joints, formed of material which is rammed, cast, gunned, or sintered into place.

mortar (refractory) = a finely ground refractory material that becomes plastic when mixed with water, and is suitable for use in laying refractory brick.

mp = **melting point** = fusion point.

mph = **miles per hour.**

M-R valve = **manual reset valve** = manual reset automatic fuel shutoff valve. See that definition.

MSHA = **United States Mine Safety and Health Administration.**

MSW = municipal solid waste.

muffle = an enclosure in a furnace protecting the load from the flame and products of combustion.

mullite refractories = refractory products consisting predominantly of mullite ($3Al_2O_3 \cdot 2SiO_2$) crystals formed either by conversion of one or more of the sillimanite group of minerals or by synthesis from appropriate materials employing either melting or sintering processes.

MW = **megawatt.** 1 MW = 3.412 million Btu/hr.

N = atomic nitrogen; newton, unit of force = 1 kg · m/s².

N_2 = molecular nitrogen.

n (prefix, pronounced nano) = × 10⁻⁹.

naphtha = a term applied to many petroleum, coal tar, and natural gas liquid by-products, usually to a narrow boiling range fraction somewhere between kerosene and gasoline. For one sample, the minimum ignition temperature was 531 F or 227 C. For another sample, the molecular weight was 98 and the dew point was 150 F. See H-fuel. Many special purpose naphthas have narrow ranges of specifications for use as solvents in particular industries. Although usable as a fuel, it is difficult to pump because of vapor lock and lack of lubricity. (Naphtha is not a naphthenic hydrocarbon.)

naphthenes = saturated hydrocarbons having a cyclic, or ring structure with the general formula C_nH_{2n}, such as cyclopropane (C_3H_6), cyclohexane (C_6H_{12}). Naphthenics are chemically stable and offer no unusual combustion problems.

nat = natl = natural.

natural convection = free convection = transfer of heat by currents set up in fluids by differences of density resulting from differences in temperature.

natural draft = a difference in pressure resulting from the tendency of hot gases to rise up a chimney, thus creating a partial vacuum in the furnace.

natural gas = any gas found in the earth, as opposed to manufactured gas.

NC = numerical control.

nc = normally closed.

n-C_4H_{10} = normal butane, as opposed to iso-butane.

NEMA = National Electrical Manufacturers Association; Annapolis, MD.

net heating value = lower heating value, the gross heating value minus the latent heat of vaporization of the water vapor formed by the combustion of the hydrogen in the fuel. For a fuel with no hydrogen, net and gross heating values are the same.

neutral atmosphere = an atmospheric condition in firing a furnace or kiln that is neither oxidizing nor reducing.

NFPA = National Fire Protection Association; Quincy, MA.

nhv = net heating value = lower heating value. (See those definitions.)

Ni = nickel.

nine-inch equivalent = a brick volume equal to that of a standard 9 in. × 4½ in. × 2½ in. straight brick; the unit of measurement of brick quantities in the refractories industry.

NIOSH = United States National Institute for Occupational Safety and Health.

nm³ = normal cubic metre, a measure of gas volume at a pressure of 760 mm of mercury column and temperature of 0 C. sm³ = standard cubic metre is measured at 15 C and 760 mm Hg. scf (US standard cubic foot) is measured at 60 F and 29.92"Hg. 1 nm³ = 1.06 sm³ = 35.31 scf.

NO = nitric oxide = nitrogen monoxide, formed in flames, electric arcs, and many other places; colorless gas, readily reacts with oxygen to form nitrogen dioxide.

NO$_2$ = nitrogen dioxide, formed from NO in the presence of sunlight and VOCs; red-brown gas, causes smog, acid rain.

no = normally open.

noise = unwanted sound.

nonane (C$_9$H$_{20}$) = a major constituent in kerosene, diesel fuel, #1 fuel oil and #2 fuel oil.

nonoxidizing = not capable of oxidizing. Usually refers to the atmosphere in a furnace or kiln. Also used to describe a burner flame when insufficient O$_2$ is present to complete combustion (reducing).

normal thread engagement = effective thread length = the amount of overlap, usually measured in inches, necessary to ensure a tight connection between threaded pipes and fittings. (See Table D.1 in the Appendix.)

NOx = NO$_\mathbf{x}$ = nitrogen oxides, specifically defined by the USEPA as NO plus NO$_2$.

nozzle = an opening, port, orifice, or jet tube through which a fluid flows. For a burner, the part that delivers air, fuel, or an air-fuel mixture to a combustion chamber.

nozzle mixing burner = a burner in which the gas and air are kept separate until discharged from the burner into the combustion chamber or tunnel. Generally used with low pressure gas (up to ½ psig or 14"wc) and low pressure air (up to 5 psig).

NTIS = National Technical Information Service, US Dept. of Commerce; Springfield, VA, USA.

null balanced system = a method for controlling a variable by balancing it against a reference quantity. The control is such that the difference between the variable and the reference is held to zero.

O = atomic oxygen.

O$_2$ = oxygen (molecular).

od, OD = outside diameter or **outside dimension**, in inches, in feet.

oec = oxygen-enriched combustion (as opposed to pac).

oil pressure atomizing burner = mechanical pressure atomizing burner, a burner in which oil under pressure is allowed to expand through a small orifice, causing the oil to break into a spray of fine droplets.

oil temperature limit switch = a device that monitors the temperature of oil between preset limits and shuts down the burner, by closing an automatic fuel shutoff valve, should improper oil temperature be detected.

oil-to-gas converter = a device used to vaporize distillate oils. The vaporized oil-air mixture can then be burned through premix gas burners. The system consists of an oil-fired air heater, atomizers, and ratio controls.

oil train = a fuel train for oil. It may include heating and recirculating systems for heavier oils.

olefins = unsaturated hydrocarbons with the general formula C$_n$H$_{2n}$, such as propylene (C$_3$H$_6$), butene (C$_4$H$_8$). They tend to polymerize. They occur in large amounts in cracked fuel oils.

ONGA = **Ontario Natural Gas Association.**

on-off control = a control scheme that turns the input on or off, but does no proportioning or throttling of the flow rates as is the case with a modulating control.

on-ratio = slang for operating at chemically correct air/fuel ratio.

open port burner = open-type burner = open burner = a burner that fires across a gap into an opening in the furnace or combustion chamber wall and is not sealed into the wall. Sometimes surrounded by an air register through which controlled secondary air or tertiary air can enter the combustion chamber. By contrast, see sealed-in burner.

optimum air supply = the quantity of air that gives greatest thermal efficiency under actual conditions. With perfect mixing of fuel and air, the optimum air supply is equal to the chemically correct (stoichiometric) amount of air.

orifice = literally, any opening, but used in this book to designate a deliberate construction in a passage, circular in shape unless otherwise specified.

Orsat analyzer = an absorption apparatus used to determine the percentage (by volume, dry basis) of CO_2, O_2, and CO in flue gases.

OSHA = **Occupational Safety and Health Administration**, a part of the U.S. Department of Labor.

osi = ounces per square inch; a measure of pressure = 1/16 psi.

overall boiler efficiency = the ratio of the useful heat in the delivered steam to the gross heat in the fuel supplied, expressed as a %.

overall coefficient of heat transfer, U = the coefficient relating the heat transferred from one point to another, to the temperature difference between the two points and the cross sectional area of heat transfer, and including the combined effects of several resistances in series as in composite walls with surface and film resistances. See formula 4/5, Volume I.

overfiring = a heat treatment that causes deformation or bloating of ceramic products, or damages any product.

overload of motors = excess amperage drawn by an electric motor resulting from application of a torque greater than that for which it was designed. This slows the motor, and the excess current breaks down the insulation, resulting in a burned-out motor.

overrate firing = operating a boiler or furnace at an input rate in excess of its rated capacity. The rated capacity of a boiler is based upon its heat transfer surface area.

oxidizing atmosphere = a furnace atmosphere with an oversupply of oxygen, thus tending to oxidize materials placed in it.

oxidizing flame = oxidizing fire, a lean flame or fire; that is, one resulting from combustion of a mixture containing too much air and too little fuel. This kind of flame produces an oxidizing atmosphere.

Oxy-fuel combustion = a system for operating a burner with 100% oxygen instead of air.

oxygen concentration = "%e" in Part 13 = % oxygen concentration (by volume) in an air-oxygen mixture or in the total of separate air and oxygen streams being fed to a burner or combustion chamber. (Normal air has a %e of 20.9; oxy-fuel has a %e of 100.)

Oxygen-enriched combustion = oec = oxygen enrichment = burning fuel with a mixture of air and pure oxygen (anywhere between e = 20.9% and 100% oxygen) to improve efficiency or to produce a higher flame temperature. The degree of oxygen enrichment is designated by the % **oxygen concentration** in the air-oxygen mixture.

oxygen process = a method for making steel in which oxygen is blown upon or through molten pig iron, whereby most of the carbon and impurities are removed by oxidation.

oxygen sensor = a device for measuring the oxygen content of a gas.

Oxygen trim (O_2 trim) = an air/fuel ratio control system that uses an oxygen sensor in the flue gas as a feedback signal to control the amount of excess oxygen (excess air).

oz = ounce(s), (avdp).

P = pressure.

ρ – fluid flow rate, lb/hr (gph × lb/gal or cfh × lb/ft³).

Pa = pascal, an SI unit of pressure or stress = 1 N/m^2 = 0.102 mm H_2O = 0.002 32 osi. 1 atm = 4976 pascals. (See Table C.5 in the Appendix.)

pac = preheated air combustion (as opposed to oec) = burning a fuel with air that has been heated (upstream of the burner) to an elevated temperature, as by a recuperator or regenerator, for the purpose of improving fuel efficiency or raising the flame temperature.

PAC = polycyclic aromatic compound (a pollutant), containing four or more closed rings, usually of the benzenoid type.

paraffins = straight chain or branched, saturated hydrocarbons with the general formula C_nH_{2n+2}, such as methane (CH_4), propane (C_3H_8), octane (C_8H_{18}). Paraffinic fuel oils are stable and easy to handle and burn.

parallel flow = co-current flow. See Figure 9.11.

PCC = primary combustion chamber; SCC = secondary combustion chamber.

PCE = pyrometric cone equivalent, the temperature corresponding to the sagging point of a ceramic cone of prescribed shape.

PEL = permissible exposure limit.

pentane = C_5H_{12}.

percent air = the actual amount of air supplied to a combustion process, expressed as a percentage of the amount theoretically required for complete combustion.

percent excess air = the percentage of air supplied in excess of that required for complete combustion. For example, 120% air equals 20% excess air. See ideal combustion. (See Table C.10 in the Appendix.)

perfect combustion = the combining of the chemically correct proportions of fuel and air in combustion so that both the fuel and oxygen are totally consumed.

perfect gas = any gas for which the quantity Pv/T is a constant over a wide range of conditions (where P is the absolute pressure, v is the specific volume, and T is the absolute temperature). There is no "perfect gas," but many gases such as oxygen, nitrogen, and hydrogen, obey this law very closely over a wide range of temperatures and pressures, and so are often called perfect gases, as opposed to vapors which do not obey Boyle's law so closely.

periodic dryer = a dryer in which ware is placed, dried, and then removed; in contrast to a continuous dryer.

periodic kiln = a batch-type kiln.

periodic time (in acoustics) = period = reciprocal of frequency = time for one cycle.

permeability = the property of porous materials which permits the passage of gases and liquids under pressure. The permeability of a body is largely dependent upon the number, size, and shape of the open connecting pores, and is measured by the rate of flow of a standard fluid under definite pressure.

pH = a chemical term used to specify the degree of acidity of a solution. A pH of 0 to 7 is acid; 7 to 14 is alkaline.

photocell flame detector = a device that generates or rectifies an electric current while exposed to the light from a flame. Failure of the current or lack of rectification may be used to close an automatic fuel shutoff valve.

PIC = pressure indicating controller.

pic = products of incomplete combustion, such as CO, OH, and aldehydes. As usually used, the pic also include some poc.

pico, abbreviated p = prefix for submultiple 10^{-12}.

p&id = process and instrumentation diagram.

PID = proportional/integral/derivative control (gain/rate/reset).

pilot = a small flame used to light a burner. An *interrupted pilot* (sometimes called *ignition pilot*) is automatically spark-ignited each time that the main burner is to be lighted. It burns during the flame-establishing period and/or trial-for-ignition period, and is automatically cut off (interrupted) at the end of the main burner flame-establishing period, while the main burner remains on. Interrupted pilots are usually preferred/required for industrial heating operations.
    Two types of pilot control <u>not suitable</u> for industrial furnaces, ovens, kilns, incinerators, nor boilers are:
    A *continuous pilot* (sometimes called a *constant pilot*, *standby pilot*, or *standing pilot*) burns without turndown throughout the entire time that the burner assembly is in service, whether or not the main burner is firing.
    An *intermittent pilot* is automatically ignited each time there is a call for heat, and maintained throughout the entire run period. It is shut off with the main burner at the end of heat demand.

piod = temperature sensing probe, usually encased.

pipe burner = any burner made in the form of a tube or pipe with ports or tips spaced over its length.

pitch = rarely-used fuel products by distillation of coal tar. Very heavy, with a high carbon/hydrogen ratio.

Pitot tube = impact tube, a tube with an open end immersed in a stream, and pointed upstream. The other end of the tube is connected to a pressure gauge, which then indicates the total pressure due to the "ram effect" of the flow. Named after a French physicist, Henry Pitot (1695-1771).

P/I transducer = a device to convert a pressure signal to electrical current.

plasma jet = ionized gas produced by passing an inert gas through a high-intensity arc, causing temperatures up to tens of thousands of degrees Celsius.

plastic refractory = a blend of ground refractory materials in plastic form, suitable for ramming into place to form monolithic linings.

PLC = programmable logic controller.

plc = **pressure loss coefficient** = pressure loss ratio = ΔP net orifice loss after recovery $\div \Delta P_{max}$ at the vena contracta.

plenum or plenum chamber = windbox = part of a ducting or piping system having a cross sectional area considerably larger than that of any connecting pipes or openings so that the velocity pressure within such section is essentially zero. (Velocity pressure is largely converted to static pressure.) A manifold should be generously sized to serve as a plenum so that flow distribution will be equal among all of its downcomers.

plug fan = a high temperature recirculating fan that is inserted through the roof of a kiln or furnace to recirculate gases within the chamber.

plug flow = a reaction situation in which the products are pushed, as a plug, ahead of the raw reactants without mixing or reacting with the incoming products or those in the reaction zone. Theoretical opposite of a well-stirred reaction.

plug valve = a gas valve = a device for shutting off the flow of a fluid (not for throttling control), consisting of a rotatable cylinder or conical element with a passage through it to permit flow in one position but block flow when rotated to other positions.

PM-10 = particulate emissions less than 10 microns in size (1 micron = 1 millionth of a metre = 0.000 039 4").

pna. See polynuclear aromatics.

poc = **products of combustion** (assumed complete combustion) or combustion gases in a combustion chamber or on their way through a flue, heat recovery device, pollution reduction equipment, or stack. Usually consist of CO_2, H_2O, N_2, but may also include O_2, CO, H_2, aldehydes, and other complex hydrocarbons; and sometimes particulates, sulfur compounds, and nitrogen compounds. May be termed flue gas, stack gas, exit gas--depending on position. Should not be called waste gas because of confusion with by-product fuels. See also pic.

POHC = principal organic hazard constituent.

poise = a metric unit for measuring absolute viscosity. One poise equals one dyne-second per square centimetre, or one gram per centimetre sound. From a physicist named Poiseuille who, in 1842, laid the groundwork for the deduction of coefficients of viscosity.

polymerization = the formation of larger, heavier molecules.

polynuclear aromatic (pna) = a petrochemical compound, usually gaseous, the molecule of which consists of three or more benzine rings. Also known as a polycyclic hydrocarbon.

port = an orifice or opening.

ported manifold type burner = a burner in which a manifold supplies fuel to a number of small ports, as for example the burners in domestic gas stoves.

positive displacement pump = any type of pump in which leakage or slip is negligible, so that the pump delivers a constant volume of fluid, building up to any pressure necessary to deliver that volume (unless the motor stalls or the pump breaks).

post-purge = an acceptable method for scavenging the combustion chamber, boiler passes and breeching to remove all combustible gases after flame failure controls have sensed pilot and main burner shutdown and fuel shutoff valves are closed.

potential flow = a movement of fluid, heat, or electrical change, the rate of which is directly proportional to the net driving force and inversely proportional to the resistance. For example, per Ohm's Law for electricity, $I = E/R$; for heat transfer $q = Q/A = \Delta T/R$ or $U\Delta T$; and per the Bernoulli equation for fluids, flow rate per unit area, $Q/A = \overline{K} \sqrt{2g.\Delta P/\rho}$. ($\overline{K}$ and U are measures of conductance, which is the reciprocal of resistance.)

pouring-pit refractory = a refractory associated with the transfer or flow control of molten steel between furnace and mold.

pouring temperature = the customary temperature at which molten metals are poured in industrial processes.

pour point = the lowest temperature at which an oil will flow when cooled and examined under conditions specified in ASTM test method D 97.

power = the rate of transfer of energy.

ppb = **parts per billion.**

ppm = **parts per million**, referring to concentration, usually of a pollutant or contaminant in air, water, or food, although also used in specifying chemical formulations. The "parts" may be volumes in a million volumes or weights in a million weights. For gaseous pollutants it is usually by volume in 1 000 000 volumes of dry air. $1 \text{ ppm} = \dfrac{M}{0.024\,04} \ \mu g/m^3$ where M is the molecular weight. 1 ppm = 0.0001% = 1 pound in 500 tons = 1 ounce in 7530 gallons of water.

ppm$_v$ = **parts per million by volume.**

preheated air = air heated prior to its use for combustion. Frequently the heating is done by hot flue gases.

preheated air combustion. See pac.

preignition purge = pre-purge, an acceptable method for scavenging the combustion chamber, boiler passes, and breeching to remove all combustible gases before the ignition system can be energized.

premix gas burner = usually, a burner (nozzle) supplied with gas and air from an upstream mixing device, as opposed to a nozzle-mix burner. Sometimes, a burner within which the gas and air are mixed before they reach its nozzle.

premixer = a device used to mix gas and air before delivery to a burner, such as an aspirator, an inspirator, or a fan mixer.

pre-purge = preignition purge.

pres = **press** = **pr** = **pressure.**

pressure-balanced a/f ratio control = system using regulators or pneumatic controls to automatically control air/fuel ratio.

pressure burner = blast burner = a premix burner delivering a combustible mixture under pressure, normally above 0.3"wc to the combustion zone.

pressure control = a system for control of air/fuel ratios by proportioning the pressures in the air and fuel supply lines.

pressure drop = the difference in pressure between any two points along the path of flow of a fluid.

pressure loss = pressure drop, often unintentional.

pressure recovery = the amount of pressure regained after a fluid passes through a restriction such as a thin plate orifice. Percent of recovery depends on the Beta ratio (d/D).

pressure-reducing regulator = pressure regulator = line pressure regulator, a device used to maintain a constant pressure in a fuel supply line regardless of the demand. It cannot maintain a pressure greater than its inlet pressure.

pressure-relieving regulator = pressure relief valve, a device used to maintain a constant pressure in a fuel supply line regardless of flow by bleeding off some fuel to atmosphere or to return to a tank.

primary air = the first stream of air to mix with fuel at a burner, but see secondary air.

primary element = the first system element that responds quantitatively to the measured variable and performs the initial measurement operation. A primary element performs the initial conversion of measurement energy.

producer gas = an artificial fuel gas of low Btu content made by passing a mixture of air and steam through a hot bed of coke or coal.

products of combustion. See poc.

products of incomplete combustion. See pic.

propane (C_3H_8) = an easily liquefiable hydrocarbon gas of the paraffin series. Propane is one of the components of raw natural gas, and it is also derived from petroleum refining processes. A major component of "LPG", liquefied petroleum gas.

proportional control = a mode of control in which there is a continuous linear relation between value of controlled variable and the position of the final control element.

proportioning = maintaining the desired ratio of fuel to air.

proportioning valve (pv) system for air/gas proportioning and mixing = a system using separate control of air and gas both of which are under pressure. The valves, controlling the air and gas flows are usually mechanically connected.

propylene (C_3H_6) = a hydrocarbon fuel similar to and usually associated with propane. If over 5% propylene is found in propane, poor combustion with some soot and carbon formation may occur.

PSD = prevention of significant deterioration (of the environment).

psf = pound(s) per square foot. (See Table C.5 in the Appendix.)

psi = pounds per square inch. (See Table C.5 in the Appendix.) **psig** = psi gauge pressure. **psia** = psi absolute pressure = gauge pressure + 14.696 psia (barometric pressure).

psychrometry = the science or techniques of measurement of the water vapor content of air or other gases.

pulse combustion = ram-jet-like combustion that provides pressure, flow, and high convection heat transfer in limited configurations for domestic and commercial applications involving large numbers of identical heating units.

pulse-controlled firing = a control scheme for furnaces with multiple industrial burners wherein all burners are operated on/off or high to very low, and modulated by varying their ratio of time on to time off. This enhances convection heat transfer because burners operate only at full fire. Individual burner cycle times are usually "stepped" to start at slightly different times so as to increase furnace temperature uniformity. See Step-Fire™.

pump lift = the net vertical difference in elevation between an open liquid surface and the suction side of a pump.

purge, post = an acceptable method for scavenging the combustion chamber, boiler passes and breeching to remove all combustible gases after flame failure controls have sensed pilot and main burner shutdown and fuel shutoff valves are closed.

purging = eliminating an undesirable substance from a pipe, piping system, or furnace by flushing it out with another substance, as in purging a furnace of unburned gas by blowing air through it.

purple peeper = (slang for) ultraviolet flame detector, a device that energizes an electronic circuit when it "sees" the small amount of ultraviolet radiation that is present in all industrial burner flames.

pv = proportioning valves, a system for air/fuel ratio control using air and fuel valves that are opened and closed in proportion to one another by a common shaft, a mechanical linkage (sometimes cam-biased), or an electronic "linkage".

PV = process variable.

PWL = L_W = sound power level.

pyrometric cone = one of a series of pyramidal-shaped pieces consisting of mineral mixtures, and used for measuring time-temperature effect. Each numbered cone is of a definite mineral composition so that when heated under standard conditions, it bends at a definite temperature, the PCE.

pyrometric cone equivalent (PCE) = the number of that Standard Pyrometric Cone whose tip would touch the supporting plaque simultaneously with a cone of the refractory material being investigated, when tested in accordance with the Method of Test for Pyrometric Cone Equivalent (PCE) of Refractory Materials (ASTM) Designation C24.

Q = volume flow rate or heat flow rate.

Q' = stp volume flow rate (weight flow rate).

Q₁, Q₂ = fluid flow rate (volume) at conditions 1 and 2, or heat transfer rate at conditions 1 and 2.

q = heat flow rate or heat flux (Btu/hr ft²).

q_c = convection heat transfer rate.

q_k = conduction heat transfer rate.

q_r = radiation heat transfer rate.

quad = 10^{15} Btu ~ 10^9 thousand cf of natural gas.

quarl = burner tile, combustion refractory.

quarl angle = tile angle, usually the included angle, but it is better to always specify "quarl included angle" or "quarl half angle".

quenching of flames = the process in which the reactants in a flame are rapidly cooled. This usually prevents the reaction from reaching completion in the localized area where quenching occurs. Large scale quenching may result in incomplete combustion.

quenching of metals = a heat treatment of metal wherein the heated piece is rapidly cooled by immersing it in water or oil for the purpose of hardening the material.

R or °R = Rankine or degrees Rankine (absolute temperature level = F + 460); or resistance, the reciprocal of conductance. (See formulas 4/5 and 4/6, Volume I.)

RACT = reasonably available control technology; RACM = ... measure.

radiant burner = radiation burner = infrared burner, a burner designed to transfer a significant part of the combustion heat in the form of radiation from surfaces of various refractory shapes or alloy screens.

radiant tube burner = a burner designed to operate inside a tube, which heats the load indirectly by radiation, or by air or gas convection.

radiation = a mode of heat transfer in which the heat travels very rapidly in straight lines without heating the intervening space. Heat can be radiated through a vacuum, through many gases, and through a few liquids and solids.

radiation burner = infrared burner = a burner that heats its load primarily by radiation--either by long burner flame, or by convection transfer to a surface that becomes a radiant emitter.

ramming mix = a ground refractory material mixed with water and rammed into place for patching shapes or for forming monolithic furnace linings.

rangeability = the ratio of maximum operating capacity to minimum operating capacity within a specified tolerance and operating condition.

ratio regulator = a proportional control device that regulates the downstream pressure in the pipeline in which it is located to maintain proportional pressures in fuel and air lines in a pressure control system, thus producing proportional flow of fuel and air.

Ratiotrol™ = North American Manufacturing Company's trademark for an air/oil ratio regulator.

RCRA = Resource Conservation and Recovery Act (USA).

rdf = refuse-derived fuel.

recirculating type oil system = a closed circuit oil system in which the oil not used at the burners is returned to the storage tank. It is good practice to use only a relatively small proportion of the oil pumped, so that turning on additional burners will not cause an excessive drop in pressure or temperature.

reclaimed oil = motor, cutting, quenching, or transformer oils that have been treated for use as fuel oil. Generally satisfactory substitutes for heavy fuel oils, but sometimes requiring unusual preparation.

rectification = the process of converting an alternating current to a pulsing direct current.

recuperator = a piece of equipment that makes use of hot flue gases to preheat air for combustion. The flue gases and air flow are in adjacent passageways so that heat is transferred from the hot gases, through the separating wall, to the cold air.

reducing atmosphere = a furnace atmosphere that tends to remove oxygen from substances placed in the furnace. It may be produced by supplying inadequate air to the burners, thus intentionally making the combustion incomplete.

reducing flame = reducing fire, a rich flame or fire; that is, one resulting from combustion of a mixture containing too much fuel and too little air, producing a reducing atmosphere.

refractories = highly heat resistant materials used to line furnaces and kilns.

refractory block = refractory tile = burner block = burner tile = combustion tile = combustion block = burner refractory = a piece of refractory material molded with a conical or cylindrical hole through its center. The block is mounted in such a manner that a burner flame fires through this hole. The block helps to maintain continuous combustion, and reduces the probability of flashback and blow-off with premix burners.

refuse = the ash and unburned fuel remaining after combustion of a solid fuel; any solid waste.

regenerator = a cyclic heat interchanger which alternately receives heat from gaseous combustion products and transfers heat to air before combustion.

regulator = a device that detects a change in a process variable, and automatically energizes a mechanism that will correct the deviation in that variable so as to return it to a preset value. A controller is externally powered, but a regulator uses energy from the system that it is regulating. In current practice, a controller is sometimes called a regulator.

regulator, gas pressure = a spring loaded, dead weighted or pressure-balanced device that can maintain the gas pressure to the burner supply line within ±10% of the operating pressure at any one rate from maximum to minimum firing rates, with variations in inlet pressure of ±40% of the rated inlet pressure.

relative humidity, rh = the partial pressure of water vapor in air divided by the vapor pressure of liquid water at the same temperature, expressed as a percent. For other liquids, the term "% saturation" is used.

relief valve = a valve that opens at a designated pressure and bleeds a system in order to prevent a build-up of excessive pressure that might damage regulators and other instruments.

residual oil = residuum = heavy oil, the heavy portion of a crude oil remaining after distillation and cracking. As refining methods increase yields of gasoline and distillate oils, the residual oil becomes heavier and more difficult to pump and atomize.

resistance = (as applied to fluid flow) the opposition to flow that makes it inevitable that there will be a pressure drop when a fluid is flowing.

resolution = the smallest change in input that can be detected by an instrument.

resonance = maximum vibration or sound--occurs when the frequency of the forcing vibration is the same as the natural frequency of a column of gas or solid body on which it is impressed. Sometimes referred to as "response".

return line = the low pressure side of a main circulating oil loop or branch circuit. That is, that part of an oil circulating system through which the unused oil flows back to the storage tank after having passed the burners.

reverberation = the sound that persists at a given point after the initial sound source has stopped.

Reynolds number = the ratio of inertia forces to viscous forces in a flowing fluid. When Reynolds number has been calculated, the friction factor may be determined.

RH or rh = relative humidity, usually %. (See above.)

$Ri = \dfrac{d_o}{h_i d_i}$ in ft² hr °F/Btu, the heat transfer resistance of the film or boundary layer on the inside of a pipe.

ribbon burner = a burner having many small closely spaced ports usually made by pressing corrugated metal ribbons into a slot.

rich mixture = a mixture of fuel and air in a premix burner system in which an excess of fuel is supplied in relation to the amount needed for complete combustion. More precisely, a fuel-rich mixture, or an air-lean mixture, or oxidant-lean ratio.

rich ratio = a proportion of fuel to air containing too much fuel or too little air for complete combustion of the fuel. More precisely, a fuel-rich ratio, or an air-lean ratio, or oxidant-lean ratio.

ricra = pronunciation for the acronym RCRA (above).

riddle tile = refractory tiles usually 9 to 12 in. thick used to make up hearths in periodic kilns. Tile shapes permit flue gases to travel down from the kiln to flue collection chambers.

ring burner = a form of atmospheric burner made with one or more concentric rings; or a form of burner used in firing boilers consisting of a perforated vertical gas ring with air admitted generally through the center of the ring. Combustion air may be supplied by natural, induced or forced draft.

rise of an arch = the vertical distance between the spring line and the highest point of the under surface of an arch.

rms = root mean square.

$R_o = \dfrac{1}{h_o}$ in ft² hr °F/Btu, the heat transfer resistance of the film or boundary layer on the outside of a pipe or a pipe's insulation.

ROG = reactive organic gases ~ VOC.

ROI = return on investment.

rotameter = a variable-area, constant-head, rate-of-flow volume meter in which the fluid flows upward through a tapered tube, lifting a shaped weight to a position where upward fluid force just balances its weight.

rotary cup oil burner = centrifugal atomizing burner, a burner which throws oil from a rotating cup into an air stream, breaking the oil into a fine conical spray.

rotary dryer or kiln = an inclined rotating drum usually refractory lined and fired with a burner at the lower end. Used to dry loose materials as they roll through by gravity, sometimes lifted, mixed, and exposed by flights or shelves attached to the inner wall.

rotary plug valve = a type of valve in which a ported sleeve or plug is rotated past an opening in the valve body.

rowlock course = a course of brick laid on edge with their longest dimensions perpendicular to the face of a wall.

RPM flange = a low pressure pipe flange design standardized by the Riveted Pipe Manufacturers.

rpm = revolutions per minute (rotational velocity) = Hz.

rtd = resistance temperature detector.

S = specific gravity, usually relative to water; also sulfur.

s = second(s).

SAE = **Society of Automotive Engineers; Warrendale, PA.**

sagger = a fired ceramic container to hold ware during firing and to protect ware from the flames.

Sankey diagram = a pictoral method for analyzing how heat is spent in a furnace, boiler, oven, kiln, incinerator, heater. See Part 3, Volume I and Part 9, Volume II.

saturated air = air containing all the water vapor it can normally hold under existing conditions.

saturated steam = steam at the boiling point for water at the existing pressure.

saturation pressure = the pressure at which a vapor confined above its liquid will be in stable equilibrium with it. Below saturation pressure, some of the liquid will change to vapor, and above saturation pressure, some of the vapor will condense to liquid.

saturation temperature = the boiling point of a liquid for the existing pressure.

Saybolt Furol = a scale used for measuring the viscosity of heavy oils. The instrument has a larger orifice and is used at a higher temperature than the Saybolt Universal instrument used for lighter oils.

Saybolt Universal = a scale used for measuring the viscosity of oil, expressed in seconds required for a specified amount of oil to flow through an orifice; hence the larger the number of seconds, Saybolt Universal (SSU), the more viscous the oil.

SCC = secondary combustion chamber.

scfh = standard (stp) cubic feet per hour.

schlieren = strata of air having densities sufficiently different from surrounding air to permit flow pattern studies by refracted light.

SCR = selective catalytic reduction, a form of de-NOx.

sealed-in burner = a burner mounted on a furnace in an airtight manner so that there can be no inflow of secondary air around the burner.

sec = **second(s)**, also abbreviated s.

secondary air = the second stream of air to mix with fuel at or near a burner. (See also tertiary air.) In an air atomizing burner, the atomizing air might be considered to be primary air and the main or combustion air to be secondary air. In an open burner, all air through the burner (atomizing and main) may be considered to be primary and all air through the register to be secondary.

semimuffle furnace = a furnace with a partial muffle, in which the products of combustion can eventually contact the load, but the temperature uniformity is improved by protecting the load from intense direct heat transfer from the flames.

sensible heat = heat, the addition or removal of which results in a change in temperature, as opposed to latent heat.

Sensitrol™ = North American Mfg. Company's trademark for a limiting orifice oil valve that is also usable for manual shutoff.

sg = **specific gravity** with respect to water with both substances at 60 F. (This abbreviation is peculiar to the combustion industry and may not be widely recognized in other fields. "sg" and "sp gr" are also used for specific gravity of a gas relative to air, a term identified as "gas gravity" in this book.)

SGA = specific gravity relative to standard dry air (0.076 32 lb/ft³ or 1.222 kg/m³).

SGW = specific gravity relative to water at 39.2 F or 4 C (62.4 lb/ft³ or 1000 kg/m³).

shaft furnace = a vertical cylindrical heating chamber, such as a cupola or blast furnace, in which lump materials are heated by convection of a rising stream of hot gases.

shaft kiln = a furnace for heating lump material, consisting of a vertical refractory-lined shaft.

shape factor = the fraction of radiation from one surface that falls upon another.

shielded cable = a single wire or multiple conductors surrounded by a separate conductor (the shield) to minimize the effects of outside electrical disturbances.

shock loss = if in converting fluid velocity to static pressure the fluid is stopped or slowed too rapidly, some of the energy of fluid velocity is converted to useless heat instead of the desired static pressure. This is called shock loss.

shutdown = the total process of terminating operation of a system.

shuttle kiln = a type of periodic kiln in which the ware is loaded on cars rather than directly on the hearth. Operating sequence is: load car, push car into kiln, fire, cool, remove car.

SI = Le Systeme International d'Unités, the international system of units, the new metric system.

Si = silicon.

silicon carbide = a compound of silicon and carbon; formula, SiC, used for making refractories that are good heat conductors.

simultaneous gas-oil burner = a burner that burns gas and oil at the same time.

single port burner = a burner having only one discharge opening or port.

sintering = a heat treatment that causes adjacent particles of material to cohere at a temperature below that of complete melting.

skewback = the course of brick, having an inclined face, from which an arch is sprung.

sl = sea level.

slagging of refractories = destructive chemical action between refractories and external agencies at high temperatures, resulting in the formation of a liquid.

sludge = heavy materials found at the bottom of fuel oil storage tanks, including oil-water emulsions, heavy chemicals, oxidation products, dirt; also sewage sludge.

slurry = a suspension of finely pulverized solid material in water, of creamy consistency.

sm³ = standard cubic metre (at 15 C and 760 mm Hg).

sm³/h = standard cubic metres per hour (volume flow rate). See also nm³.

small port burner = ported manifold burner = a manifold containing many small holes through which an air-gas mixture flows, and outside of which the mixture burns, such as a domestic gas stove burner.

SME = Society of Manufacturing Engineers; Dearborn, MI.

Sn = tin.

SNCR = selective non-catalytic reduction.

SNG = substitute natural gas or synthetic natural gas, usually made from coal or petroleum products. Trade names include hi-gas, bi-gas.

snubber = any device or method for slowing or easing a reaction, as a brake on a drum, a dampening device on a spring, or an orifice in a pneumatic signal line.

SO₂ = sulfur dioxide.

SO₃ = sulfur trioxide.

soak (soaking) = to hold the load in a kiln or furnace at one temperature for a time to allow equalization of temperature throughout the load.

soldier course = a course of brick set on end.

sonic velocity = velocity of sound = critical velocity, V_C.

soot = a black substance, consisting of very small particles of carbon or heavy hydrocarbons, which appears in smoke resulting from incomplete combustion.

sound = any cyclical pressure variation in an elastic medium (gas, liquid, or solid) that is perceived and interpreted by the ear.

sound absorption = conversion of sound energy into another form of energy, usually heat, when the sound enters an acoustic medium.

sound power = acoustic source energy rate = acoustic power, PWL, or L_w, usually measured in watts.

sound pressure = sound intensity, in microbars or micropascals.

sound pressure level, SPL = L_p = ratio of a measured sound pressure to that at some arbitrary base, usually in decibels.

sound power level = L_w or PWL, is the total acoustic energy rate radiating from a point source of origin of a sound, in watts.

sour gas, sour oil = fuel containing a large proportion of sulfur or sulfur compounds. Low sulfur or treated fuels are termed sweet gas, sweet crude.

SOx = SO_x = sulfur oxides, e.g. SO_2, SO_3.

space heating = heating large volumes of air to the temperatures desired for human occupancy or for storage.

spalling of refractories = the loss of fragments (spalls) from the face of a refractory structure, through cracking and rupture, with exposure of inner portions of the original refractory mass.

spanner tile = a piece of refractory that bridges an opening--as to provide support above a burner tile.

sparging = sprinkling over, scattering across, or mixing into.

spark-ignited pilot = an electrically-ignited small flame used to light a main burner. See pilot.

specific gravity of a gas = gas gravity = the ratio of the density of the gas to the density of dry air at standard temperature and pressure.

specific gravity of a liquid = the ratio of the density of the liquid to the density of water. ("sp gr 60/60 F" means specific gravity when both the liquid and the water are at 60 F.)

specific heat = the amount of heat required to raise a unit weight of a substance through one degree temperature rise. 1 Btu/lb °F = 1 cal/gm °C.

specific volume = the volume occupied by a unit weight of a substance under any specified conditions of temperature and pressure; the reciprocal of density; ft³/pound or m³/kg.

specific weight = density, the weight per unit volume of a substance.

sp gr (60/60 F) = **specific gravity** with respect to water with both substances at 60 F.

sp ht – specific heat.

SPL = sound pressure level = L_p.

spring line or spring of a refractory arch = the line of contact between the inside surface of an arch and the skewback; usually used to specify the height of an arched refractory chamber.

sprung arch = a curved structure supported only by abutments at the side or ends.

spuds = a gas orifice, a small drilled hole for the purpose of limiting gas flow to a desired rate; a flame holder; a small-port premix nozzle.

sq ft = square foot (or feet); ft^2 is preferred.

sq in. = square inch (or inches); in.^{2} is preferred.

SR1 = seconds, Redwood No. 1.

SSF or SFS = seconds, Saybolt Furol, a unit of kinematic viscosity. See Figure 2.7, Volume I.

SSU or SUS = seconds, Saybolt Universal; a measure of viscosity. See Part 2, Volume I.

stability, combustion = that quality of a burner enabling it to remain lighted over a wide range of air/fuel mixture ratios and input rates without benefit of a pilot or spark.

stability, oil = the resistance of an oil to breakdown. Results of instability are sludge formation, carbon or coke formation, sooting, and gummy or waxy deposits.

stack gases. See flue gases and poc.

stackbonded = method of installing ceramic fiber, perpendicular to inner furnace wall surface.

stack loss = flue gas loss, the sensible heat carried away by the dry flue gas plus the sensible and latent heat carried away by the water vapor in the flue gas.

staged combustion = a physical arrangement of burner parts and surrounding injector nozzles that adds the air to the fuel or the fuel to the air in stages.

standard air = air at standard temperature and pressure, namely 60 F (15.56 C) and 29.92 inches of mercury. (14.696 lb per in.^{2} or 760 mm Hg.) See Table 1.1, Volume I.

standard atmosphere = the accepted normal atmospheric pressure at sea level, equal to 29.92 inches of mercury column, 14.696 lb per sq. inch absolute, or 760 mm Hg.

standard barometer = the reading of a barometer for standard atmospheric pressure; equal to 29.92 inches of mercury column or 14.696 psia or 760 mm Hg.

standard cubic foot = a measure of gas volume, at 60 F and 29.92"Hg.

standard cubic metre = a measure of gas volume, at 15 C and 760 mm Hg, = sm^3.

standard pressure = standard atmosphere, equal to a pressure of 29.92"Hg. 14.696 lb per sq. in., or 760 mm of mercury.

standard temperature = 60 F (15.56 C) in this book and in engineering. For the fan industry, 70 F (21.1 C). In scientific work, 32 F (0 C) or 39.2 F (4 C).

standard volume = the volume of a gas or air measured at 60 F (15.56 C) temperature and 29.92 inches of mercury pressure (760 mm Hg or 14.696 psia).

standby pilot--(explained under pilot).

standing pilot--(explained under pilot).

static pressure = the force per unit area exerted by a fluid upon a surface at rest, or across a surface parallel to the direction of fluid flow. For flow in a pipe, it is measured by a gauge connected to the side of the pipe, perpendicular to the direction of flow. (See Figure 5.3, Volume I.)

steady state = a condition in which the variables of a process are constant with respect to time.

StepFire™ system = North American Mfg. Company's trademark for its pulse-controlled firing system which converts the temperature demand input signal to the on-time and rotational-time of burner group firing.

Stephan-Bolzman law = total energy radiated from a body is proportional to the 4th power of its absolute temperature = 4th power law.

std = standard.

steam atomizing burner = a burner that uses high pressure steam to tear droplets of oil from an oil stream and propel them into the combustion space so that they vaporize quickly.

stoichiometric combustion – See stoichiometric ratio.

stoichiometric ratio = chemically correct ratio of air to fuel, or oxygen to fuel, that will leave no unused fuel nor oxygen after combustion.

stoke – the cgs unit of kinematic viscosity. One stoke equals one centimetre squared per second. From Sir George Gabriel Stokes (1819-1903), British mathematician and physicist who is noted for his work on the friction of fluids in motion.

stp = **standard temperature and pressure** (60 F and 14.696 psia or 15.56 C and 760 mm Hg)

stp flow rate = the rate of flow of a fluid, by volume, corrected to standard temperature and pressure.

stp volume = the volume that a quantity of gas or air would occupy at standard temperature and pressure.

strainer = a fine mesh screen or filter used to separate foreign particles from an oil or steam stream.

street elbow = street ell = a 90 degree pipe fitting with a male thread on one end and a female thread on the other. Not recommended because of higher pressure drop than a standard elbow with female threads on both ends.

stretcher = a brick laid flat with its length parallel to the face of the wall.

sublimation = change of state directly from solid to gas, or from gas to solid.

suction line = that part of an oil circulating system between the oil storage tank and the first pump.

suction type mixer = aspirator mixer, an air/gas proportioning device that uses the venturi principle to cause the combustion air to induce the proper amount of gas into the air stream. It is used with low pressure air and zero gas.

superduty fireclay brick = fireclay brick having a PCE not lower than cone 33, and that meets certain other requirements outlined in ASTM Standard C 27-58T.

superheated steam = water vapor at a temperature above the saturation temperature for the existing pressure, e.g. at 250 F and 14.696 psia. The water vapor in the atmosphere is superheated (above the saturation temperature for its partial pressure) and changes from superheated to saturated when lowered to its dew point.

supervising gas valve = supervising gas cock = a valve that sends a signal indicating that its main passage is closed. When such a signal is received from all fuel valves on a furnace, the operator may proceed to purge the furnace and then light the pilots. See Supervising Valve System, Part 7, Volume II.

surface coefficient = h_r = radiation heat transfer coefficient = the rate of heat flow per unit area (heat flux) for each degree of temperature difference between the emitting surface and the absorbing surface. (See formula 4/2, Volume I.)

surface tension = molecular attraction or cohesion of molecules on the surface of a liquid, which gives it the appearance of having an elastic skin.

suspended arch = a furnace roof consisting of refractory shapes suspended from overhead supporting members. By contrast, see sprung arch.

swirl = spinning or spiral motion of a fluid; usually an aid to combustion stability. See Burner Characteristics, Part 6, Volume II.

swp = steam working pressure.

synthetic gas = any man-made gaseous fuel; substitute gas; manufactured gas; usually made from coal, petroleum by-products, or waste materials by complex molecular reforming processes.

taconite = a compact ferruginous chert or slate in which the iron oxide is so finely dispersed that substantially all of the iron-bearing particles are smaller than 20 mesh. Typical analysis of the ore grade shows 32.0% total Fe.

take-off lines = pipelines leading from the high pressure side of a main circulating oil loop to the branch circuits.

TAPPI = Technical Association of the Pulp and Paper Industry; Atlanta, GA.

tc = t/c = T/C = thermocouple = temperature of casing or cold face.

TEFC = totally enclosed fan cooled; refers to electric motors.

temp = temperature.

Tempest® = North American Mfg. Company's trademark for small high velocity (high momentum) burners.

tera, abbreviated T = prefix for multiple 10^{12}.

tertiary air (pronounced tur'-she-ary) = a third supply of air introduced downstream from the secondary air.

TH = total heat (psychrometric chart).

THC = total hydrocarbons, or total hydrocarbon emissions.

theoretical air = stoichiometric air = on-ratio air = correct air, the chemically correct amount of air required for complete combustion of a given quantity of a specific fuel.

theoretical draft = the calculated difference between the pressure in a furnace and the atmospheric pressure (caused by buoyancy of hot gases in the furnace and chimney). This does not include the effects of friction or nonstandard conditions.

theoretical flame temperature = adiabatic flame temperature. See also hot mix temperature.

therm = 100 000 Btu.

thermal conductance = the amount of heat transmitted by a material divided by the difference in temperature of the surfaces of the material. Also known as conductance. Where heat is transferred by more than one mechanism through a structure of mean cross sectional area Am, conductance, C = gross rate of heat transfer divided by temperature drop between its faces. C = $\Sigma q/\Delta t$ = KmAm/X.

thermal conductivity, k = the ability of a material to conduct heat, measured in flow of Btu per hour through a square foot of cross sectional area and one foot (or inch) of thickness with 1°F of temperature difference across this thickness. The refractory and insulation industries use Btu in./ft² hr °F. Most others use Btu ft/ft² hr °F.

thermal de-NOx = an after treatment to remove nitrogen oxides from products of combustion, using NH_3 (ammonia), or urea, with or without catalysts. See SCR.

thermal expansion = an increase in volume and linear dimensions resulting when a substance is heated.

thermal NO = nitric oxide formed by temperature effects as opposed to that formed by fixed nitrogen from the fuel (fuel NO).

thermal shock = a sudden temperature change.

thermie (French therm) = 1000 kcal.

thermistor = semi-conductor elements that have a large negative temperature coefficient of resistivity. These are convenient for measuring temperature changes over short spans very accurately.

thermopile = many thermocouples assembled in series to provide a signal strong enough to actuate a valve without amplification by an external power source.

throttling valve = a valve used to control the flow rate of a fluid. A throttling valve does not necessarily provide tight shutoff.

thrust controlled flame = turbulent diffusion flame in which aspiration of combustion air into the combustible gas is controlled by thrust forces.

TIC = temperature indicating controller.

tile – See burner tile.

timed trial-for-ignition = that period of time during which the programming flame supervising controls permit the burner fuel valves to be open before the flame sensing device is required to detect the flame.

T/I transducer = a device to convert a temperature signal to electric current.

T/I transmitter = a transducer that converts the millivoltage from a thermocouple to an electrical current signal.

TLV = **threshold limit value** (highest allowable concentration).

TMS = **The Metallurgical Society; Warrendale, PA.**

toe = tons of oil equivalent, i.e. for equivalent heating ability. (1 toe = 40.66 gigajoules; See Appendix for unit conversions.)

ton, short = 2000 lb = 907.2 kg; ton, long = 2240 lb = 1015.9 kg.

tonne = metric ton = 1000 kg = 2205 lb.

torque = bending moment; a rotational form of doing work. Torque multiplied by rotational speed equals power. A torque of 1 pound(force)·inch = 0.112 984 8 Newton·metre of torque.

TORR = 1 mm Hg, a unit of pressure.

tosca = pronunciation for the acronym TSCA. (See below.)

totalizer = a counter that totals or accumulates a total count, such as an odometer. Like a positive displacement gas or oil meter, its reading must be divided by elapsed time to determine rate. Sometimes erronoeously called an "integrator". Not to be confused with a "summer", which adds two or more flow rates.

total pressure = impact pressure = stagnation pressure, the pressure measured by an impact tube, the sum of static pressure and velocity pressure. The total pressure of a motionless fluid is equal to its static pressure.

town gas = city gas, artificial gas (usually made from coal, as producer gas, or low Btu gas).

tpy = tons per year.

T_r = temperature of a radiation-receiving surface, in degrees Rankine (Fahrenheit + 459.67).

tramp air = infiltrated air, usually undesirable added excess air.

transducer = an element or device which receives information in the form of one physical quantity and converts it to information in the form of the same or another physical quantity, e.g. pressure to milliamps.

transmitter = a device that conveys information from one location to another.

triatomic gases = gases having three atoms in each molecule, such as CO_2, H_2O, SO_2. These gases are capable of gas radiation.

T_s = temperature of a radiation source, in degrees Rankine (Fahrenheit + 459.67).

TSC = two stage combustion.

TSCA = Toxic Substances Control Act (USA).

tunnel burner = a burner sealed into the furnace wall in which combustion takes place mostly in a refractory tunnel or tuyere which is part of the burner.

tunnel kiln = a tunnel-shaped furnace through which ware is pushed on cars passing through preheating, firing, and cooling zones consecutively, enabling continuous operation.

turbo blower = a centrifugal blower in which the air leaving the blade tips passes through a narrow slot into a large volume chamber, thus efficiently converting velocity energy to static pressure.

turbulence = a state of being highly agitated. Turbulent flow is fluid flow in which the velocity of a given particle changes constantly both in magnitude and direction.

turndown = the ratio of maximum to minimum input rates; turndown ratio, abbreviated t/d = high fire rate/low fire rate.

tuyere (pronounced tweer) = a refractory shape containing one or more holes through which air and other gases are introduced into a furnace.

TWA = time-weighted average.

TwinBed® = North American Mfg. Company's trademark for its integral burner-regenerator system.

TY = temperature control split range device.

UBC = used beverage containers, usually aluminum.

UEL = upper explosive limit.

UHC = unburned hydrocarbons.

UL = Underwriters Laboratories; Northbrook, IL.

ULC = Underwriters Laboratories (Canada).

ultimate analysis = a statement of the quantities of the various elements of which a substance is composed, usually expressed in percentage by weight.

ultimate %CO_2 = the percentage by volume of carbon dioxide that appears in the dry flue gases when a fuel is completely burned with its chemically correct air/fuel ratio. The theoretical maximum %CO_2 possible.

Universal, Saybolt. See Saybolt Universal.

upper limit of flammability (formerly inflammability) = the maximum percentage of fuel in an air-fuel mixture which can be ignited. Above this percentage, the mixture will be too rich to burn.

USEPA = United States of America Environmental Protection Agency; Washington, DC.

USgal = United States gallon.

UST = underground storage tank (*vs.* AST).

UV = ultraviolet.

V = volt, a unit of electromotive force.

VA = volt-ampere.

vac = vacuum.

valve control = a system for control of air/fuel ratio by mechanical linkage of valves having the same characteristic.

vanadium = a metal, the oxide of which appears in the ash resulting from the combustion of fuel oils. Vanadium oxide has a destructive effect on refractories.

vapor = a gas that is near the condensing or liquid state, whose molecules are so close together that the forces between them significantly affect their behavior so that $P_1v_1/T_1 \neq P_2v_2/T_2$.

vaporizing oil burner = a burner in which the oil is vaporized in a single step by direct heating of the liquid.

vapor lock = an obstruction to the flow of a liquid in a pipe caused by vapor from the liquid, or by air.

vapor pressure = the pressure of the vapor of a liquid or solid in equilibrium with the liquid or solid.

V_c = critical velocity = sonic velocity.

velocity head = velocity pressure expressed in feet of column of the flowing fluid.

velocity pressure = the difference between total pressure and static pressure; that is, the difference between the pressure which a fluid flowing in a pipe exerts upon the upstream face of an obstruction in the pipe and the pressure which it exerts upon the walls of the pipe.

velocity profile = a diagram of vector arrows (length proportional to velocity) superimposed across a sectional view of a duct or pipe.

vena contracta taps = the pressure taps (located upstream and downstream of a thin metering orifice) positioned to give the highest possible pressure differential readings.

vent = a hole or opening for the escape of a fluid.

venturi = a section in a pipe or passageway that converges to a narrow constriction, then smoothly flares out again. Named for the Italian physicist G. B. Venturi (1762-1822), who first noted the effect of constriction.

v_g = specific volume of a saturated vapor.

visc = **viscosity** = the tendency of a fluid to resist flow. A measure of resistance to flow. Very significant in design and selection of oil burners and oil handling systems. See kinematic viscosity and absolute viscosity.

viscosity, absolute = dynamic viscosity = a measure of a fluid's tendency to resist flow, without regard to its density. Absolute viscosity is kinematic viscosity multiplied by density or specific gravity.

viscosity, kinematic = the relative tendency of a fluid to resist flow, including the effect of the fluid's density. Kinematic viscosity is equal to absolute viscosity ÷ density or specific gravity.

viscous friction = resistance to flow of fluids caused by energy dissipation and generation of stresses by distortion of fluid elements; flow resistance; internal friction.

vitiated air (pronounced vish-ee-ate-ed air) = air with <20.9% O_2.

vitrification = a process of permanent chemical and physical change (reduced porosity) at high temperatures in a ceramic body, with the development of a substantial proportion of glass.

VOC = volatile organic compound = any of a group of hydrocarbons that react in the atmosphere with nitrogen oxides, heat, and sunlight to form ozone; thus aggravating smog and global warming problems. Some enforcement agencies exclude CO, CO_2, carbonic acid, carbonates, metallic carbides, and methane from their definition of VOCs.

vol = **volume.**

volatile = easily vaporized; the more easily vaporized component of a liquid.

volume, combustion = the space occupied by the fuel while it is actually burning, including both the flame and invisible combustion zone.

volume flow rate = the quantity (measured in units of volume) of a fluid flowing per unit of time, as cubic feet per minute or gallons per hour.

volume, specific = the volume occupied by one pound of a substance under any specified conditions of temperature and pressure.

volumetric analysis = a statement of the various components of a substance (usually applied to gases only), expressed in percentages by volume.

vortex shedding = in the flow of fluids past an object, the eddying or formation of alternating whirlpools downstream of the object. A form of burble.

vs = **versus**, as compared to.

W = tungsten, or watt(s), a unit of power or heat flow rate. 1W = 1J/s.

wall loss = the heat lost from a furnace or tank to or through its walls.

warmup time = bring-up time = the time required to bring a furnace and its charge, if any, up to operating temperature.

waste gases = by-product fuel. See discussion under poc and flue gases.

water and sediment = bsw (bottom sediment and water). Impurities and foreign material found in fuel oils. See Part 2, Volume I.

water (blue) gas = an artificial fuel made by forcing steam over incandescent carbon to form a mixture of hydrogen and carbon monoxide. $C + H_2O \leftrightarrow CO + H_2$. The gas is poisonous because of its high CO content.

water column = the leg(s) of a manometer that uses water as the measuring fluid for pressure or pressure drop. The difference in height of the two water columns is usually stated in inches or millimetres of water column. See "wc below, and Table C.5 in the Appendix.

watt density = a measure of the concentration (in relation to surface area) of heat input, particularly critical in electrically heated oil heaters. Usually measured in watts per square inch, abbreviated wsi.

watt-seconds = a unit of work or heat equal to that resulting from the use of electricity at the rate of 1 watt for one second. One watt-second equals one joule.

wavelength (λ, lambda) = the distance traveled by a sound as its pressure wave varies through one complete cycle.

wax = the large hydrocarbon molecules that precipitate out of a liquid fuel when it is gradually cooled. This is the major factor controlling the pour point of a fuel oil.

wb = wet bulb temperature. (See below.)

"wc = "H₂O = inches of water column = inches of water gauge ("wg), a measure of pressure. See Table C.5 in the Appendix.

weight flow rate = the quantity (measured in units of weight) of a fluid flowing per unit of time, as pounds per second.

weight, specific = density, the weight per unit volume of a substance.

well-stirred reaction = a chemical combining of atoms or compounds in which the products of the reaction are thoroughly mixed with the incoming raw reactants. Theoretical opposite of a plug flow reaction.

wet basis = flue gas analysis by an analyzer that does not scrub or dry the sample before analysis; or an analysis in which all dry basis readings have been multiplied by (1 − %moisture/100%).

wet bulb temperature, wb = the temperature indicated by a thermometer, the bulb of which is covered with a wet wick; the temperature at which heat and mass transfer are in equilibrium. "Wet bulb depression" is the difference between wet bulb and dry bulb readings. When the wb depression is zero, there will be no evaporative cooling, and the relative humidity will be 100%; that is, the air is saturated with all the water vapor that it can hold.

wg = **water gauge**. When preceded by inches or millimetres, a measure of pressure. (See "wc.) "wg = "wc. (See Table C.5 in the Appendix.)

whb = **waste heat boiler.**

Wobbe Index = Wobbe Number, an index used to show fuel interchangeability. Wobbe Index = gross heating value in Btu/ft³ divided by √gas gravity. See Interchangeability of Fuels in Part 2, Volume 1.

wsi = **watts per square inch.** See watt density.

wt = **weight.**

W-T-E boiler/incinerators = waste-to-energy boiler/incinerators.

wwp = **water working pressure.**

× = times, or multiplied by.

XSAir = **excess air**; usually measured in % excess above the stoichiometrically correct amount of air.

xylene (C_8H_{10}) = a colorless, flammable, toxic liquid of the benzene series. Obtained mostly from coal tar.

zero gas = gas at atmospheric pressure (zero gauge pressure).

Zn = zinc.

zone, control = that section of a furnace within which temperature is controlled by one temperature measurement (and usually with one control valve).

Greek letters and symbols:

α (alpha) = proportional to; quarl angle; tile angle; angle.

β (beta) = d/D = ratio of orifice diameter to inside diameter of a pipe (both in same units).

Δ (delta) = differential, difference in, change in, increment. See ΔP, Δp, ΔT.

ΔP = pressure drop, differential, or loss in any consistent units, but sometimes used for lb/ft² (psf) to differentiate from Δp in lb/in.² (psi).

Δp = pressure drop, differential, or loss usually in psi, but sometimes osi or "wc.

ΔT = difference in temperature, change in temperature, drop in temperature.

ε = (epsilon) = absolute roughness, ft. (Relative roughness, e/D is used on Figures 5.13a and 5.14.)

λ = (lambda) = wavelength.

μ = (mu) = absolute viscosity; (sometimes) = microampere; or micron = 1 millionth of a metre; (as a prefix, pronounced micro) = one millionth or × 10^{-6}.

ν = (nu) = kinematic viscosity.

π (pi) = 3.141 592 654 = ratio of circumference of a circle to the diameter of that circle.

ρ (rho) = density.

Σ (sigma) = sum, summation, total of.

φ (phi) = equivalence ratio = a means of expressing air/fuel ratio = the actual amount of fuel expressed as a decimal ratio of the stoichiometrically correct amount of fuel. (See Table C.10 in the Appendix.)

Ω (omega) = ohm, unit of electric resistance = 1 V/A.

+   plus.

−   minus.

±   plus or minus, tolerance.

·   times, or multiplied by.

÷   divided by.

/   divided by, per, for each.

√   the square root of.

=   equals, or is equal to.

$\overset{m}{=}$ measured by.

≠   not equal to.

<   (is) less than.

≤   smaller than or equal to.

>   (is) greater than.

≥   greater than or equal to.

≪   much smaller than.

≫   much larger than.

~   proportional to, similar to.

≈   approximately equal to.

≡   congruent to.

∴   therefore.

∠   angle.

‖   parallel to.

⊥   perpendicular to, at right angles to, normal to.

∞   infinity, infinite.

°   degrees. (Used only to specify the size of an angle or a temperature difference or temperature drop--not a temperature level; e.g. the temperature drop through a wall is 250°, from 310 F to 60 F.)

°C   temperature difference or change in degrees Celsius.

°E   degrees Engler, a measure of viscosity. See Figure 2.7.

°F   temperature difference or change in degrees Fahrenheit.

pound(s), or number.

%   percent.

'   foot (or feet), minutes (1/60 of 1° angle).

"   inch(es), seconds (1/60 of 1' angle = 1/3600 of 1° angle); ditto.

INDEX

---

* See also Volume I.

---

* See also Volume I.

---

* See also Volume I.

* See also Volume I.

---

* See also Volume I.

---

* See also Volume I.

---

* See also Volume I.

---

* See also Volume I.

---

* See also Volume I.

----

* See also Volume I.

* See also Volume I.

---

* See also Volume I.

---

* See also Volume I.

---

* See also Volume I.

---

* See also Volume I.

* See also Volume I.

* See also Volume I.